515.7246 DeV
DeVito.
Functional analysis and
 linear operator theory.

**The Lorette Wilmot Library
Nazareth College of Rochester**

FUNCTIONAL ANALYSIS AND LINEAR OPERATOR THEORY

CARL L. DeVITO
UNIVERSITY OF ARIZONA, TUCSON

ADDISON-WESLEY PUBLISHING COMPANY
The Advanced Book Program
Redwood City, California • Menlo Park, California • Reading, Massachusetts
New York • Don Mills, Ontario • Wokingham, United Kingdom • Amsterdam
Bonn • Sydney • Singapore • Tokyo • Madrid • San Juan

Publisher: *Allan M. Wylde*
Production Manager: *Jan V. Benes*
Marketing Manager: *Laura Likely*
Cover Design: *Iva Frank*
Electronic Production: *Lori McWilliam Pickert*

Library of Congress Cataloging-in-Publication Data

DeVito, Carl L.
Functional analysis and linear operator theory/Carl DeVito.
　　p. cm.
　Includes index.
　1. Linear operators. 2. Functional analysis. I. Title.
QA 329.2.D48　1990　515′.7246-dc　　90-693
ISBN 0-201-11941-2

This book was prepared using the TEX typesetting language.

Copyright © 1990 by Addison-Wesley Publishing Company, The Advanced Book Program, 350 Bridge Parkway, Redwood City, CA 94065.

All rights reserved. No part of this publication may be reproduced, stored in a retrieval system, or transmitted in any form or by any means, electronic, mechanical, photocopying, recording, or otherwise, without the prior permission of Addison-Wesley Publishing Company.

ABCDEFGHIJ-MA-943210

PREFACE

A course in functional analysis and the theory of linear operators on a Hilbert Space is usually given to students of mathematics in their second year of graduate study. There are, however, many other students who would clearly benefit from knowing something about this subject. Among these are first-year graduate students and some advanced undergraduate students in engineering, applied mathematics, and the physical sciences. Operator-theoretic methods are already being used to solve problems which arise in these areas, and the potential for future applications is very great. The difficulty is that few of these students have the background needed to take a course in this subject as it is usually taught. Some years ago I decided to try to design a course in operator theory and the relevant parts of functional analysis which would meet the needs of these students, but one that required as background only the kind of mathematics courses that most of them had had. My main problem was to help the members of my class make the transition from the "methods of mathematics" courses they had taken to the more sophisticated realm of abstract analysis. My experiences over the years led me to write this book.

About the Course

Was the course successful? The answer, of course, depends on how one measures success, and I measured it in two ways. Toward the end of the course I would assign to each student some outside reading, usually some latter section of one of the more sophisticated operatory theory books or some research paper applying operator-theoretic methods, and have them present this material to the class. Almost invariably the students were able to do this; i.e., they were able to read difficult material, present it in an organized way, and respond intelligently to the questions which I, or other members of the class, asked them. If you consider the sophistication of the subject and the background of my students I think this is a very good measure of success. The second measurement was student feedback. I was told by many students, sometimes a year or two after they had taken the course, that they found the class invaluable when they studied things like quantum physics, quantum chemistry, signal processing, etc., and most felt that they had a much firmer grasp of mathematics after taking this class. On the basis of these two tests I feel that teaching this course was worthwhile. I hope that others who teach from this book will have a similar experience.

About the Book

This book tries to bridge the gap between abstract analysis and methods of mathematics. In chapter one I put a great deal of effort into trying to explain what was going on. I tried to motivate the abstract ideas I was presenting and I tried to show how they related to the more concrete ideas my students were used to. I also tried to show that there were advantages to abstraction. This opens up the instructor to the charge of condescension, and so I carefully checked with my students over the years to see what they thought of these efforts. They were emphatic and surprisingly unanimous in stating that the instructor who follows this practice is the one who really respects his or her students and that such an instructor is always preferable to one who "dumps a bunch of theorems in front of us and calls that teaching." Now, of course, as the students become more familiar with the subject matter, the need for explanations decreases and I rapidly devote less and less space to them.

Another thing I did was to use a spiral approach to textbook writing. The style in writing mathematics for mathematics students is to say something once and only once and we train our students to be aware of this. This book, however, is written for students who are primarily interested in using mathematics. As important as mathematics is to their course of study, they forget material that hasn't been discussed for awhile and appreciate a brief review. So I do repeat myself and go over some topics more than once. I also took some pains to give references to where definitions, theorems, etc., being used in a discussion first appeared in the book. This aids the student in following the discussion because it enables him or her to quickly locate the relevant background material. Finally, the exercises have been designed to get the student involved in the text's mathematical development. This is accomplished by giving as problems some small results which are referred to later on in the text. Any such problem is marked with a star. For the most part these are straightforward exercises which help build the student's confidence and test his or her understanding of the material. However, all of the starred problems in chapters 1–4, the introductory part of the book, are solved in detail in appendix B. Some instructors may want to present some of these results in class. This may be true, in particular, of the properties of projection operators which I put in problem 2(d) of Exercises 1.4.

Applications

In a book aimed at people from engineering and the sciences, the conventional wisdom holds that a great many real-world problems should be discussed in detail. Curiously enough, my own experience does not support this idea. My students told me that they want to know that what they are learning has applications but they don't want to see the details. To do so would mean learning the concepts and terminology of the application's subject. It is simply not true that every engineering student knows, for example, signal processing well enough to appreciate, or even follow, an application to this area, and it is not true that every physical science student knows enough about quantum mechanics to understand how some of its subtleties can be best expressed in operator-theoretic terms. Furthermore, engineering students do not all have the same interests and neither do all science students. I never found a single physical application that interested more than half the class, and usually much less than that. However, every engineer and every physical scientist knows the importance of differential equations, integral equations, and Fourier series. Any application of operator theory to one of these areas always evoked a strong positive response from the class. There are many such applications in this book.

Topic Coverage

The first four chapters of the book form the basis for a one-semester course in operator theory and functional analysis. Chapter one contains a discussion of finite and infinite dimensional vector spaces, norms and inner products, orthogonal sets and Fourier series, and the spaces ℓ^2 and $L^2[a,b]$. In order to define this last space we need the Lebesgue integral and a discussion of this topic is given in section 1.8. Here I had to be heuristic in my arguments and very selective in my choice of subjects and, no doubt, I have said too much for some tastes and not enough for others. Chapter two is concerned with linear operators and their properties. The norm of an operator and the spectrum of an operator are defined, equivalent norms on a vector space are discussed, and examples of linear operators are given. In chapter three we discuss compact operators and the Riesz theory. My students had had very little experience in working with sequences and so I tried to give a very detailed discussion of the way this important tool is used. Chapter four concludes the introductory section with a discussion of projection operators, complementary subspaces, invariant subspaces, and Hilbert–Schmidt operators.

Chapters five and six contain a very detailed discussion of the spectral theorem for various classes of operators. Chapter five begins with a review of linear algebra and a proof of the spectral theorem for compact, self-adjoint operators. The theory is then extended to the class of compact normal operators. These results are next applied to Fredholm integral equations of both the first and the second kind. The last section of this chapter contains some examples of integral operators which are not compact and a discussion of (nonlinear) contraction mappings and their application to differential equations. In chapter six we give a proof of the spectral theorem for self-adjoint operators. All of the details are given, including a brief review of the Riemann–Stieltjes integral (the spectral theorem for unitary operators is discussed in chapter seven and this theorem, for the class of normal operators, is stated in appendix three). When I taught a full-year course in this subject, I often covered chapters one through five in some detail. I then went rather quickly over chapter six, concentrating on the statements of the results. After this I covered as much of chapters seven and eight as time permitted. These last two chapters deal with unbounded operators.

Chapter seven begins by looking at a specific differential equation and solving this equation by operator-theoretic methods. The natural operator associated with this equation is unbounded and this leads to an investigation of the properties of such operators. A discussion of closed operators, adjoints, differentiation and $L^2[a,b]$, and the Cayley transform of an operator follows. Some of these ideas are used in section 7.6 where the spectral theorem for unitary operators is discussed. The last section of this chapter treats the spectral theorem for unbounded operators.

Chapter eight begins by looking at the same differential equation considered in chapter seven, but this time it is worked over an infinite interval. Operator-theoretic methods are again used to solve this equation, and again, the operator is unbounded. In this case, however, the eigenvectors of the operator are not in the space in which the operator is defined. Still, using these functions, a Fourier transform can be defined and used to solve the problem. The chapter includes a discussion of L^2-spaces on unbounded intervals and unbounded operators on such a space. The Fourier transform is treated in some detail and it is used to show the equivalence of some important operators. Finally, one method of giving some meaning to the so-called "non-normalizable" eigenvectors which arise in connection with so many very useful operators is given.

Acknowledgements

While I must take credit, or blame, for all of the writing and any factual errors the book may contain, I had a great deal of valuable advice and assistance from two colleagues: Professor Donald Dudley of Electrical and Computer Engineering, and Professor William Faris of Mathematics and Applied Mathematics. I'd like to thank Don for his advice and enthusiastic support, and Bill for reading an early version of this book and making many valuable comments on it. They understood the many difficulties involved in designing and teaching this course, they encouraged their students to take the course, and they were invariably helpful in suggesting ways to deal with the problems of teaching operator theory. My students were also very helpful, and I'd like to thank them for their many useful suggestions over the years. Many of them told me, in private, about difficulties they had which they would never bring up before a class. These are dealt with in the book, usually after the phrase "experience impels us to state." I would also like to express my appreciation to the people at Addison-Wesley Publishing Company, especially Allan M. Wylde, Jan Benes, and Roger Vaughn, for the help they gave me while I was writing this book. Finally, I would like to thank Lori McWilliam Pickert for her excellent typing.

CONTENTS

Preface	iii

1 Linear Algebra and Analysis — 1
- 1.1 Vector Spaces — 1
- 1.2 Norms on a Vector Space — 6
- 1.3 Convergence in and Completeness of Normed Spaces — 9
- 1.4 Inner Products — 14
- 1.5 An Infinite Dimensional Hilbert Space — 19
- 1.6 Orthogonal and Orthonormal Sets — 23
- 1.7 Fourier Series — 37
- 1.8 The Space $L^2[a,b]$ — 45
- 1.9 More About the Space $L^2[a,b]$ — 57

2 Linear Mappings and Operators — 65
- 2.1 Bounded Linear Maps — 65
- 2.2 Comparing Norms — 73
- 2.3 The Spectrum of a Linear Operator — 80
- 2.4 Some Operators and their Spectra — 88

3 Compact Operators — 99
- 3.1 Compact Sets — 101
- 3.2 More about Compact Sets — 108
- 3.3 The Spectrum of a Compact Operator — 114

4 Spectral Theory, The Basic Tools — 127
- 4.1 Some Infinite Dimensional Geometry Projections — 127
- 4.2 Projections on a Hilbert Space — 132
- 4.3 The Adjoint Exists! — 138
- 4.4 The Adjoint of a Compact Operator — 145

5 The Spectral Theorem, Part I — 153
- 5.1 Matrices — 153

5.2	The Spectral Theorem, Finite-dimensional Case	157
5.3	Invariant Subspaces	163
5.4	The Spectral Theorem: Compact Operators	169
5.5	The Spectral Theorem: Compact Operators (continued)	182
5.6	Fredholm Integral Equations	186
5.7	Some Further Remarks on Integral Operators	193

6 The Spectral Theorem, Part II — 199

6.1	The Spectrum of a Self-adjoint Operator	200
6.2	An Important Function Space	205
6.3	Functions of a Bounded Self-adjoint Operator	212
6.4	The Resolution of the Identity	221
6.5	The Riemann-Stieltjes Integral	224
6.6	The Spectral Theorem for Self-Adjoint Operators	231

7 Unbounded Operators — 242

7.1	Closed Operators	242
7.2	The Adjoint and the Graph of an Operator	248
7.3	Differentiation in the Space $L^2[a,b]$	252
7.4	Some Differential Operators	255
7.5	Cayley Transform, Deficiency Indices	259
7.6	The Spectral Theorem for Unitary Operators	265
7.7	The Spectral Theorem for Unbounded Operator	272

8 Non-normalizable Eigenvectors — 284

8.1	The Space $L^2(\mathbf{R})$	287
8.2	The Fourier Transform on $L^2(\mathbf{R})$	292
8.3	More About the Fourier Transform	304

A The Axioms for a Vector Space — 313

B Solutions to Starred Problems — 315

C The Spectral Theorem for Normal Operators — 346

Bibliography — 350

List of Symbols — 351

Index — 352

1

LINEAR ALGEBRA AND ANALYSIS

Our purpose here, as the title indicates, is to collect together some facts from linear algebra and analysis which will be useful later on. At the same time, we shall establish a common notation and terminology which we will use throughout the book.

1.1   Vector Spaces

In all that follows, $\mathbb{R}$ will denote the real numbers and $\mathbb{C}$ will denote the complex numbers. These sets have a rich (and very well-known) algebraic structure and because of this they are called fields. We assume that the reader is familiar with the concept of a vector space V over a field $\mathcal{K}$ (a precise definition is given in Appendix I). In all that follows we will, unless the contrary is explicitly stated, understand that $\mathcal{K}$ is either $\mathbb{R}$ or $\mathbb{C}$; in most cases it does not matter which of these fields we are working over and when it does matter, the field we have in mind will be clear from the context. Most of the vector spaces we shall consider will be infinite dimensional and so the familiar definitions from linear algebra, given below, have been slightly modified so as to apply to this case. The important thing to remember is that a vector space, whatever its dimension, is an algebraic object. By this we mean that we can apply our operations of addition and scalar multiplication only a finite number of times. So in a general vector space the symbol $\sum_{j=1}^{\infty} v_j$ has no meaning since such a sum involves the (non-algebraic) notion of convergence.

Definition 1

Let V be a vector space over $\mathcal{K}$. A non-empty subset W of V is said to be a linear manifold in V if:

   a. For any two vectors u, w in W the vector $u + w$ is in W;

   b. For any vector w in W and any scalar α in K, the vector αw is in W.

The two simplest examples of linear manifolds in a vector space V are the space V itself and the set consisting of the zero vector alone. These are sometimes called the trivial linear manifolds.

Definition 2

Let V be a vector space over $\mathcal{K}$ and let S be a non-empty subset of V. A vector v is said to be a linear combination of vectors in S if there is a *finite* set $u_1, u_2, \ldots, u_k$ in S and scalars $\alpha_1, \ldots, \alpha_k$ such that $v = \sum_{j=1}^{k} \alpha_j u_j$. The set of all vectors in V which are linear combinations of vectors in S will be denoted by lin S.

It is easy to see that lin S is a linear manifold which contains S (see exercise 2a below). We call it the linear manifold spanned by, or generated by, S. In particular, if lin $S = V$ we say that S spans V.

Definition 3

Let V be a vector space over $\mathcal{K}$. A non-empty subset S of V is said to be a linearly independent subset of V if the only member of lin S which is equal to the zero vector is the one whose scalar coefficients are all zero.

It is customary to say that the empty set is a linearly independent subset of V. The reason for this convention will be seen very soon. We can restate Definition 3 as follows: S is linearly independent in V if and only if (i) $\sum_{j=1}^{p} \alpha_j v_j \in linS$ and (ii) $\sum_{j=1}^{p} \alpha_j v_j = 0$ implies (iii) $\alpha_1 = \alpha_2 = \ldots = \alpha_p = 0$. Perhaps we should stress that the zero appearing in (ii) is very different from the one in (iii).

At this point an example might be helpful. Let $C_r[0,1]$ denote the set of all real-valued functions which are defined and continuous in $[0,1]$. For f, g in $C_r[0,1]$ we define $f + g$ to be the function which takes each $t \in [0,1]$

to the number $f(t) + g(t)$; i.e., $(f + g)(t) = f(t) + g(t)$ for all $t \in [0, 1]$. Similarly, if $f \in C_r[0, 1]$ and $\alpha \in \mathbb{R}$, note that we need real scalars here since our functions are real-valued, we define $(\alpha f)(t)$ to be $\alpha f(t)$ for all $t \in [0, 1]$. Clearly, $C_r[0, 1]$, with the algebraic operations just defined, is a vector space over $\mathbb{R}$. Now consider the set $S = \{1, x, x^2, x^3, \ldots, x^n, \ldots\}$ in this space. What is lin S? Well an element $f \in C_r[0, 1]$ is in lin S if, and only if, we can find a *finite* set $x^{n_1}, x^{n_2}, \ldots, x^{n_k}$ in S and scalars $\alpha_1, \ldots, \alpha_k$ such that $f(x) = \alpha_1 x^n 1 + \ldots + \alpha_k x^n$ (equality here means equality as functions; i.e., the two sides are equal for each $t \in [0, 1]$). Thus lin S is simply the set of all polynomial functions on $[0, 1]$. Since there are many continuous functions which are not polynomials we see that S does not span $C_r[0, 1]$. To put it another way, lin S is a non-trivial linear manifold in $C_r[0, 1]$.

Let us show that S is linearly independent in $C_r[0, 1]$. Suppose that some element of lin S is equal to the zero vector. This means that we have a *finite* set $x^{n_1}, x^{n_2}, \ldots, x^{n_k}$ in S, and scalars $\alpha_1, \ldots, \alpha_k$ such that

$$\sum_{j=1}^{k} \alpha_j x^{n_j} = 0 \tag{1}$$

Now the right-hand side of equation (1) is the zero vector of $C_r[0, 1]$. This is the function on $[0,1]$ which takes each $t \in [0, 1]$ to zero. So our equation says that the function on the left-hand side of (1) is zero for every $t \in [0, 1]$. Since this left-hand side is a polynomial function it can be zero for only finitely many values of t (depending on its degree) unless it is the zero polynomial; i.e., unless $\alpha_1 = \alpha_2 = \ldots = \alpha_k = 0$.

One can show that any linearly independent subset of a vector space V can be 'extended' to linearly independent set which spans V; i.e., given a linearly independent subset S of V we can find a linearly independent subset T of V such that $S \subseteq T$ and lin $T = V$. Clearly any set which properly contains T could not be linearly independent and so we may say that T is a maximal linearly independent subset of V. It turns out that any two maximal linearly independent subsets of the same vector space can be placed in one-to-one correspondence, and this gives us a nice way to define the dimension of a vector space.

Definition 4

Let V be a vector space over $\mathcal{K}$. A linearly independent subset of V which also spans V will be called a Hamel basis for V. If V has a Hamel basis consisting of a finite number of vectors, then we shall say that V is a finite dimensional vector space. In this case the number of vectors in any Hamel basis for V is called the dimension of V and is denoted by dim. V. A vector space which is not finite dimensional is said to be infinite dimensional.

We have not defined the dimension of an infinite dimensional vector space although we could. The space $C_r[0,1]$ is, as we saw above, an infinite dimensional vector space. Many other examples will be given below. We shall also sketch the process by which one can construct 'all' finite dimensional spaces over $\mathcal{K}$. The extreme case is that of the vector space consisting of one (necessarily zero) vector. It is customary to say that the empty set is a basis for this space, and hence that this is a vector space of dimension zero.

Definition 5

Let V, W be two vector spaces over the same field $\mathcal{K}$. A map T from V into W (we sometimes abbreviate this last phrase by the symbol $T : V \to W$) is said to be a linear map if: (i) $T(u+v) = T(u) + T(v)$ for all u, v in V; (ii) $T(\alpha u) = \alpha T(u)$ for all α in $\mathcal{K}$ and all u in V. A linear map which is both one-to-one and onto will be called an isomorphism. Finally, a linear map from V to itself will be called a linear operator on V.

Linear mappings are of fundamental importance in all that follows. Even in calculus one studies those functions which can be approximated locally (i.e., at each point) by linear functions. The reason that this fact is so often overlooked is that the linear mappings from $\mathbb{R}$ to $\mathbb{R}$ are so trivial. It becomes more apparent when one studies functions of several variables.

Corresponding to every linear map T from V into W are two linear manifolds. They are the kernel or null space of T, $\mathrm{Ker}T = \{V \in V | T(v) = 0\}$ and the range of T, $R(T) = \{w \in W | w = T(v) \text{ for some } v \in V\}$. We close this paragraph with two more pieces of terminology.

Definition 6

Two vector spaces over the same field are said to be isomorphic if there is an isomorphism from one of these spaces onto the other.

Many of the examples of vector spaces given in this book consist of families of functions defined on a set S, say, and taking their values in $\mathcal{K}$ or in some fixed vector space over $\mathcal{K}$. If f, g are in such a family then their sum, $f + g$, is defined to be the function which takes each $s \in S$ to $f(s) + g(s)$. So $(f+g)(s) = f(s) + g(s)$. Similarly, for $\alpha \in K$ the function αf is defined by $(\alpha f)(s) = \alpha(f(s))$ for all $s \in S$. Whenever we say that a family of functions is a vector space these are the algebraic operations that we have in mind. In fact, whenever we speak of the sum of two functions, or of the product of a function and a scalar, the functions just defined are what we mean (see that space $C_r[0,1]$ defined above and problems 6 and 7).

Exercises 1

A number of useful facts, facts that will be referred to later on in the text, are scattered among the exercises. Any problem that is referred to later on is marked with a star. Solutions to the starred problems are given in Appendix II.

1. Let V be a vector space over $\mathcal{K}$.

   a. Show that every linear manifold in V contains the zero vector.

   b. Show that the intersection of any non-empty family of linear manifolds in V is a linear manifold in V.

2. Let V be a vector space over $\mathcal{K}$ and let S be a non-empty subset of V.

   a. Verify that lin S is a linear manifold in V (this is one of those problems which is so easy that it may be hard).

   b. Show that any linear manifold of V which contains S must also contain lin S. In fact, if $m = \{M \subseteq V | M$ is a linear manifold, and $M \supseteq S\}$, then $\cap\{M | M \in m\} = linS$.

   c. What condition on a set $S \subseteq V$ is necessary and sufficient to ensure that $linS = S$?

* 3. Let V be a vector space over $\mathcal{K}$. Choose any non-zero vector v_1 in V. Next choose a vector v_2 in $V \setminus lin\{v_1\}$ (i.e., in V but not in the linear span of v_1). Then choose v_3 in $V \setminus lin\{v_1, v_2\}$, and continue in this way. Clearly there are two cases: (i) There is first an integer p such that $V \setminus lin\{v_1, v_2, \ldots, v_p\} = 0$; (ii) No such integer exists.

   a. Assuming (i) show that $v_1, \ldots, v_p$ is a Hamel basis for V.

   b. Assuming (ii) show that the sequence $v_1, v_2, \ldots$ is linearly independent.

4. Let V be a vector space over $\mathcal{K}$ and suppose that $S \subseteq V$ spans V. Show that if no proper subset of S spans V, then S must be linearly independent.

5. Let V be a vector space over $\mathcal{K}$ and let $T \subseteq V$ be a linearly independent set. Suppose that any subset of V which properly contains T is not linearly independent. Show that T spans V.

6. Let Ω be a non-empty set and let $\mathcal{F}_r(\Omega)$ be the set of all real-valued functions on Ω. This is a vector space under the usual definitions of

addition and scalar multiplication for functions (see the discussion of $C_r[0,1]$ above).

    a. If $\Omega = \{1, 2, \ldots, n\}$ show that $\mathcal{F}_r(\Omega)$ is isomorphic to $\mathbf{R}^n$, the space of all n-tuples of real numbers.

    b. If $\Omega = \{1, 2, 3, \ldots, n \ldots\}$ show that $\mathcal{F}_r(\Omega)$ is isomorphic to the vector space of all sequences of real numbers; call this space S_r.

    c. Referring to (b) define, for each fixed n, $e_n(j) = 0$ if $n \neq j$, $e_n(n) = 1$. So $e_n \in \mathcal{F}_r(\Omega)$ and the image of e_n under the isomorphism of (b) is the sequence $(0, 0, \ldots, 0, 1, 0, \ldots)$, the one being in the nth place. Show that $\{e_n\}_{n=1}^{\infty}$ is linearly independent but that it is not a Hamel basis for S_r.

  7. Let Ω be a non-empty set and let $\mathcal{F}_c(\Omega)$ be the set of all complex valued functions on Ω. This is a vector space over $\mathbf{C}$. If $\Omega = \{1, 2, \ldots, n\}$ show that $\mathcal{F}_c(\Omega)$ is isomorphic with $\mathbf{C}^n$ the space of all n-tuples of complex numbers.

* 8. Let V, W be two isomorphic vector spaces. Show that either V and W have the same dimension or they are both infinite dimensional.

* 9. Let V be an n-dimensional vector space over $\mathcal{K}$, and let $v_1, \ldots, v_n$ be a Hamel basis for V.

    a. Show that the representation of any $v \in V$ in the given Hamel basis is unique; i.e., if $v = \sum_{j=1}^{n} \alpha_j v_j$ and $v = \sum_{j=1}^{n} \beta_j v_j$ then $\alpha_j = \beta_j$ for each $j = 1, 2, \ldots, n$.

    b. Now suppose that W is another n-dimensional vector space over the same field as V, and let $w_1, \ldots, w_n$ be a Hamel basis for W. By (a) we can define at map T from V onto W in two steps. First set $Tv_j = w_j$ for each $j = 1, 2, \ldots, n$. Next we extend T to all of V by linearity. More explicitly, given $v \in V$ we write $v = \sum \alpha_j v_j$, where the scalars α_j are uniquely determined, and set $Tv = \sum_{j=1}^{n} \alpha_j w_j$. Show that T is an isomorphism from V onto W.

1.2 Norms on a Vector Space

In elementary physics, one learns that a vector is a quantity that has both magnitude and direction. For us, a vector is just an element of a vector space and so we can say nothing, in general, about its nature. In particular, n-tuples of real or complex numbers, continuous functions, polynomials, etc.

are examples of vectors, because each of these objects is in some vector space. This is the basis for the often repeated remark, usually attributed to B. Russell, that mathematics is the subject in which we don't know what we are talking about. However, many of the vector spaces that arise in applications have certain special functions defined on them that enable us to assign a length, and sometimes even a direction, to the vectors in that space. One such function, the one that we shall study here, is called a norm. The importance of norms, and they are very important, will become clear as we proceed but, for now, the reader may want to think of a norm on a vector space as simply a function that enables us to assign a length to each of the vectors in that space. It is instructive, in this connection, to recall how three-, and then n-, dimensional space is usually treated.

Most discussions begin by assuming, sometimes without explicit mention, that the space of our experience is three-dimensional and Euclidean. Next we set up a rectangular coordinate system at some point thereby assigning to each point an ordered triple of real numbers, (x, y, z), called the coordinates of the point. A vector $\vec{v}$ is then defined to be the directed line segment from $(0, 0, 0)$ to, say, (x, y, z) and we often write $\vec{v} = (x, y, z)$. The length of $\vec{v}$, $\|\vec{v}\|$, is then computed by using the Pythagorean theorem. One finds

$$\|\vec{v}\| = \sqrt{x^2 + y^2 + z^2} \tag{1}$$

In this way we obtain a simple mathematical model that we usually identify with the space of our experience. Our model consists of the vector spaces of all ordered triples of real numbers, $\mathbb{R}^3$, and the functions $\|\cdot\|$ defined by (1); note that the vector space operations in $\mathbb{R}^3$ are defined in the usual coordinate-wise fashion.

Let us turn now to n-dimensional Euclidean space. Here, of course, when $n \geq 4$ our geometric intuition deserts us and so we fall back on algebra. More explicitly: Let n be an arbitrary but, in any discussion, fixed positive integer; so, experience impells us to stress, n is never "infinity." The vector space $\mathbb{R}^n$ consists of all ordered, n-tuples of real numbers (a purely abstract object nothing like the space of our experience) with addition and scalar multiplication defined in the usual coordinate-wise fashion. When $n = 3$ we identified $\mathbb{R}^3$ with the space of our experience and since the latter is Euclidean we freely used this geometry, via the Pythagorean theorem, to calculate the length of any vector in $\mathbb{R}^3$. This led us to formula (1). When $n > 3$ we use formula (1) to *define* the length of a vector. This is the preferable way to go even when $n \leq 3$. So we define the Euclidean norm $\|\cdot\|$ on $\mathbb{R}^n$ as follows: For every $\vec{v} = (x_1, \ldots, x_n)$

$$\|\vec{v}\| = \Big(\sum_{j=1}^{n} x_j^2\Big)^{\frac{1}{2}} \tag{2}$$

In a similar way we can assign a length to every vector in the space $\mathbb{C}^n$. This is the vector space of all ordered n-tuples of complex numbers with algebraic operations defined coordinate-wise and, by convention, complex scalars. It is, of course, possible to use real scalars but except for problem 3 below, we shall never do that. The Euclidean norm on $\mathbb{C}^n$ is defined as follows: For any $\vec{v} = (\xi_1, \ldots, \xi_n)$

$$\|\vec{v}\| = \left(\sum_{j=1}^{n} |\xi_j|^2\right)^{\frac{1}{2}} \qquad (3)$$

The presence of the absolute value on the right-hand side of equation (3) is necessary if non-zero vectors are to have non-zero lengths; e.g., the vector $(1, i) \in \mathbb{C}^2$ is non-zero and its length is $(|1|^2 + |i|^2)^{1/2} = \sqrt{2}$, not zero which is what formula (2) would assign to it.

Let us turn now to the formal definition of a norm.

Definition 1

Let E be a vector space over $\mathcal{K}$. A non-negative, real-valued function p on E is said to be a norm on E if:

a. $p(x + y) \leq p(x) + p(y)$ for all x, y in E;

b. $p(\alpha x) = |\alpha| p(x)$ for all x in E and all α in $\mathcal{K}$;

c. $p(x) = 0$ if, and only if, x is the zero vector.

If p is a norm on E it is customary to denote, for each x in E, the number $p(x)$ by $\|x\|$. A vector space on which a norm is defined will be called a normed space. If we want to emphasize the norm, say $\|\cdot\|$, on E, we shall speak of the normed space $(E, \|\cdot\|)$.

We have already mentioned, without proof, that the function $p(\vec{v}) = (\sum_{j=1}^{n} |x_j|^2)^{1/2}$, $\vec{v} = (x_1, \ldots, x_n)$, is a norm, we called it the Euclidean norm, on $\mathbb{R}^n$ (or $\mathbb{C}^n$). It is easy to see that this function has properties (b) and (c) of Definition 1, and the proof that it also has property (a) will be obtained below as a special case of a more general result. One should note that in $\mathbb{R}^3$ (and also in $\mathbb{R}^2$) property (a) says that the sum of the lengths of any two sides of a triangle always exceeds the length of the third side. For this reason, (a) is referred to as the triangle inequality.

An important example of an infinite dimensional normed space is the following: Take the vector space of all continuous functions on $[0, 1]$, $C[0, 1]$, and for each f in this space, define $\|f\|_\infty$ by

$$\|f\|_\infty = \max\{|f(t)| \mid o \leq t \leq 1\} \qquad (4)$$

We have not specified, because it does not matter, whether are functions are real or complex valued and what our field of scalars is. It is very easy to check that $\|\cdot\|_\infty$ has properties (a), (b), and (c) and hence $(C[0,1], \|\cdot\|_\infty)$ is a normed space. We call $\|\cdot\|_\infty$ the sup (for supremum) norm.

Exercises 2

1. For this problem we work with the space $\mathbb{R}^2$.

   a. Show that the function $\|\cdot\|_M$, defined by $\|(x,y)\|_M = \max\{|x|, |y|\}$ for all $(x,y) \in \mathbb{R}^2$, is a norm on $\mathbb{R}^2$.

   b. Show that the function $\|\cdot\|_S$, defined by $\|(x,y)\|_S = (|x|+|y|)$ for all $(x,y) \in \mathbb{R}^2$, is a norm on $\mathbb{R}^2$.

   * c. In any normed space $(E, \|\cdot\|)$ the set $\mathcal{B}_1$, also denoted by $\mathcal{B}_1(\|\cdot\|)$, is defined to be $\{x \in E \mid \|x\| \leq 1\}$ and is called the unit ball of E. Draw the unit ball in the space $\mathbb{R}^2$ with the Euclidean norm and in each of the spaces $(\mathbb{R}^2, \|\cdot\|_M)$, $(\mathbb{R}^2, \|\cdot\|_S)$.

* 2. For any fixed n let $\vec{e}_j$ be the n-tuple consisting of zeroes except for a one in the jth place, $1 \leq j \leq n$. Show that $\{\vec{e}_j\}_{j=1}^n$, regarded as a subset of $\mathbb{R}^n$, is a Hamel basis for this space and that, regarded as a subset of $\mathbb{C}^n$ it is a Hamel basis for $\mathbb{C}^n$ as well. We call this the standard basis in $\mathbb{R}^n$ or $\mathbb{C}^n$.

3. Suppose, just for this problem, that we regard $\mathbb{C}^n$ as a vector space over $\mathbb{R}$. Find a Hamel basis for this space and calculate its dimension.

* 4. Let n be a fixed, positive integer. Show that any n-dimensional vector space over $\mathbb{R}$ is isomorphic (Section 1, Definition 6) to $\mathbb{R}^n$, and that any n-dimensional vector space over $\mathbb{C}$ is isomorphic to $\mathbb{C}^n$.

* 5. Let E, F be two vector spaces over the same field and suppose that $\phi: E \to F$ is an isomorphism (Section 1, Definition 5) from E onto F. Suppose that we have a norm $\|\cdot\|_F$ on F. For each $x \in E$ define $\|x\|_E$ to be the number $\|\phi(x)\|_F$. Show that $\|\cdot\|_E$ is a norm on E.

1.3 Convergence in and Completeness of Normed Spaces

The presence of a norm on a vector space enables us to talk about convergence in that space and this makes a great deal of analysis possible. The key concept is that of a convergent sequence.

Definition 1

Let $(E, \|\cdot\|)$ be a normed space. A sequence $\{x_k \mid k = 1, 2, \ldots\}$ of points of E is said to be convergent to the point x_0 of E if, given any $\varepsilon > 0$, there is an integer K such that $\|x_0 - x_k\| < \varepsilon$ whenever $k \geq K$. When this is the case we shall write $x_0 = \lim x_k$.

In the space $\mathbb{R}^n$ convergence for the Euclidean norm simply means component-wise convergence. More precisely, the sequence $\{\vec{v}_k \mid k = 1, 2, \ldots\}$ of vectors in $\mathbb{R}^n$ is convergent to the vector $\vec{v}_0$ for the Euclidean norm if, and only if,

$$\lim(\vec{v}_k)_j = (\vec{v}_0)_j \tag{1}$$

for each $j = 1, 2, \ldots, n$.

Here $(\vec{v}_k)_j$ denotes the jth component of $\vec{v}_k$.

In order to prove this, we just write $\vec{v}_k = (x_{1k}, x_{2k}, \ldots, x_{nk})$ for each k including $k = 0$. Next, we consider separately the necessity and the sufficiency parts of our statement.

a. If $\lim \vec{v}_k = \vec{v}_0$ for the Euclidean norm then for each fixed j, $1 \leq j \leq n$, we have

$$|x_{jk} - x_{j0}| \leq (|x_{jk} - x_{j0}|)^{1/2} \leq (\sum_{j=1}^{n} |x_{jk} - x_{j0}|^2)^{1/2} = \|\vec{v}_k - \vec{v}_0\|$$

and since the right-hand side tends to zero as k tends to infinity we see that $\lim x_{jk} = x_{j0}$ for each j.

b. Suppose now that (1) holds; i.e., that $\lim x_{jk} = x_{j0}$ for each $j = 1, 2, \ldots, n$. Then, given $\varepsilon > 0$ we can choose K_j so that $|x_{jk} - x_{0k}| < \varepsilon/\sqrt{n}$ whenever $k \geq K_j$ and we do this for each $j = 1, 2, \ldots, n$. Let K be the largest of the numbers $K_1, \ldots, K_n$. Then for all $k \geq K$ we have

$$\|\vec{v}_k - \vec{v}_0\| = (\sum_{j=1}^{n} |x_{jk} - x_{j0}|^2)^{1/2} \leq (n\frac{\varepsilon^2}{n})^{1/2} = \varepsilon$$

A similar argument shows that convergence in $\mathbb{C}^n$ for the Euclidean norm is just coordinate-wise convergence also. Let us turn now to the space $(C[0, 1], \|\cdot\|_\infty)$. First recall that a sequence of functions on $[0,1]$, that after all is what vectors in $C[0, 1]$ are, can converge in at least two ways: pointwise on $[0,1]$ or uniformly over $[0,1]$. The sequence $\{f_k | k = 1, 2, \ldots\}$ converges to f_0 pointwise on $[0, 1]$ if, given $\varepsilon > 0$ and $t_0 \in [0, 1]$, we can find K such that

$$|f_0(t_0) - f_k(t_0)| < \varepsilon \tag{2}$$

whenever $k \geq K$.

In general, the integer K depends on both ε and t_0, so it will change if t_0 is changed. This same sequence is said to converge uniformly to f_0 if, given $\varepsilon > 0$, we can find K such that

$$|f_0(t) - f_k(t)| < \varepsilon \tag{3}$$

whenever $k \geq K$ for all $t \in [0,1]$.

Since we are dealing here with continuous functions on the closed, bounded interval $[0,1]$, line 3 may be written

$$\max_{0 \leq t \leq 1} |f_0(t) - f_k(t)| = \|f_0 - f_k\|_\infty < \varepsilon \tag{4}$$

for all $k \geq K$.

Hence, if $\{f_k\}$ converges to f_0 uniformly over $[0,1]$, then this sequence converges to f_0 for the sup norm. The converse is also true. If $\{f_k\}$ converges to f_0 for the sup norm, then given $\varepsilon > 0$ we can find K such that (4) is true and this clearly implies (4). Thus we may state: The sequence $\{f_k\}$ of vectors in $C[0,1]$ converges to the vector $f_0 \in C[0,1]$ for the sup norm if, and only if, the sequence of functions $\{f_k\}$ converges to the function f_0 uniformly over $[0,1]$.

The answer to the question of whether or not an integral equation has a solution in a particular function space often depends on whether or not the space is complete (Section 8). Let us turn now to a discussion of this concept.

Definition 2

Let $(E, \|\cdot\|)$ be a normed space and let $\{x_k\}$ be a sequence of points of E. We shall say that $\{x_k\}$ is a Cauchy sequence if, given $\varepsilon > 0$ there is an integer K such that $\|x_n - x_m\| < \varepsilon$ whenever both $m, n \geq K$. We shall say that the normed space $(E, \|\cdot\|)$ is a complete normed space, or that it is a Banach space, if every Cauchy sequence of points of E converges to a point of E.

Since the space $\mathbb{R}$ with the Euclidean norm, $\|x\| = \sqrt{x^2} = |x|$, is known to be complete, it follows easily from our discussion of convergence in $\mathbb{R}^n$ with the Euclidean norm that the latter is a Banach space for each fixed n. The same is, of course, true of $\mathbb{C}^n$ with the Euclidean norm. Let us show now that $(C[0,1], \|\cdot\|_\infty)$ is a Banach space.

Let $\{f_k\}$ be a sequence in $C[0,1]$ which is a Cauchy sequence for $\|\cdot\|_\infty$. Given $\varepsilon > 0$ we may choose K such that

$$|f_n(t) - f_m(t)| \leq \|f_n - f_m\|_\infty < \varepsilon \tag{5}$$

whenever both $m, n \geq K$.

Several things follow from this. First, we note that for any fixed t the sequence $\{f_k(t)\}_{k=1}^{\infty}$ is a Cauchy sequence for $\|\cdot\|$ (in $\mathbb{R}$ or in $\mathbb{C}$ depending on where our functions take their values). Thus, since either of these is compete, we see that $\lim f_k(t)$ exists for all $t \in [0,1]$ and we shall call this limit $f_0(t)$. Secondly, letting $n \to \infty$ in (5) gives us

$$|f_0(t) - f_m(t)| \leq \varepsilon \tag{6}$$

for all $t \in [0,1]$ and any $m \geq K$.

It follows that $\{f_m\}$ converges to f_0 uniformly over $[0,1]$ and so, by a well-known theorem [12], f_0 must be continuous on this interval. Finally, since uniform convergence is equivalent to convergence for $\|\cdot\|_\infty$ we see that $f_0 = \lim f_k$ for this norm.

We can, of course, also talk about infinite series in a normed space and this will be important later on. Given a normed space $(E, \|\cdot\|)$ and elements $x_1, x_2, \ldots, x_n, \ldots$ of E we consider the sequence

$$x_1, x_1 + x_2, x_1 + x_2 + x_3, \ldots, \sum_{j=1}^{n} x_j, \ldots \tag{7}$$

We shall say that the series $\sum_{j=1}^{\infty} x_j$ is convergent in E if, and only if, the sequence (7) converges to a point of E. We call (7) the sequence of partial sums of our series and the limit of (7), when there is one, is called the limit or the sum of $\sum_{j=1}^{\infty} x_j$. Recall that a series $\sum_{j=1}^{\infty} c_j$ of complex numbers is convergent if it converges absolutely; i.e., if $\sum_{j=1}^{\infty} |c_j|$ is convergent. This leads us to ask if the convergence of the series $\sum_{j=1}^{\infty} \|x_j\|$ implies that of $\sum_{j=1}^{\infty} x_j$. When $(E, \|\cdot\|)$ is a Banach space the answer is yes. Ths property actually characterizes those normed spaces which are complete. More precisely:

Theorem 1

A normed space $(E, \|\cdot\|)$ is a Banach space if, and only if, it satisfies the following condition: For any sequence $\{x_j\}$ of points of E the series $\sum_{j=1}^{\infty} x_j$ is convergent to a point of E whenever the series of real numbers $\sum_{j=1}^{\infty} \|x_j\|$ is convergent.

Proof. Assume that $(E, \|\cdot\|)$ is a Banach space, let $\{x_j\}$ be a sequence of points of E, and assume that $\sum_{j=1}^{\infty} \|x_j\|$ is convergent. For any two positive integers m, n, with $m > n$, we have

$$\|\sum_{j=1}^{m} x_j - \sum_{j=1}^{n} x_j\| \leq \sum_{j=n+1}^{m} \|x_j\|$$

Since this last sum tends to zero as m and n tend to infinity, $\{\sum_{j=1}^{n} x_j | n = 1, 2, \ldots\}$ is a Cauchy sequence in the Banach space $(E, \|\cdot\|)$ and hence the series $\sum_{j=1}^{\infty} x_j$ is convergent.

Now assume that $(E, \|\cdot\|)$ satisfies our condition. Let $\{y_n\}$ be a Cauchy sequence in this space. We may choose a subsequence $\{z_n\}$ of $\{y_n\}$ such that $\|z_i - z_j\| < 2^{-i}$ for all $j \geq i$. Let $x_1 = z_1$ and, for $i > 1$, let $x_i = z_i - z_{i-1}$. Clearly $\sum_{i=1}^{n} x_i = z_n$ and $\sum_{i=1}^{n} \|x_i\| \leq \|z_1\| + 1$ for all n. It follows that $\lim \sum_{i=1}^{n} x_i = \lim z_n$ exists in E. But since $\{z_n\}$ is a subsequence of $\{y_n\}$ and the latter is a Cauchy sequence, $\{y_n\}$ must converge to a point in E.

Exercises 3

* 1. Let $(E, \|\cdot\|)$ be a normed space and let $\{x_k\}$ be a sequence of points of E which converges to the point $x_0 \in E$. Show that $\lim \|x_k\| = \|x_0\|$.

* 2. Let E be a vector space and let $\|\cdot\|_1$, $\|\cdot\|_2$ be two norms on E. We shall say that $\|\cdot\|_1$ is weaker than $\|\cdot\|_2$, and we shall write $\|\cdot\|_1 \leq \|\cdot\|_2$, if there is a positive number λ such that $\lambda \mathcal{B}_1(\|\cdot\|_2) \subseteq \mathcal{B}_1(\|\cdot\|_1)$ (Exercises 2, problem 1c). We shall discuss this more systematically in Chapter 2 (Section 2).

    a. Show that $\|\cdot\|_1 \leq \|\cdot\|_2$ if, and only if, there is a positive number λ such that $\lambda \|x\|_1 \leq \|x\|_2$ for all x in E. Hint: This is clearly true if x is the zero vector, so we may assume that x is non-zero.

    b. Suppose that $\|\cdot\|_1 \leq \|\cdot\|_2$ and that the sequence $\{x_k\} \subseteq E$ converges for $\|\cdot\|_2$ to $x_0 \in E$. Show that this sequences converges to x_0 for $\|\cdot\|_1$ also.

    c. We shall say that $\|\cdot\|_1$ and $\|\cdot\|_2$ are equivalent, and we shall write $\|\cdot\|_1 \equiv \|\cdot\|_2$, if $\|\cdot\|_1 \leq \|\cdot\|_2$ and $\|\cdot\|_2 \leq \|\cdot\|_1$. Show that $\|\cdot\|_1 \equiv \|\cdot\|_2$ if, and only if, there are positive numbers λ, μ such that $\lambda \|x\|_1 \leq \|x\|_2 \leq \mu \|x\|_1$, for all x in E.

    d. If $\|\cdot\|_1$ and $\|\cdot\|_2$ are equivalent norms on E, show that $(E, \|\cdot\|_1)$ and $(E, \|\cdot\|_2)$ have the same convergent sequences. More precisely, show that the sequence $\{x_k\} \subseteq E$ converges to $x_0 \in E$ for $\|\cdot\|_1$ if, and only if, it converges to x_0 for $\|\cdot\|_2$.

    e. We refer to Exercise 2, problem 1. Show that the norms $\|\cdot\|_S$ and $\|\cdot\|_M$ defined on $\mathbb{R}^2$ are each equivalent to the Euclidean norm on this space.

    f. Suppose that we have three norms, $\|\cdot\|$, $\|\cdot\|_1$, $\|\cdot\|_2$, on a vector space E and that $\|\cdot\|_1 \equiv \|\cdot\|$ and $\|\cdot\|_2 \equiv \|\cdot\|$. Show that $\|\cdot\|_1 \equiv \|\cdot\|_2$.

* 3. Let $(E, \|\cdot\|)$ be a normed space and let $\{y_n\}$ be a Cauchy sequence of points of E. Suppose that $\{y_n\}$ has a subsequence $\{y_{k(n)}\}_{n=1}^\infty$ which is convergent to a point $y_0 \in E$. Show that $\{y_n\}$ also converges to y_0.

1.4 Inner Products

We turn now to the other property which we intuitively associate with vectors. We mean, of course, direction. Again we shall begin the discussion informally by recalling the space $\mathbb{R}^3$ and some elementary physics. Recall that the dot or inner product of two vectors arises often in mechanics and elsewhere. Given $\vec{v}_1 = (x_1, y_1, z_1)$ and $\vec{v}_2 = (x_2, y_2, z_2)$ we define their dot product, denoted $\vec{v}_1 \cdot \vec{v}_2$ or $\langle \vec{v}_1, \vec{v}_2 \rangle$, to be

$$\langle \vec{v}_1, \vec{v}_2 \rangle = \|\vec{v}_1\| \|\vec{v}_2\| \cos \theta \tag{1}$$

where θ is the angle between these vectors in the plane they determine (if the vectors do not determine a plane then the angle between them is either zero or π depending on whether they have the same or opposite directions). An immediate consequence of (1) is the useful formula

$$\langle \vec{v}, \vec{v} \rangle = \|\vec{v}\|^2 \tag{2}$$

for any vector $\vec{v}$.

If we want to define the inner product of two vectors in $\mathbb{R}^n$ and we try to use (1) to do so, then we must consider the meaning of the angle θ in this, higher dimensional, space. We prefer to handle these matters in a different way. Returning to $\mathbb{R}^3$ and the vectors $\vec{v}_1, \vec{v}_2$ above we may apply the law of cosines to the triangle whose sides are $\vec{v}_1, \vec{v}_2$ and $(\vec{v}_2 - \vec{v}_1)$. We have:

$$\begin{aligned} \|\vec{v}_2 - \vec{v}_1\|^2 &= \|\vec{v}_1\|^2 + \|\vec{v}_2\|^2 - 2\|\vec{v}_1\|\|\vec{v}_2\| \cos \theta \\ &= \|\vec{v}_1\|^2 + \|\vec{v}_2\|^2 - 2\langle \vec{v}_1, \vec{v}_2 \rangle \end{aligned} \tag{3}$$

We find, after some algebra, that

$$\langle \vec{v}_1, \vec{v}_2 \rangle = x_1 x_2 + y_1 y_2 + z_1 z_2 \tag{4}$$

and this formula does not contain θ. Hence for any two vectors $\vec{v}_1 = (x_1, \ldots, x_n)$ and $\vec{v}_2 = (y_1, \ldots, y_n)$ in $\mathbb{R}^n$ we define

$$\langle \vec{v}_1, \vec{v}_2 \rangle = \sum_{j=1}^n x_j y_j \tag{5}$$

Note that with this as our definition of dot product equation 2 is valid for any vector in $\mathbb{R}^n$. At this point we could, if we wanted, define angles in $\mathbb{R}^n$ and hence give our vectors direction. What we might do is, provided our vectors are non-zero, combine (5) and (1) to define θ as follows:

$$\theta = \arccos\left[\frac{\langle \vec{v}_1, \vec{v}_2 \rangle}{\|\vec{v}_1\|\|\vec{v}_2\|}\right] \tag{6}$$

But observe that, if (6) is to have any meaning, we must know that the quantity in brackets is between minus one and plus one, i.e., for any two vectors $\vec{v}_1, \vec{v}_2$ in $\mathbb{R}^n$ we must be able to show that

$$|\langle \vec{v}_1, \vec{v}_2 \rangle| \leq \|\vec{v}_1\| \cdot \|\vec{v}_2\| \tag{7}$$

This important inequality is correct as we shall see soon. Hence our scheme for introducing angles into $\mathbb{R}^n$ can be carried out as outlined above. We shall not pursue this however since, as far as the author knows, it is rarely useful to do so. The important thing about an inner product is not that it enables us to define angle, but that it enables us to define "right angle."

The situation in $\mathbb{C}^n$ is slightly more complicated. If we are to have equation (2) valid here (recall that is the relation $\langle \vec{v}, \vec{v} \rangle = \|\vec{v}\|^2$) then we must modify (5). For any $\vec{v}_1 = (z_1, z_2, \ldots, z_n)$ and $\vec{v}_2 = (w_1, w_2, \ldots, w_n)$ in $\mathbb{C}^n$ we define $\langle \vec{v}_1, \vec{v}_2 \rangle$ by the formula

$$\langle \vec{v}_1, \vec{v}_2 \rangle = \sum_{j=1}^n z_j \bar{w}_j \tag{8}$$

where the bar over the w_j denotes complex conjugation. Observe that equation (6) could be used to define angle in this space but it is not very intuitively satisfying because the quantity in brackets (when $\vec{v}_1, \vec{v}_2$ are in $\mathbb{C}^n$) is, in general, a complex number. However, as we mentioned above, this is no great loss.

The entire discussion just given can be regarded as motivation for the following:

Definition 1

Let V be a vector space over the field $\mathcal{K}$. A function σ from $V \times V$ into $\mathcal{K}$ will be called an inner product on V if it has the three following properties:

    i. $\sigma(u,v) = \overline{\sigma(v,u)}$ for all u, v in V;

    ii. $\sigma(\alpha u + \beta v, w) = \alpha \sigma(u,w) + \beta \sigma(v,w)$ for all u, v, w in V and all α, β in $\mathcal{K}$;

    iii. $\sigma(u,v) \geq 0$ for all v in V with equality holding if, and only if, v is the zero vector.

Given an inner product σ on V, it is customary to write $\langle u,v\rangle$ in place of $\sigma(u,v)$ for all u, v in V. We call $\langle u,v\rangle$ the inner product of the vector u and v, and we call the pair $(V,\langle\rangle)$ an inner product space. It is very easy to check that the functions defined in (5) and (8) are inner products on the spaces $\mathbb{R}^n$ and $\mathbb{C}^n$ respectively. We observed above that the inner products gave us the Euclidean norms on these spaces via the relation $\langle \vec{v},\vec{v}\rangle = \|\vec{v}\|^2$. It is in fact true that an inner product always gives us a norm in this way. More explicitly: Let $(V,\langle\rangle)$ be an inner product space and for each $v \in V$ define $p(v)$ to be the number $(\langle v,v\rangle)^{1/2}$. Then the function p is a norm on V. Let us sketch the proof of this fact.

We must show that (a) $p(u+v) \leq p(u)+p(v)$ for all u, v in V; (b) $p(\alpha v) = |\alpha|p(v)$ for all v in V and all α in $\mathcal{K}$; (c) $p(v) \geq 0$ with equality if, and only if, v is the zero vector (Section 2, Definition 1). The last of these, (c), is immediate from part (iii) of our definition of inner product. To prove (b) let $v \in V$, $\alpha \in \mathcal{K}$ then:

$$\begin{aligned} p(\alpha v) &= (\langle \alpha v, \alpha v\rangle)^{1/2} && \text{by definition;} \\ &= (\alpha\langle v, \alpha v\rangle)^{1/2} && \text{by (ii);} \\ &= (\alpha, \overline{\langle \alpha v, v\rangle})^{1/2} && \text{by (i);} \\ &= (\alpha\overline{\alpha}\langle v,v\rangle)^{1/2} && \text{because by (iii) } \langle v,v\rangle \text{ is a real number;} \\ &= (|\alpha|^2\langle v,v\rangle)^{1/2} = |\alpha|p(v) \end{aligned}$$

We have only to prove (a). In order to do that, we shall first prove another important inequality from which (a) can be deduced. Various special cases of this result were obtained by Cauchy, by Schwarz, and by Bunyakowski, and we shall refer to it as the C.S.B. inequality.

Theorem 1

Let $(V,\langle\rangle)$ be an inner product space. Then for any two vectors u, v in V we have: $|\langle u,v\rangle|^2 \leq \langle u,u\rangle\langle v,v\rangle$.

Proof. If we happen to have $\langle u,v\rangle = 0$ then the result is immediate from part (iii) of Definition 1. So let us suppose that $\langle u,v\rangle \neq 0$. We leave it to the reader to show (see problem 1) that this implies that neither u nor v is the zero vector. Let us set $\alpha = \langle u,v\rangle/\langle v,v\rangle$. Then:

$$\begin{aligned} \frac{|\langle u,v\rangle|^2}{\langle v,v\rangle} &= \alpha\overline{\langle u,v\rangle} = \alpha\langle v,u\rangle \text{ by (i);} && (9) \\ &= \overline{\alpha}\langle u,v\rangle \text{ because the first quantity in (9) is real;} \\ &= \overline{\alpha}\alpha\langle v,v\rangle \text{ from the definition of } \alpha; \\ &= |\alpha|^2\langle v,v\rangle \end{aligned}$$

Now we also have

$$0 \leq \langle u - \alpha v, u - \alpha v \rangle = \langle u, u \rangle - \alpha \langle v, u \rangle - \overline{\alpha}\langle u, v \rangle + |\alpha|^2 \langle v, v \rangle \quad (10)$$
$$= \langle u, u \rangle - \frac{|\langle u, v \rangle|^2}{\langle v, v \rangle} \text{ by (9)}$$

Our inequality follows from this.

Let us return now to the proof that the function $p(v) = (\langle v, v \rangle)^{1/2}$ is a norm. We must still prove the triangle inequality and we shall use the C.S.B. inequality to do this. We have, for any u, v in V,

$$\begin{aligned}
p(u+v)^2 &= \langle u+v, u+v \rangle \text{ by definition;} \\
&= \langle v, u+v \rangle + \langle u, u+v \rangle \text{ by (ii);} \\
&= \overline{\langle u+v, v \rangle} + \overline{\langle u+v, u \rangle} \text{ by (i);} \\
&= \langle u, u \rangle + \overline{\langle u, v \rangle} + \overline{\langle v, u \rangle} + \langle v, v \rangle \\
&= p(u)^2 + \overline{\langle u, v \rangle} + \overline{\langle v, u \rangle} + p(v)^2 \\
&\leq p(u)^2 + 2|\langle u, v \rangle| + p(v)^2 \text{ (because the middle terms add up} \\
&\qquad \text{to twice the real-part of } \langle u, v \rangle); \\
&\leq p(u)^2 + 2p(u)p(v) + p(w)^2 \text{ (by the C.S.B. inequality);} \\
&= (p(u) + p(v))^2
\end{aligned}$$

and this is the triangle inequality for p.

We have now shown that on any inner product space $(V, \langle \rangle)$ we have the norm $\|\cdot\| = (\langle , \rangle)^{1/2}$. We call this the norm associated with, or defined by, the inner product. Whenever we talk about a norm on an inner product space we shall mean, unless the contrary is explicitly stated, the norm associated with the inner product. With this understanding we can write the C.S.B. inequality, for any inner product space $(V, \langle \rangle)$ as:

(∗) $|\langle u, v \rangle| \leq \|u\|\|v\|$ for all u, v in V.

Note that the norm associated with the inner product on $\mathbf{R}^n$ (equation (5)) is just the Euclidean norm, and the same is true of the norm associated with the inner product on $\mathbf{C}^n$ (equation (8)). Thus we have finally shown that these functions really are norms on $\mathbf{R}^n$ and $\mathbf{C}^n$ (recall that we never did prove the triangle inequality for these functions until now).

Finally we should point out that there are norms which are not associated with an inner product. One can show that the sup norm on $C[0,1]$ (Section 2, equation (4)) is such a norm.

The fact that any inner product space has a norm means that we can talk about convergent sequences and series in such a space (see Section 3). In particular, we may ask whether or not an inner product space is complete

18 CHAPTER 1 LINEAR ALGEBRA AND ANALYSIS

(Section 3, Definition 2) for the norm associated with the inner product. Those that are, are called Hilbert spaces. More formally:

Definition 2

Let $(H, \langle\rangle)$ be an inner product space and let $\|\cdot\|$ be its associated norm. If $(H, \|\cdot\|)$ is a complete normed space, then we shall say that $(H, \langle\rangle)$ is a Hilbert space.

The inner product spaces $(\mathbb{R}^n, \langle\rangle)$ and $(\mathbb{C}^n, \langle\rangle)$ defined above are Hilbert spaces and, at the moment, these are the only examples that we have. Others will be given later. Incidentally, it is said that David Hilbert once asked a colleague what a Hilbert space was and when he was told he remarked "Oh, is that all."

Exercises 4

* 1. Let $(V, \langle\rangle)$ be an inner product space and let u, v, w be elements of V.

    a. Show that $\langle u, 0\rangle = \langle 0, u\rangle = 0$. Hint: $u + 0 = 0$.

    b. Show that $\langle u, \alpha v + \beta v\rangle = \overline{\alpha}\langle u, v\rangle + \overline{\beta}\langle u, w\rangle$ for any scalars α, β.

    c. Derive the "parallelogram law" $\|u+v\|^2 + \|u-v\|^2 = 2\left(\|u\|^2 + \|v\|^2\right)$.

    d. Prove that $\langle u, v\rangle = \frac{1}{4}[\|u+v\|^2 - \|u-v\|^2 + i\|u+iv\|^2 - i\|u-iv\|^2]$ where $i = \sqrt{-1}$. This is called the polar identity.

    e. Prove the Pythagorean theorem in V; i.e., if $\langle u, v\rangle = 0$ then $\|u+v\|^2 = \|u\|^2 + \|v\|^2$.

    f. If $\{v_n\}$ is a sequence in V which converges to the vector v, show that $\lim_{n\to\infty}\langle u, v_n\rangle = \langle u, v\rangle$.

    g. If $\{u_n\}, \{v_n\}$ are sequences in V which converge to u, v respectively, show that $\lim_{n\to\infty}\langle u_n, v_n\rangle = \langle u, v\rangle$.

* 2. Let $a < b$ be two fixed, real numbers and let $C[a, b]$ be the vector space of all continuous functions on $[a, b]$. For any two vectors f, g in this space define

    $$\langle f, g\rangle = \int_a^b f(t)\overline{g(t)}dt$$

    a. Show that $\langle, \rangle$ is an inner product on $C[a, b]$.

b. Let $a = -\pi$, $b = \pi$ and let n be an integer. Recall that $\exp(int) = \cos nt + i\sin nt$. Compute $\|\exp(int)\|$ and, for any integer m different from n, compute $\langle \exp(int), \exp(imt)\rangle$.

c. Show that the norm defined on $C[-\pi, \pi]$ by this inner product is not the sup norm. Hint: Choose some vector in this space and compute its length for each of these norms.

3. Let $(V, \langle,\rangle)$ be an inner product space and let u, v be elements of V.

   a. Show that $\langle u, v\rangle = \|u\|\|v\|$ if, and only if, u and v ar linearly independent.

   b. Show that $\|u + v\| = \|u\| + \|v\|$ if, and only if, $u = av$ for some non-negative scalar a.

1.5 An Infinite Dimensional Hilbert Space

We shall present here the space ℓ^2. It is simple enough so that one can "see into" its structure and this may give the impression that the space is a kind of curiosity, of little practical importance. We shall see below that this is not the case.

To begin, ℓ^2 is the set of all sequences $\{z_j | j = 1, 2, \ldots\}$ of complex numbers such that $\sum_{j=1}^{\infty} |z_j|^2 < \infty$. It is not clear that this is a vector space under the coordinate-wise operations, but one thing is clear. If $\vec{z} = \{z_j\}$ and $\vec{w} = \{w_j\}$ are in ℓ^2 then for any fixed n, $\{z_j\}_{j=1}^{n}$ and $\{w_j\}_{j=1}^{n}$ are in $\mathbb{C}^n$, hence by the C.S.B. inequality (Section 4, Equation (*))

$$|\sum_{j=1}^{n} z_j \overline{w_j}| \leq (\sum_{j=1}^{n} |z_j|^2)^{1/2} (\sum_{j=1}^{n} |w_j|^2)^{1/2} \tag{1}$$

Let us digress briefly here to make a useful observation. Suppose that each z_j and each w_j happens to be a real number. Let $sgn(z_j w_j) = 1$ if the product is positive, minus one if it is negative, and zero if $z_j w_j = 0$. Then $z' = \{sgn(z_j w_j)z_j\}$ is still in ℓ^2 because $|sgn(z_j w_j)z_j| = |z_j|$, and

$$\sum_{j=1}^{n} sgn(z_j w_j)z_j w_j = \sum_{j=1}^{n} |z_j w_j| \tag{2}$$

In the case of complex z_j, w_j we can let α_j be, for each fixed j, a complex number with modulus one such that $\alpha_j z_j \overline{w_j} = |z_j \overline{w_j}|$; just set α_j equal to

$\overline{z_j \overline{w}_j}/|z_j \overline{w}_j|$. Again we note that $\{\alpha_j z_j\}$ is still in ℓ^2 and

$$\sum_{j=1}^n \alpha_j z_j \overline{w}_j = \sum_{j=1}^n |z_j \overline{w}_j| \tag{3}$$

The point of this digression is that for any $\vec{z} = \{z_j\}$ and any $\vec{w} = \{w_j\}$ in ℓ^2 we have, for each fixed n, not only inequality (1) but also the stronger inequality

$$\sum_{j=1}^n |z_j \overline{w}_j| \le (\sum_{j=1}^n |z_j|^2)^{1/2} (\sum_{j=1}^n |w_j|^2)^{1/2} \tag{4}$$

Note that the right-hand side of (4) converges as $n \to \infty$. Thus we may state (we really do not even need the bar over the w_j):

$$\text{For any } \{z_j\}, \{w_j\} \text{ in } \ell^2, \sum_{j=1}^\infty |z_j \overline{w}_j| \text{ is convergent.} \tag{5}$$

Using (5) it is easy to see that ℓ^2 is a vector space under the coordinate-wise operations. For any $\vec{z} = \{z_j\}$, and any $\vec{w} = \{w_j\}$ in ℓ^2 we may write, with n fixed,

$$\sum_{j=1}^n |z_j + w_j|^2 \le \sum_{j=1}^n (|z_j| + |w_j|)^2 = \sum_{j=1}^n |z_j|^2 + 2\sum_{j=1}^n |z_j w_j| + \sum_{j=1}^n |w_j|^2 \tag{6}$$

The first and last sums on the right-hand side of (6) are convergent because $\vec{z}$ and $\vec{w}$ are in ℓ^2. Also, again since $\vec{z}$ and $\vec{w}$ are in ℓ^2 (5) tells us that the middle sum on the right-hand side of (6) is convergent. Thus $z + w = \{z_j + w_j\}$ is in ℓ^2, and clearly $\lambda \vec{z} = \{\lambda z_j\}$ is in ℓ^2 for any fixed scalar λ.

We may define an inner product on the vector space ℓ^2 as follows; again using (5) as justification,

Definition 1

For any $\vec{z} = \{z_j\}$, $\vec{w} = \{w_j\}$ in ℓ^2 we let

$$\langle \vec{z}, \vec{w} \rangle = \sum_{j=1}^\infty z_j \overline{w}_j$$

As we mentioned above, we may justify this definition by using (5) since the latter implies the convergence of $\sum z_j \overline{w}_j$ for any $\{z_j\}$ $\{w_j\}$ in ℓ^2. It is very easy to check that $\langle , \rangle$ is an inner product on ℓ^2. Observe that for any $\vec{z} = \{z_j\}$, our inner product gives us

$$\|\vec{z}\| = (\sum_{j=1}^\infty |z_j|^2)^{1/2} \tag{7}$$

and that, if we define the distance from $\vec{z}$ to $\vec{w}$ to be $\|\vec{z} - \vec{w}\|$ then

$$\|\vec{z} - \vec{w}\| = (\sum_{j=1}^{\infty} |z_j - w_j|^2)^{1/2} \tag{8}$$

which is the natural generalization of the distance formula from analytic geometry.

Let us now show that $(\ell^2, \langle,\rangle)$ is a Hilbert space; i.e., we shall show that $(\ell^2, \|\cdot\|)$ is a complete normed space (Section 4, Definition 2). Unfortunately, this is a little bit messy because each vector in ℓ^2 is a sequence and so a sequence of vectors will be a sequence of sequences.

Suppose that $\{\vec{z}_n | n = 1, 2, \ldots\}$ is a Cauchy sequence in ℓ^2. Then for each fixed n we have

$$\vec{z}_n = \{z_{n1}, z_{n2}, z_{n3}, \ldots\} \tag{9}$$

Now given $\varepsilon > 0$ there is an integer N such that for $m, n \geq N$

$$\|\vec{z}_n - \vec{z}_m\| < \varepsilon \tag{10}$$

Referring to (8) and (9) we may rewrite (10) as

$$\sum_{j=1}^{\infty} |z_{nj} - z_{mj}|^2 < \varepsilon^2 \text{ for } m, n \geq N \tag{11}$$

It follows from this that, for each fixed j, the sequence $\{z_{nj} | n = 1, 2, \ldots\}$ is a Cauchy sequence of complex numbers; for

$$|z_{nj} - z_{mj}|^2 \leq \sum_{j=1}^{\infty} |z_{nj} - z_{mj}|^2 \leq \varepsilon \tag{12}$$

for all $m, n \geq N$.

Let us set $z_{0j} = \lim_{n \to \infty} z_{nj}$ for each $j = 1, 2, 3, \ldots$ and consider $\vec{z}_{0j} = \{z_{0j} | j = 1, 2, \ldots\}$. We shall show that $\vec{z}_{0j} \in \ell^2$ and that the given sequence $\{\vec{z}_n\}$ converges to this vector for the ℓ^2-norm.

Choose $n > N$ and write, for any fixed integer k,

$$\begin{aligned}(\sum_{j=1}^{k} |z_{0j}|^2)^{1/2} &\leq (\sum_{j=1}^{k} |z_{0j} - z_{nj}|^2)^{1/2} + (\sum_{j=1}^{k} |z_{nj}|^2)^{1/2} \\ &\leq \varepsilon + \|\vec{z}_n\|\end{aligned} \tag{13}$$

Here we have used the fact that $(z_{01}, z_{02}, \ldots, z_{0k})$ and $(z_{n1}, z_{n2}, \ldots, z_{nk})$ are in $\mathbb{C}^k$ hence we may apply the triangle inequality in $\mathbb{C}^k$ to write the first inequality in (13). The second line follows from (11) and the fact that the norm of $\vec{z}_n$ in ℓ^2 must exceed the norm of $(z_{n1}, \ldots, z_{nk})$ in $\mathbb{C}^k$. Now by suppressing the middle terms in (13) we have

$$\left(\sum_{j=1}^{k} |z_{0j}|^2\right) \leq \varepsilon + \|\vec{z}_n\| \tag{14}$$

Here n is fixed and k is arbitrary. Letting $k \to \infty$ we see that $\vec{z}_n$ is in ℓ^2 as claimed.

Finally, let us return to (11). Recall:

$$\sum_{j=1}^{\infty} |z_{nj} - z_{mj}|^2 < \varepsilon^2 \text{ for any } m, n \geq N$$

We may take the limit as $m \to \infty$ to get

$$\sum_{j=1}^{\infty} |z_{nj} - z_{0j}|^2 \leq \varepsilon^2 \text{ for all } n \geq N \tag{15}$$

Clearly this just says

$$\|\vec{z}_n - \vec{z}_0\| \leq \varepsilon \text{ for all } n \geq N, \tag{16}$$

and this means $\{\vec{z}_n\}$ converges to $\vec{z}_0$ in ℓ^2.

The reader will have noted that we could carry out the above discussion for the case of "real" ℓ^2; i.e., the set of all sequences of real numbers $\{x_j\}$ such that $\sum |x_j|^2 < \infty$. As just indicated we shall, when necessary, call this real ℓ^2 and refer to the space defined earlier as complex ℓ^2.

Exercises 5

* 1. For each positive integer n let $\vec{e}_n$ be the sequence $\{\delta_{nj} | j = 1, 2, \ldots\}$ where $\delta_{nj} = 0$ if $n \neq j$ and $\delta_{nn} = 1$. Clearly each $\vec{e}_n$ is a unit vector (i.e., a vector with norm one) in ℓ^2.

    a. Show that $\langle \vec{e}_n, \vec{e}_m \rangle = 0$ if $n \neq m$ and 1 if $n = m$.

    b. Compute $\|\vec{e}_n - \vec{e}_m\|$ for all m, n.

    c. Show that $lin\{e_j | j = 1, 2, \ldots\} \notin \ell^2$.

  2. Let us define two mappings from ℓ^2 to ℓ^2 as follows:

    * i. $A_r(\vec{z}) = A_r(\{z_1, z_2, z_3, \ldots, z_n, \ldots\}) = \{0, z_1, z_2, z_3, \ldots\}$

ii. $A_\ell(\vec{z}) = A_\ell(\{z_1, z_2, z_3, \ldots\}) = \{z_2, z_3, z_4, \ldots\}$

Not surprisingly A_r is called the right-shift operator, and A_ℓ is called the left-shift operator, on ℓ^2.

a. Show that A_r and A_ℓ are linear operators on ℓ^2 (Section 1, Definition 5).

b. Show that the composition $A_\ell \circ A_r$, recall $A_\ell \circ A_r(\vec{z}) = A_\ell[A_r(\vec{z})]$ is equal to the identity operator but that $A_r \circ A_\ell$ is not (the identity operator I on any space V, say, is defined by $Iv = v$ for all $v \in V$).

1.6 Orthogonal and Orthonormal Sets

We begin our discussion by taking a closer look at the space $\mathbb{R}^n$ and its Hamel basis $\{\vec{e}_j | j = 1, 2, \ldots, n\}$ mentioned before (Exercises 2, problem 2). Suppose that a vector $\vec{v} \in \mathbb{R}^n$ is given and that we wish to find scalars $\alpha_1, \ldots, \alpha_n$ such that

$$\vec{v} = \sum_{j=1}^{n} \alpha_j \vec{e}_j \tag{1}$$

Since $\langle \vec{e}_j, \vec{e}_k \rangle$ is zero when $j \neq k$, and is equal to one when $j = k$, we can find α_k, for any fixed k, as follows:

$$\langle \vec{v}, \vec{e}_k \rangle = \langle \sum_{j=1}^{n} \alpha_j \vec{e}_j, \vec{e}_k \rangle = \sum_{j=1}^{n} \alpha_j \langle \vec{e}_j, \vec{e}_k \rangle = \alpha_k \tag{2}$$

This calculation not only shows us how to find the α_k's but it shows us that they are unique. So each $\vec{v} \in \mathbb{R}^n$ can be written as in (1) in exactly one way. Let us notice one more consequence of the fact that $\langle \vec{e}_j, \vec{e}_k \rangle$ is zero, except when $j = k$ when the inner product is one. If $\vec{u}, \vec{v}$ ar two vectors in $\mathbb{R}^n$ and if

$$\vec{u} = \sum_{j=1}^{n} \alpha_j \vec{e}_j, \quad \vec{v} = \sum_{j=1}^{n} \beta_j \vec{e}_j \tag{3}$$

then

$$\begin{aligned} \langle \vec{u}, \vec{v} \rangle &= \langle \sum \alpha_j \vec{e}_j, \sum \beta_k \vec{e}_k \rangle = \sum_j \alpha_j \langle \vec{e}_j, \sum_k \beta_k \vec{e}_k \rangle \\ &= \sum_j \alpha_j \{ \sum_k \beta_k \langle \vec{e}_j, \vec{\ell}_k \rangle \} \\ &= \sum_j \alpha_j \beta_j \end{aligned} \tag{4}$$

So we can compute the inner product of two vectors from a knowledge of their representations in the basis $\{\vec{e}_j\}$. Similar remarks apply to the space $\mathbf{C}^n$ and its basis $\{\vec{e}_j | j = 1, \ldots n\}$. In this case, however, the β's in (4) would have to be conjugated.

Let us give names to the properties of the set $\{\vec{e}_j\}$ which make all this happen.

Definition 1

Let $(V, \langle,\rangle)$ be an inner product space. Two vectors u, v in V are said to be orthogonal if $\langle u, v \rangle = 0$. A non-empty subset S of V is called an orthogonal set if any two distinct vectors in S are orthogonal. Finally, an orthogonal set S is called an orthonormal set if every vector in S has norm one.

The basis $\{\vec{e}_j | j = 1, \ldots, n\}$ is also an orthonormal subset of $\mathbf{R}^n$ and, as we have seen, this makes it convenient to work with. Let us observe another remarkable fact: Suppose that $(V, \langle,\rangle)$ is an inner product space over $\mathbf{R}$, that V has dimension n, and that $v_1, \ldots, v_n$ is a basis for V which is also an orthonormal subset of V; so $\langle v_j, v_k \rangle = 0$ when $j \neq k$ and is one when $j = k$. Define a map T from V onto $\mathbf{R}^n$ as follows: First, for any j, $1 \leq j \leq n$, set $T(v_j)$ equal to $\vec{e}_j \in \mathbf{R}^n$. Next extend T to all of V by linearity; i.e., given $v \in V$ write

$$v = \sum_{j=1}^{n} \alpha_j v_j \tag{5}$$

(this can be done in one and only one way) and set

$$T(v) = \sum \alpha_j T(v_j) = \sum \alpha_j \vec{e}_j \in \mathbf{R}^n \tag{6}$$

This gives us an isomorphism from V onto $\mathbf{R}^n$ (Exercises 1, problem 9). But now, since our bases $\{v_j\}$ and $\{\vec{e}_j\}$ are orthonormal sets, T has an addition property. For all u, v in V

$$\langle u, v \rangle = \langle T(u), T(v) \rangle \tag{7}$$

Let us prove this. First write

$$u = \sum \alpha_j v_j, \quad v = \sum \beta_j v_j \tag{8}$$

then

$$\begin{aligned}
\langle u, v \rangle &= \langle \sum \alpha_j v_j, \sum \beta_k v_k \rangle = \sum \alpha_j \langle v_j, \sum \beta_k v_k \rangle \\
&= \sum \alpha_j \{ \sum \beta_k \langle v_j, v_k \rangle \} = \sum \alpha_j \beta_j
\end{aligned} \tag{9}$$

and

$$\begin{aligned}\langle T(u),T(v)\rangle &= \langle T(\sum \alpha_j v_j), T(\sum \beta_j v_j)\rangle \quad (10)\\ &= \langle \sum \alpha_j \vec{e}_j, \sum \beta_j \vec{e}_j\rangle\\ &= \sum \alpha_j \langle \vec{e}_j, \sum \beta_k \vec{e}_k\rangle\\ &= \sum \alpha_j \{\sum \beta_k \langle \vec{e}_j, \vec{e}_k\rangle\} = \sum \alpha_j \beta_j\end{aligned}$$

In both (9) and (10) we used orthonormality. In (9) that of the set $\{v_j\}$ and in (10) that of the set $\{\vec{e}_j\}$.

This simple calculation proves a rather remarkable theorem. However, before we can say just what it is that we have proved we must introduce some terminology.

Definition 2

Let $(V, \langle,\rangle_1)$ and $(V, \langle,\rangle_2)$ be two inner product spaces over the same field. A linear map T from V_1 into V_2 is called an isometry if

$$\langle u, v\rangle_1 = \langle T(u), T(v)\rangle_2$$

for all vectors u, v in V_1. An isomorphism from V_1 onto V_2 which is also an isometry will be called an isometric isomorphism. Finally, we shall say that two given inner product spaces over the same field are isometrically isomorphic if there is an isometric isomorphism from one of these spaces onto the other.

The argument given in the paragraph just before Definition 2 shows that any inner product space over $\mathbb{R}$ that has an orthonormal basis (i.e., an orthonormal set which is a Hamel basis) containing n elements is isometrically isomorphic to $\mathbb{R}^n$. A similar argument shows that if the space is defined over $\mathbb{C}$, then it is isometrically isomorphic to $\mathbb{C}^n$. Now which inner product spaces have such a basis? To answer this question, we need only recall the Gram-Schmidt process familiar from linear algebra. It is useful to state the result in some generality:

Lemma 1

Let $(V, \langle,\rangle)$ be an inner product space and let $\{u_j | j = 1, 2, \ldots, k\}$ be any finite, linearly independent subset of V. Then there is an orthonormal set $\{v_j | j = 1, 2, \ldots, k\}$ in V such that $lin\{u_j | j = 1, \ldots, k\} = lin\{v_j | j = 1, 2, \ldots, k\}$.

Proof. The notation lin was defined above (Section 1, Definition 2). We shall simply sketch the argument and leave the details to the reader. Since

$\{u_j\}$ is linearly independent it cannot contain the zero vector. Thus we may define v_1 as follows:

$$v_1 = u_1/\|u_1\| \tag{11}$$

Observe that $lin\{u_1\} = lin\{v_1\}$. Next define v_2 in two steps. First let:

$$w_2 = u_2 - \langle u_2, v_1\rangle v_1 \tag{12}$$

so that $\langle v_1, w_2\rangle = 0$. Next set

$$v_2 = w_2/\|w_2\| \tag{13}$$

From (12) and (13) we see that $u_2 \in lin\{v_1, v_2\}$. We already have $u_1 \in lin\{v_1, v_2\}$. Also, $v_1 \in lin\{u_1, u_2\}$ and, by (12) and (13), v_2 is in here as well. Hence

$$lin\{v_1, v_2\} = lin\{u_1, u_2\} \tag{14}$$

Continue. To define v_3 first let

$$w_3 = u_3 - \langle u_3, v_1\rangle v_1 - \langle u_3, v_2\rangle v_2 \tag{15}$$

so that $\langle w_3, v_1\rangle = \langle w_3, v_2\rangle = 0$, and then set

$$v_3 = w_2/\|w_3\| \tag{16}$$

It is clear from (15) and (16) that $u_3 \in lin\{v_1, v_2, v_2\}$ and $v_3 \in lin\{u_1, u_2, u_3\}$ so again we have

$$lin\{v_1, v_2, v_3\} = lin\{u_1, u_2, u_3\} \tag{17}$$

This process will end when we have constructed v_k.

Note that if the set $u_1, \ldots, u_k$ happens to be a Hamel basis for $(V, \langle,\rangle)$, then the same is true of the orthonormal set $v_1, \ldots, v_k$; for it spans V by construction and is easily seen to be linearly independent (problem 1). Combining Lemma 1 with our observations made earlier we have the following:

Theorem 1

Any n-dimensional inner product space over $\mathbb{R}$ (resp. $\mathbb{C}$) is isometrically isomorphic to $\mathbb{R}^n$ (resp. $\mathbb{C}^n$).

Corollary 1

Any finite dimensional inner product space over $\mathbb{R}$ or $\mathbb{C}$ is a Hilbert space (Section 4, Definition 2).

ORTHOGONAL AND ORTHONORMAL SETS

Proof. Let $(V, \langle , \rangle)$ be an inner product space over $\mathbb{R}$ having finite dimension n, say. Let T be an isometric isomorphism from V onto $\mathbb{R}^n$. We must show that V is complete for the norm given by $\langle , \rangle$ (Section 4, Definition 2). To do this we must show that any sequence $\{v_k\} \subseteq V$ which is a Cauchy sequence (Section 3, Definition 2) is convergent to a vector $v_0 \in V$. Consider the sequence $\{T(v_k)\} \subseteq \mathbb{R}^n$. Since T is an isometry we have:

$$\|T(v_p) - T(v_q)\|^2 = \langle T(v_p) - T(v_q), T(v_p) - T(v_q) \rangle = \langle v_p - v_q, v_p - v_q \rangle = \|v_p - v_q\|^2 \tag{18}$$

Thus $\{T(v_k)\}$ is a Cauchy sequence in the Hilbert space $\mathbb{R}^n$, and so it must converge to some point, say $\vec{w}$, in $\mathbb{R}^n$. But T maps V onto $\mathbb{R}^n$ so there is a unique $v_0 \in V$ for which $T(v_0) = \vec{w}$.

Finally,

$$\begin{aligned} \|v_0 - v_k\|^2 &= \langle v_0 - v_k, v_0 - v_k \rangle = \langle T(v_0 - v_k), T(v_0 - v_k) \rangle \\ &= \|T(v_0) - T(v_k)\|^2 = \|\vec{w} - T(w_k)\|^2 \end{aligned}$$

which shows that the sequence $\{v_k\}$ converges to v_0.

Let us turn now to the discussion of infinite dimensional inner product spaces. As one might expect, there are some complications here which must be dealt with. First, let us get down a few positive facts. Suppose that $(V, \langle , \rangle)$ is an inner product space and let $\{v_n | n = 1, 2, \ldots\}$ be an infinite sequence of vectors in V which is also an orthonormal set. Suppose further that, for some vector $u \in V$, there are scalars $c_1, c_2, \ldots$ such that

$$u = \sum_{j=1}^{\infty} c_j v_j \tag{19}$$

where convergence is for the norm of V. Then, as in the finite case, we can determine the c_j's in terms of the known vectors u and v_j. In fact:

$$\begin{aligned} \langle u, v_n \rangle &= \langle \sum_{j=1}^{\infty} c_j v_j, v_n \rangle = \langle \lim_{m \to \infty} \sum_{j=1}^{m} c_j v_j, v_n \rangle \tag{20} \\ &= \lim_{m \to \infty} \langle \sum_{j=1}^{m} c_j v_j, v_n \rangle \text{ (Exercises 4, problem 1g)} \\ &= c_n \end{aligned}$$

for each fixed n; showing that the coefficients in (19) are unique.

Note that we started here with the assumption that the series in (19) converged to u and that led us to (20). The fact is that if we take any vector in V and compute the sequence of numbers given by (20) we get a sequence in ℓ^2. This remarkable fact is a consequence of a useful inequality due to Bessel.

Theorem 2

Let $\{v_n | n = 1, 2, \ldots\}$ be an orthonormal sequence in the inner product space $(V, \langle,\rangle)$, let u be any vector in V and, for each fixed j, let $c_j = \langle u, v_j \rangle$. Then $\sum_{j=1}^{\infty} |c_j|^2 \leq \|u\|^2$.

Proof. It is clear that $|c_1|^2 \leq |c_1|^2 + |c_2|^2 \leq |c_1|^2 + |c_2|^2 + |c_3|^2 \leq \ldots$. In other words, the sequence

$$\{\sum_{j=1}^{m} |c_j|^2 \mid m = 1, 2, \ldots\} \tag{21}$$

is monotonic. Hence, in order to prove that the series

$$\sum_{j=1}^{\infty} |c_j|^2 \tag{22}$$

converges it suffices to show that (21) is bounded above. Now

$$0 \leq \langle u - \sum_{j=1}^{m} c_j v_j, u - \sum_{j=1}^{m} c_j v_j \rangle = \|u\|^2 - \sum_{j=1}^{m} |c_j|^2 \tag{23}$$

Here the inequality comes from the third property of inner products (Section 4, Definition 1) and the equality comes from a direct calculation using the orthonormality of the v_j's. It follows from (23) that

$$\sum_{j=1}^{m} |c_j|^2 \leq \|u\|^2 \tag{24}$$

for any fixed m and this, as we have already observed, proves the convergence of the series (22). Finally, letting $m \to \infty$ in (24) gives us the desired inequality.

Now at this point there are two questions which arise. First, does Theorem 2 show that for any $u \in V$ the series

$$\sum_{j=1}^{\infty} \langle u, v_j \rangle v_j \tag{25}$$

is convergent to a point of V? The answer is "yes" if $(V, \langle,\rangle)$ is a Hilbert space and "no" otherwise. We will prove this, in connection with some other useful facts, soon (also, see Section 8). Second, if it happens that (25) does converge must its limit be u? The answer is "no." Perhaps the easiest way to demonstrate this, although it may seem a little contrived to the reader, is to assume that (25) does converge to u and that we have a second orthonormal sequence $\{u_k\}$ in V such that $u_k = v_{k+1}$ for all k. So we are supposing that we, inadvertently, left v_1 out. Then the series

$$\sum_{j=1}^{\infty} \langle u, u_j \rangle v_j$$

would still converge, but we would not expect it to converge to u again. Somehow we must guarantee that our orthonormal set is not missing my vectors; i.e., that it is "complete." We can do this as follows:

Definition 3

Let $(V, \langle,\rangle)$ be an inner product space and let $\mathcal{S}$ be an orthonormal subset of V. We shall say that $\mathcal{S}$ is a maximal orthonormal set if there is no orthonormal subset of V which properly contains $\mathcal{S}$; i.e., if $\mathcal{T}$ is an orthonormal subset of V and if $\mathcal{S} \subset \mathcal{T}$, then $\mathcal{S} = \mathcal{T}$.

Lemma 2

Let $(V, \langle,\rangle)$ be an inner product space. Then:

    a. V contains a maximal orthonormal subset, and any two such sets can be placed in one-to-one correspondence;

    b. An orthonormal subset $\mathcal{S}$ of V is maximal if, and only if, for any $v \in V$ the condition $\langle v, w \rangle = 0$ for all $w \in \mathcal{S}$ implies that v is the zero vector.

Proof. The proof of (a) requires some sophisticated set theory (e.g., Zorn's lemma or the Hausdorff maximal principle) and so we omit it. Let us prove (b). First suppose that $\mathcal{S}$ is a maximal orthonormal set and that for some vector $v \in V$ we have $\langle v, w \rangle = 0$ for all $w \in \mathcal{S}$. If v is not the zero vector we may set $v' = v/\|v\|$ and then have a set $\mathcal{S} \cup \{v'\}$, which is both orthonormal and properly contains $\mathcal{S}$. This contradiction shows that v must be the zero vector.

Next suppose that S is an orthonormal subset of V with the property that $\langle v, w \rangle$ for all $w \in S$ implies that v is the zero vector. If S' is an orthonormal set which properly contains S then any $v \in S' \setminus S$ would satisfy $\langle v, w \rangle = 0$ all $w \in S$. This forces us to conclude that v is the zero vector which is a contradiction since, as an element of the orthonormal set S', v must have norm one.

It is easy to see that a maximal orthonormal subset of $\mathbb{R}^n$ or $\mathbb{C}^n$ must contain n vectors (problem 2b below), but there spaces do not concern us here. Here we are interested in spaces which contain infinite orthonormal sets, hence infinite, maximal orthonormal sets. Recall that infinite sets are of two types. First we have those which can be placed in one-to-one correspondence with the positive integers (i.e., those that can be "counted") and then we have those which cannot. The former are called countably, or denumerably, infinite and the latter are called uncountably infinite or, simply, uncountable. Now the most convenient spaces to work with are those Hilbert spaces whose maximal orthonormal sets are countably infinite. This includes most of the spaces which arise in applications. There is a very nice way of characterizing these spaces but, at the moment, we cannot even state what that is since we lack the concepts necessary to do so (see section 9). Instead, we shall prove a theorem which lists some of the properties of countably infinite, maximal orthonormal sets which make them so convenient to work with. The reader who has studied Fourier series will find some of these familiar and we shall see why in the next section.

Theorem 3

Let $(H, \langle, \rangle)$ be a Hilbert space and let $\{v_n | n = 1, 2, \ldots\}$ be an orthonormal sequence in H. If this sequence has any one of the four following properties, then it has all four of them.

   a. $\{v_n | n = 1, 2, \ldots\}$ is a maximal orthonormal set (i.e., the condition $\langle v, v_n \rangle = 0$ for all n holds if, and only if, v is the zero vector);

   b. For each $v \in H$ the series $\sum_{j=1}^{\infty} \langle v, v_j \rangle v_j$ converges to v for the norm of H;

   c. (Parseval's relation) For each $v \in H$, $\|v\|^2 = \sum_{j=1}^{\infty} |\langle v, v_j \rangle|^2$;

   d. (also called Parseval's relation) For each u and each v in H, $\langle u, v \rangle = \sum_{j=1}^{\infty} \langle u, v_j \rangle \overline{\langle v, v_j \rangle}$.

Proof. Assume that our orthonormal sequence has property (a). Then for any fixed vector $v \in H$ and any integers m, n with $m > n$, we have by direct calculation:

$$\|\sum_{j=1}^{m}\langle v,v_j\rangle v_j - \sum_{j=1}^{n}\langle v,v_j\rangle v_j\|^2 = \|\sum_{j=n+1}^{m}\langle v,v_j\rangle v_j\|^2 \qquad (26)$$

$$= \sum_{j=n+1}^{m}|\langle v,v_j\rangle|^2$$

$$= \sum_{j=1}^{m}|\langle v,v_j\rangle|^2 - \sum_{j=1}^{n}|\langle v,v_j\rangle|^2$$

and by Theorem 2 these last two sums converge, to the same limit, as m and n tend to infinity. Thus the sequence

$$\{\sum_{j=1}^{m}\langle v,v_j\rangle v_j | m=1,2,\ldots\}$$

is a Cauchy sequence in the Hilbert space H (Section 2, Definition 3). It follows that this sequence converges to a point of H which we shall denote by

$$\sum_{j=1}^{\infty}\langle v,v_j\rangle v_j.$$

What we must do now is show that this last sum is actually equal to the vector v chosen above. This will prove that (a) implies (b). We make use of (a):

$$\langle v - \sum_{j=1}^{\infty}\langle v,v_j\rangle v_j, v_n\rangle = \langle v - \lim_{m\to\infty}\sum_{j=1}^{m}\langle v,v_j\rangle v_j, v_n\rangle$$

$$= \langle \lim_{m\to\infty}(v - \sum_{j=1}^{m}\langle v,v_j\rangle v_j), v_n\rangle = \lim_{m\to\infty}\langle v - \sum_{j=1}^{m}\langle v,v_j\rangle, v_n\rangle$$

$$= \langle v,v_n\rangle - \lim_{m\to\infty}\sum_{j=1}^{m}\langle v,v_j\rangle\langle v_j,v_n\rangle = \langle v,v_n\rangle - \langle v,v_n\rangle = 0$$

for any fixed n; we have used Exercise 4, problem 1g repeatedly.

Thus, by (a),

$$v - \sum_{j=1}^{\infty}\langle v,v_j\rangle v_j = 0$$

which shows that (a) implies (b).

Now let us suppose that the orthonormal sequence $\{v_j\}$ has property (b).

Then for each $v \in H$ we may write

$$\begin{aligned}
\|v\|^2 &= \langle v, v \rangle = \langle \sum_{j=1}^{\infty} \langle v, v_j \rangle v_j, v \rangle = \sum_{j=1}^{\infty} \langle v, v_j \rangle \langle v_j, v \rangle \\
&= \sum_{j=1}^{n} \langle v, v_j \rangle \overline{\langle v, v_j \rangle} = \sum_{j=1}^{\infty} |\langle v, v_j \rangle|^2
\end{aligned}$$

and this shows that (b) implies (c). But clearly (c) implies (b) because, for any $v \in H$ a direct calculation gives

$$\|v - \sum_{j=1}^{m} \langle v, v_j \rangle v_j\|^2 = \|v\|^2 - \sum_{j=1}^{m} |\langle v, v_j \rangle|^2$$

and, if we assume (c), the right-hand side of this equation tends to zero as m tends to infinity, giving (b). Thus (b) and (c) are equivalent.

Next we assume that our sequence has property (c), and hence property (b) as well, and consider property (d). For any u and v in H we can use (b) to write:

$$\begin{aligned}
\langle u, v \rangle &= \langle u, \lim_{m \to \infty} \sum_{j=1}^{m} \langle v, v_j \rangle v_j \rangle \\
&= \lim_{m \to \infty} \langle u, \sum_{j=1}^{m} \langle v, v_j \rangle v_j \rangle = \lim_{m \to \infty} \sum_{j=1}^{m} \overline{\langle v, v_j \rangle} \langle u, v_j \rangle \\
&= \sum_{j=1}^{\infty} \langle u, v_j \rangle \overline{\langle v, v_j \rangle}
\end{aligned}$$

which is (d).

Finally, let us show that (d) implies (a). If we assume (d) then, for any $v \in H$, we have

$$\|v\|^2 = \langle v, v \rangle = \sum_{j=1}^{\infty} \langle v, v_j \rangle \overline{\langle v, v_j \rangle}$$

Now suppose that $\langle v, v_j \rangle = 0$ for every j. Then our last equation clearly shows that $\|v\|^2 = 0$ and this says that v must be the zero vector (Section 2, Definition 1c).

Recall that a set which is either finite or denumerably infinite is called a countable set. At this point we have shown that the four properties list in

Theorem 3 are equivalent for any countable orthonormal subset of a Hilbert space. This leads us to:

Definition 4

Let $(H, \langle, \rangle)$ be a Hilbert space. A countable orthonormal subset of H is said to be an orthonormal basis for H if it has any, and hence all, of the four properties listed in Theorem 3.

Perhaps a word of caution is in order here. In a finite dimensional space an orthonormal basis is just an orthonormal set which is also a Hamel basis for the space. In an infinite dimensional Hilbert space an orthonormal basis is certainly an orthonormal set, but it is not a Hamel basis for the space; the sums involved in the definition of a Hamel basis are always finite while the sums involved in Theorem 3 may be, and in many cases are, infinite. As a practical matter an orthonormal basis in an infinite dimensional space is a valuable, important tool while a Hamel basis in the same space is of little, if any, importance.

An example of an infinite dimensional Hilbert space which has a countable orthonormal basis is easily given. Take the space ℓ^2 (Section 5) and the subset $\{\vec{e}_j | j = 1, 2, \ldots\}$ (Exercises 5, problem 1). This set is clearly orthonormal and if $\vec{v} = \{a_n\} \in \ell^2$ then $\langle \vec{v}, \vec{e}_n \rangle = a_n$ for every n and so $\langle \vec{v}, \vec{e}_n \rangle = 0$ for all n implies that every $a_n = 0$; i.e., that $\vec{v}$ is the zero vector in ℓ^2. In some sense this is the only example. More precisely:

Theorem 4

Any Hilbert space over $\mathbb{R}$ (resp. $\mathbb{C}$) which has a countably infinite orthonormal basis is isometrically isomorphic to real (resp. complex) ℓ^2.

Proof. The proof is straightforward but rather long and tedious. It closely resembles the proof in the finite dimensional case, but there are some subtle points which require special care here.

Suppose that H is a Hilbert space over $\mathbb{C}$ (for a change) and suppose that $\{v_1, v_2, \ldots\}$ is a countably infinite orthonormal basis for H. For each j define $T(v_j)$ to be the vector $\vec{e}_j \in \ell^2$. Note that "extending T by linearity" as we did in the finite dimensional case would only define T on the set lin $\{v_1, \ldots, v_n, \ldots\}$ which is not all of H (see the discussion just after Definition 4). What we do instead is recall that each $v \in H$ can be written as

$$v = \sum_{j=1}^{\infty} \langle v, v_j \rangle v_j$$

and this is the only way (see the discussion just before Theorem 3) it can be so written. Furthermore, by (c) of Theorem 3 we have

$$\sum_{j=1}^{\infty} |\langle v, v_j \rangle|^2 < \infty$$

Thus we may define $T(v)$ to be the vector

$$\sum_{j=1}^{\infty} \langle v, v_j \rangle \vec{e}_j$$

of ℓ^2. This gives us a map from H into ℓ^2 which we must now prove is linear, one-to-one, onto, and an isometry.

Linearity is easy to prove. Given $u, v \in H$ and scalars α, β we note that $\langle \alpha u + \beta v, v_j \rangle = \alpha \langle u, v_j \rangle + \beta \langle v, v_j \rangle$. Thus

$$\alpha u + \beta v = \sum_{j=1}^{\infty} \langle \alpha u + \beta v, v_j \rangle v_j = \alpha \sum_{j=1}^{\infty} \langle u, v_j \rangle v_j + \beta \sum_{j=1}^{\infty} \langle v, v_j \rangle v_j$$

and from this we see that

$$T(\alpha u + \beta v) = \alpha \sum \langle u, v_j \rangle \vec{e}_j + \beta \sum \langle v, v_j \rangle \vec{e}_j = \alpha T(u) + \beta T(v)$$

Next we show that T is one-to-one. Since we know that T is linear $T(u) = T(v)$ is equivalent to $T(u-v) = 0$ and so to show that T is one-to-one we need only show that $T(w) = 0$ if, and only if, w is the zero vector of H. Clearly, from the definition of T, $T(0) = 0$. Suppose now that $T(w)$ is the zero vector; i.e.,

$$\sum_{j=1}^{\infty} \langle w, v_j \rangle \vec{e}_j$$

is the zero vector in ℓ^2. Since $\{\vec{e}_j\}$ is an orthonormal basis for ℓ^2 this is possible only if $\langle w, v_j \rangle = 0$ for all j; the zero vector, like every other vector of ℓ^2, can be represented in just one way and that is the sum with all coefficients zero. However, $\{v_j\}$ was given to be an orthonormal basis for H. By (a) of Theorem 3 we see that w must be the zero vector of H.

Now let us show that T maps H onto ℓ^2. This takes a little work and uses the fact that H is a Hilbert space. Suppose that $\vec{w} \in \ell^2$ is given. Then since $\{\vec{e}_j\}$ is an orthonormal basis for ℓ^2 we have

$$\vec{w} = \sum_{j=1}^{\infty} \langle \vec{w}, \vec{e}_j \rangle \vec{e}_j$$

and

$$\sum_{j=1}^{\infty} |\langle \vec{w}, \vec{e}_j \rangle|^2 < \infty$$

Let us set $\langle \vec{w}, \vec{e}_j \rangle = a_j$ where we use the letter a to remind us that each a_j is a scalar. Now we claim that there is a vector $v \in H$ such that $\langle v, v_j \rangle = a_j$ for each j and that, in fact,

$$v = \sum_{j=1}^{\infty} a_j v_j$$

What we must do, of course, is show that this last sum really does converge to an element of H. We have

$$\|\sum_{j=1}^{n} a_j v_j - \sum_{j=1}^{m} a_j v_j\|^2 = \|\sum_{j=m+1}^{n} a_j v_j\|^2 = \sum_{j=m+1}^{n} |a_j|^2$$

by the orthonormality of the set $\{v_j\}$, and the latter sum tends to zero as $m, n \to \infty$ because

$$\sum |a_j|^2 < \infty$$

Thus the sequence

$$\{\sum_{j=1}^{n} a_j v_j \,|\, n = 1, 2, \ldots\}$$

is a Cauchy sequence in the Hilbert space H and hence is convergent to a point of H which we may call v. It is clear now that $T(v)$, by construction, gives us the vector $\vec{w}$ we began with. Thus T is onto.

Finally, for any u, v in H we have, using (d) of Theorem 3

$$\langle u, v \rangle = \sum_{j=1}^{\infty} \langle u, v_j \rangle \overline{\langle v, v_j \rangle}$$

and, using (d) of Theorem 3 again but this time in ℓ^2,

$$\langle T(u), T(v) \rangle = \sum_{j=1}^{\infty} \langle u, v_j \rangle \overline{\langle v, v_j \rangle}$$

Hence T is an isometry.

We have said that every inner product space has a maximal orthonormal set and we have given a very thorough discussion of those Hilbert spaces that have countable maximal orthonormal sets. We have also said that we can given a nice characterization of these spaces and that we shall do so later (Section 9). Most of the Hilbert spaces which arise in applications are of this kind. The reader may wonder just what can be said about those Hilbert spaces H whose maximal orthonormal sets $\mathcal{H}$ are uncountable. What happens is this: For each $v \in H$ there is a countable subset $\{v_1, v_2, \ldots\}$ of $\mathcal{H}$, which depends on v and hence will change if v is changed, such that

$$v = \sum_{j=1}^{\infty} \langle v, v_j \rangle v_j$$

where convergence is in the norm of H. We also have

$$\sum_{j=1}^{\infty} |\langle v, v_j \rangle|^2 < \infty$$

This is all that we shall say about such spaces. From now on, unless the contrary is explicitly stated, an orthonormal basis in a Hilbert space will always mean a countable (i.e., finite or denumerably infinite) set.

One final remark. What we have called a countable orthonormal basis for a Hilbert space is sometimes referred to as a "complete orthonormal set." We alluded to this in our discussion following Theorem 2. This terminology involves using the word "complete" in a manner which is totally different from the way in which we used it earlier (Section 3, Definition 2) and, for this reason, we shall avoid it. The reader should be aware however that this terminology is common especially in some of the older books.

Exercises 6

* 1. Let $(V, \langle, \rangle)$ be an inner product space.

    a. Show that any orthonormal subset of V is linearly independent (Section 1, Definition 3).

    b. Let $\mathcal{S}$ be a non-empty orthogonal subset of V which does not contain the zero vector. Then $\mathcal{B} = \{\frac{v}{\|v\|} | v \in \mathcal{S}\}$ is clearly an orthonormal set. Show that $\mathcal{B}$ is a maximal orthonormal set if, and only if, $\langle u, v \rangle = 0$ for all $v \in \mathcal{S}$ implies that u is the zero vector.

* 2. a. Suppose that we apply the Gram-Schmidt process (Lemma 1) to a finite subset $\{v_1, \ldots, v_n\}$ of an inner product space. We have seen that if this set is linearly independent we will obtain an orthonormal set. Suppose that the set is already orthonormal, what do we obtain then?

    * b. Show that any maximal orthonormal subset of $\mathbb{R}^n$ or $\mathbb{C}^n$ must contain n vectors.

* 3. In the inner product space $(C[-\pi, \pi], \langle, \rangle)$ (Exercises 4, problem 2) show that the set $\{\exp(int)|_{n=0,\pm1,\pm2,\ldots}\}$ is an orthogonal set which is not an orthonormal set. Follow 1(b) above to construct an orthonormal subset from this.

* 4. Let $(H, \langle, \rangle)$ be a Hilbert space with a countably infinite orthonormal basis $v_1, v_2, \ldots$. Let $a_1, a_2, \ldots$ be a sequence of numbers. Show that there is a

vector $u \in H$ for which $\langle u, v_j \rangle = a_j$. for each $j = 1, 2, \ldots$ if, and only if, $\sum_{j=1}^{\infty} |a_j|^2$ converges. This is what enabled us to prove that the map from H into ℓ^2 defined in the proof of Theorem 4 was actually onto.

5. Show that the set $Z = \{0, \pm 1, \pm 2, \pm 3, \ldots\}$ is countably infinite by setting up a one-to-one correspondence between this set and the set $1, 2, 3, \ldots$.

1.7 Fourier Series

A fundamental result in approximation theory and numerical analysis is the theorem of Weierstrass which states that, given any continuous function on $[a, b]$ and any $\varepsilon > 0$ we can find a polynomial p such that

$$|f(t) - p(t)| < \varepsilon$$

for all t, $a \le t \le b$. To put it another way: Given any $f \in C[a,b]$ and any $\varepsilon > 0$ there is a $p \in lin\{1, x, x^2, \ldots\}$ such that

$$\|f - p\|_\infty = \max_{a \le t \le b} |f(t) - p(t)| < \varepsilon$$

We shall deduce Weierstrass' theorem from an important result in the theory of Fourier series. This result, Fejér's theorem, has some other important consequences which will provide a link between Fourier series and our discussion of orthonormal sets given in the last section.

We shall work with the space $C_p[-\pi, \pi]$ of all complex-valued, continuous functions for $[-\pi, \pi]$ which satisfy

$$f(\pi) = f(-\pi)$$

So the subscript p stands for "periodic." Clearly the function e^{int} is in this space for any fixed integer n.

Definition 1

Let $f \in C_p[-\pi, \pi]$. The numbers

$$\hat{f}(n) = \frac{1}{2\pi} \int_{-\pi}^{\pi} f(t) e^{-int} dt, \quad n = 0, \pm 1, \pm 2, \ldots$$

are called the Fourier coefficients of f and the series

$$\sum_{-\infty}^{\infty} \hat{f}(n) e^{int}$$

is called the Fourier series of f.

The question we would like to answer is this: Given the Fourier coefficients, and hence the Fourier series, of a function $f \in C_p[-\pi, \pi]$ can we calculate f? Now if the series converges to f, either uniformly (i.e., for the sup norm) over $[-\pi, \pi]$ or at each point of this interval, then of course we can. However, the Fourier series of a continuous function can diverge at infinitely many points [2]. We shall be able to give a more precise description of the set of points at which such a series can diverge in the next section. Here we want to show that, in any case, we can recapture f from its Fourier series. Of course, this requires something a little more elaborate than simply taking the limit of the series.

Given any sequence of complex numbers, say $a_0, a_1, a_2, \ldots$, we define the Cèsaro means of this sequence to be the numbers

$$\sigma_1 = a_0, \sigma_2 = \frac{a_0 + a_1}{2}, \sigma_3 = \frac{a_0 + a_1 + a_2}{3}, \ldots$$

If this new sequence $\sigma_1, \sigma_2, \sigma_3, \ldots$ converges to, say, σ, then we say that the original sequence $a_0, a_1, a_2, \ldots$ is Cèsaro convergent to σ. Now any convergent sequence is Cèsaro convergent to its (ordinary) limit (problem 1(b)), but there are many divergent series which are convergent in this, more general, sense; $1, 0, 1, 0, 1, 0, \ldots$, for example, is Cèsaro convergent to $1/2$. Consider now a series

$$\sum_{-\infty}^{\infty} a_n$$

For each $n = 0, 1, 2, \ldots$ let

$$s_n = \sum_{-n}^{n} a_k$$

If the sequence $\{s_n\}$ is Cèsaro convergent to, say, s', then we say that the original series is Cèsaro convergent, or Cèsaro summable, to s'.

Let us now apply this idea to Fourier series. In order to do that we must first derive a convenient expression for the Cèsaro means of such a series. We shall express them as convolutions. So, given

$$\sum_{-\infty}^{\infty} \hat{f}(n) e^{int}$$

let $\sigma_n(x; f)$ denote the nth Cèsaro mean of this series. Then

$$\sum_{-n}^{n} \hat{f}(k) e^{ikx} = \sum_{-n}^{n} e^{ikx} \left[\frac{1}{2\pi} \int_{-\pi}^{\pi} f(y) e^{-iky} dy \right]$$

$$= \frac{1}{2\pi} \int_{-\pi}^{\pi} f(y) \left[\sum_{-n}^{n} e^{ik(x-y)} \right] dy$$

Now if we let $K_n(x-y)$ denote the nth Cèsaro mean of the specific series appearing in the square brackets, so $K_n(x-y)$ is the nth Cèsaro mean of

$$\sum_{-n}^{n} e^{ik(x-y)}$$

then

(*) $\quad \sigma_n(x;f) = \dfrac{1}{2\pi} \displaystyle\int_{-\pi}^{\pi} f(y) K_n(x-y) dy$

We shall now prove a simple lemma about $K_n(x-y)$ and use it to obtain our main result.

Lemma 1

Let $K_n(x)$ be the nth Cèsaro mean of the series

$$\sum_{-\infty}^{\infty} e^{ikx}$$

then, for $n = 1, 2, 3, \ldots$ we have

i. $K_n(x) = \dfrac{1}{n}\left(\dfrac{1-\cos nx}{1-\cos x}\right) = \dfrac{1}{n}\left(\dfrac{\sin \frac{1}{2}nx}{\sin \frac{1}{2}x}\right)^2$;

ii. $K_n(x) \geq 0$ for all x;

iii. $\dfrac{1}{2\pi}\int_{-\pi}^{\pi} K_n(x)dx = 1$

Furthermore, if I is any open interval containing zero, then

$$\lim_{n\to\infty} \sup\{K_n(x) | x \text{ in } (-\pi, \pi] \text{ but not in } I\} = 0$$

Proof. We first observe that

$$(n+1)K_{n+1}(x) - nK_n(x) = \sum_{-n}^{n} e^{ikx} = \sum_{0}^{n} e^{ikx} + \sum_{1}^{n} e^{-ikx} \qquad (1)$$

because

$$K_{n+1}(x) = \dfrac{s_n + s_{n-1} + \ldots + s_0}{n+1}, \quad K_n(x) = \dfrac{s_{n-1} + \ldots + s_0}{n}$$

where

$$s_n = \sum_{-n}^{n} e^{ikx}$$

the last two series in (1) are geometric and their sums are

$$\frac{1-e^{i(n+1)x}}{1-e^{ix}}, \quad \frac{1-e^{-i(n+1)x}}{1-e^{ix}}-1$$

respectively. Adding these expressions and putting the result in (1) we obtain

$$(n+1)K_{n+1}(x) - nK_n(x) = \frac{\cos nx - \cos(n+1)x}{1-\cos x} \qquad (2)$$

Now

$$K_1(x) = \sum_0^0 e^{ikx} \equiv 1$$

and so, using (2), we can find $K_2(x)$. Then, knowing $K_2(x)$ we can use (2) again to get $K_3(x)$ and so on. This proves (i) and (ii) follows immediately from (i).

To prove (iii) we note that if f is identically equal to 1 then the same is true of each Cèsaro mean of the Fourier series for f; because

$$\hat{f}(n) = \frac{1}{2\pi}\int_{-\pi}^{\pi} 1 \cdot e^{-ikx}\,dx = 0$$

except when $n=0$ and so the Fourier series for f has only one term (it is identically equal to 1) and, finally, in the calculation of the nth Cèsaro mean this 1 occurs n times where n is the number we divide by. Thus $\sigma(x;f) = 1$ and using (*) we get

$$1 = \frac{1}{2\pi}\int_{-\pi}^{\pi} K_n(x-y)\,dy$$

Setting $x=0$ in this and noting that, from (i), each $K_n(x)$ is an even function, we have (iii).

Finally, if $0 < \delta < \pi$ and $\delta \le |x| \le \pi$ then

$$(\sin \tfrac{1}{2}x)^2 \ge (\sin \tfrac{1}{2}\delta)^2$$

because our inequalities guarantee that $1/2x$ and $1/2\delta$ are in that interval where the sine is monotonically increasing. So, if I is an open interval containing zero and if $\delta > 0$ is so small that $(-\delta, \delta) \subset I$ then

$$\sup\{K_n(x)|x \text{ in } (-\pi,\pi] \text{ but not in } I\} \le \sup\{K_n(x)|\delta \le |x| \le \pi\} \le \frac{(\sin \tfrac{1}{2}\delta)^{-2}}{n}$$

and, clearly, the limit of the last of these, as n tends to infinity is zero.

We come now to our main result. It is an impressive theorem, but perhaps even more impressive are the corollaries that we can deduce from it.

Theorem 1

(Fejer). The Fourier series of any $f \in C_p[-\pi,\pi]$ is Cèsaro summable to f uniformly over $[-\pi,\pi]$ (i.e., for the sup norm).

Proof. Let $f \in C_p[-\pi,\pi]$ be given and, for simplicity, let us denote the nth Cèsaro mean of the Fourier series for f, what we have called $\sigma_n(x;f)$, by $\sigma_n(x)$. Then, using (*) and (iii) of Lemma 1, we may write

$$\begin{aligned}\sigma_n(x) - f(x) &= \frac{1}{2\pi}\int_{-\pi}^{\pi} f(t)K_n(x-t)dt - \frac{1}{2\pi}\int_{-\pi}^{\pi} K_n(t)f(x)dt \\ &= \frac{1}{2\pi}\int_{-\pi}^{\pi}\{f(x-t) - f(x)\}K_n(t)dt\end{aligned}$$

where, if necessary, we may extend f beyond $[-\pi,\pi]$ by periodicity. Then

$$\begin{aligned}|\sigma_n(x) - f(x)| &\leq \frac{1}{2\pi}\int_{-\pi}^{\pi} |f(x-t) - f(x)|K_n(t)dt \quad (3) \\ &= \frac{1}{2\pi}\int_{-\delta}^{\delta} + \frac{1}{2\pi}\int_{|t|\geq\delta}\end{aligned}$$

Now

$$\begin{aligned}\frac{1}{2\pi}\int_{-\delta}^{\delta}|f(x-t)-f(x)|K_n(t)dt &\leq \sup_{-\delta\leq t\leq\delta}|f(x-t)-f(x)|\frac{1}{2\pi}\int_{-\pi}^{\pi}K_n(t)dt \\ &\leq \sup_{-\delta\leq t\leq\delta}|f(x-t)-f(x)|\end{aligned}$$

by (iii) of Lemma 1. The second integral on the right-hand side of (3) is easily bounded as follows:

$$\frac{1}{2\pi}\int_{-\delta}^{\delta}|f(x-t)-f(x)|K_n(t)dt \leq (\frac{1}{2\pi})2 \quad \max\{|f(x)|\mid -\pi\leq x\leq\pi\}$$

$$\sup\{K_n(t)\mid |t|\geq\delta\}$$

Now given $\varepsilon > 0$ we may choose $\delta > 0$ so that

$$|f(x-t)-f(x)| < \frac{\varepsilon}{2}$$

whenever $t \in [-\delta,\delta]$ and $x \in [-\pi,\pi]$ (here we are using the uniform continuity of f). Having chosen δ we next choose N so that

$$\sup\{K_n(t)\|t|\geq\delta\} \leq \frac{\pi\varepsilon}{2\max\{|f(t)|\|-\pi\leq t\leq\pi\}} \quad \text{for all } n \geq N$$

this being possible by Lemma 1. Putting these estimates into (3) gives us

$$|\sigma_n(x) - f(x)| < \varepsilon$$

for all $n \geq N$ and all $x \in [-\pi, \pi]$.

Our first corollary, when translated into the language of section 6 will tell us that a certain orthonormal set is maximal.

Corollary 1

The only function in $C_p[-\pi, \pi]$ whose nth Fourier coefficient is zero for every n is the zero function (i.e., if $\hat{f}(n) = 0$ for every n, then $f(t) = 0$ for all $t \in [-\pi, \pi]$).

Proof. It is clear that the zero function has every Fourier coefficient zero. Conversely, suppose that $f \in C_p[-\pi, \pi]$ and that $\hat{f}(n) = 0$ for every n. Then all of the sums

$$\sum_{-n}^{n} \hat{f}(k)e^{ikx}$$

are zero and so each $\sigma_n(x; f)$, which is just an average of these sums, is zero. But by the theorem

$$\sigma_1(x; f), \sigma_2(x; f), \ldots$$

converges to f for the sup norm. Thus $f \equiv 0$.

Corollary 2

Let $f \in C_p[-\pi, \pi]$. Then given any $\varepsilon > 0$ we can find a "trigonometric polynomial" (we mean an element of $lin\{e^{inx} | n = 0, \pm 1, \pm 2, \ldots\}$) $\varepsilon(x)$ such that

$$\|f - \varepsilon\|_\infty < \varepsilon$$

Proof. We need only note that the Cèsaro means of f, what we called $\sigma_n(x)$ in the proof of Theorem 1, are all trigonometric polynomials.

Corollary 3

(Weierstrass). Let f be any continuous function on a closed, bounded interval $[a, b]$. Then, given $\varepsilon > 0$, there is a polynomial $p(x)$ such that

$$|f(x) - p(x)| < \varepsilon$$

for all x in $[a, b]$.

Proof. Assume first that $[a,b] \subset (-\pi, \pi)$ and extend f, in any way, so that it is continuous on $[-\pi, \pi]$ and satisfies $f(\pi) = f(-\pi)$. Given $\varepsilon > 0$ we find, as we may by Corollary 2, a trigonometric polynomial $\varepsilon(x)$ such that

$$|f(x) - \varepsilon(x)| < \varepsilon/2$$

for all x in $[-\pi, \pi]$. Now $\varepsilon(x)$ is the sum of, say, ℓ terms of the form $c_n e^{inx}$ and each of these can be expanded in a Maclaurin's series. By truncating this series, we obtain a polynomial $p_n(x)$ for which

$$|c_n e^{inx} - p_n(x)| < \varepsilon/2\ell$$

for all x in $[-\pi, \pi]$. Combining our inequalities gives us

$$|f(x) - p(x)| < \varepsilon$$

for all x in $[-\pi, \pi]$. The case in which $[a,b]$ is not contained in $[-\pi, \pi]$ is left to the reader.

Let us now return to the discussion of orthonormal sets (Section 6) and see what the results of this section tell us. The space $(C_p[-\pi, \pi], \langle,\rangle)$ is an inner product space, recall that

$$\langle f, g \rangle = \int_{-\pi}^{\pi} f(t) \overline{g(t)} dt$$

and the set

$$\{\frac{e^{int}}{\sqrt{2\pi}} | n = 0, \pm 1, \pm 2, \ldots\} \tag{4}$$

is orthonormal (Exercises 6, problem 3). Corollary 1 above together with problem 2b of Exercises 6 shows that this is a maximal (Section 6, Definition 3) orthonormal set. We would like to conclude from this that

$$f = \sum \langle f, \frac{e^{int}}{\sqrt{2\pi}} \rangle \frac{e^{int}}{\sqrt{2\pi}} \tag{5}$$

(note this is just the Fourier series for f) for every $f \in C_p$, convergence here is for the norm induced by $\langle,\rangle$. The maximality of (4) would imply (5) if $(C_p, \langle,\rangle)$ were a Hilbert space (see Section 6, Theorem 3 the proof that (a) $\to$ (b)). Since $(C_p, \langle,\rangle)$ is not complete we must give some other proof that (4) implies (5). If we knew that the series in (5) converged to an element of C_p then it would be easy to show, using the maximality of (4), that this limit must be f. But this is exactly what we do not know. We can prove, using Bessel's inequality (Section 6, Theorem 2), that the sequence of partial sums of the series in (5) is a Cauchy sequence but since $(C_p, \langle,\rangle)$ is not complete, we cannot say that this sequence converges. As a matter of fact (5) is true! We just cannot prove it yet and this is annoying. The incompleteness of the space $(C_p, \langle , \rangle)$ will cause more trouble in the next section. There, however, we shall do something about it.

One last remark. The reader can check, by direct calculation if necessary, that the set
$$\{1, \sin nt, \cos nt | n = 1, 2, \ldots\}$$
is an orthogonal subset of $(C_p, \langle , \rangle)$ and that
$$\{\frac{1}{\sqrt{2\pi}}, \frac{\sin nt}{\sqrt{\pi}}, \frac{\cos nt}{\sqrt{\pi}} | n = 1, 2, \ldots\} \tag{6}$$
is an orthonormal set. It is actually a maximal orthonormal set but we leave the proof of this to the exercises (problem 3).

Exercises 7

1. **a.** Show in detail that $1, 0, 1, 0, 1, 0, \ldots$ is Cèsaro convergent to $1/2$.

   b. Let $\{a_n\}_{n=0}^\infty$ be a sequence of complex numbers which converges to b. Show that $a_0, \frac{a_0+a_1}{2}, \frac{a_0+a_1+a_2}{3}, \ldots$ also converges to b. Hint:
   $$\frac{a_0 + a_1 + \cdots + a_k + a_{k+1} + \cdots + a_n}{n} = \left[\frac{a_0 + \cdots + a_k}{n}\right] + \left[\frac{a_{k+1} + \cdots + a_n}{n}\right]$$
   When a_{k+1} is near b the second term on the right-hand side is near $\frac{(n-k)b}{n}$ and the first term is near zero.

* 2. **a.** Let $f \in C[a, b]$. Use Weierstrass' theorem to show that there is a sequence $\{p_n(x)\}$ of polynomials (hence elements of $C[a, b]$) which converges to f for the sup norm.

   b. Suppose that each $p_n(x)$ in (a) has real coefficients. Show that there is a sequence $\{q_n(x)\}$, each term of which is a polynomial with rational coefficients, which converges to f for the sup norm.

   c. Suppose that each $p_n(x)$ in (a) has complex coefficients. Show that there are sequences $\{r_n(x)\}, \{s_n(x)\}$ each term of which is a polynomial with rational coefficients such that $\{r_n(x) + is_n(x)\}$ converges to f for the sup norm.

3. Let $f \in C_p$ and suppose that $\langle f, 1 \rangle = 0$, $\langle f, \sin nt \rangle = 0 = \langle f, \cos nt \rangle$ for $n = 1, 2, \ldots$. Show that f must be the zero function. Hint: Recall that $e^{int} = \cos nt + i \sin nt$.

1.8 The Space $L^2[a,b]$

We want to motivate our discussion, given below, of the Lebesgue integral. Let us begin by writing down an integral equation and then presenting, in an informal and purely heuristic manner, one method of solving it. So suppose that a, b are real numbers with $a < b$, that $g(s) \in C[a,b]$ is given, that $K(s,t)$ is a known function that is continuous on the square $[a,b] \times [a,b] = \{(x,y) \in \mathbb{R}^2 | a \leq x \leq b, a \leq y \leq b\}$ and consider

$$u(s) = g(s) + \int_a^b K(s,t)u(t)dt \qquad (1)$$

in the unknown function u. We will discuss integral equations more systematically later on. For now let us first write $K(u)(s)$ for

$$\int_a^b K(s,t)u(t)dt$$

so that equation (1) becomes

$$u(s) = g(s) + K(u)(s) \qquad (2)$$

Next let $\{\sigma_j(s)\}_{j=1}^\infty$ be a maximal orthonormal set in $(C[a,b], \langle,\rangle)$ (Exercises 4, problem 2) and, for each k, use (2) to get

$$\langle u, \sigma_k \rangle = \langle g, \sigma_k \rangle + \langle K(u), \sigma_k \rangle \qquad (3)$$

Let us suppose now that we have a solution u which can be written in the form

$$u(s) = \sum_{j=1}^\infty \langle u, \sigma_j \rangle \sigma_j(s) \qquad (4)$$

Writing $u_j = \langle u, \sigma_j \rangle$, $g_k = \langle g, \sigma_k \rangle$ and putting (4) into (3) we obtain

$$\langle \sum_{j=1}^\infty u_j \sigma_j, \sigma_k \rangle = g_k + \langle K(\sum_{j=1}^\infty u_j \sigma_j), \sigma_k \rangle$$

or

$$u_k = g_k + \langle K(\sum_{j=1}^\infty u_j \sigma_j), \sigma_k \rangle \qquad (5)$$

Now, under reasonable assumptions which we need not state here, we can write

$$K(\sum_{j=1}^{\infty} u_j \sigma_j) = \int_a^b K(s,t)(\sum_{j=1}^{\infty} u_j \sigma_j(t))dt \qquad (6)$$

$$= \sum_{j=1}^{\infty} u_j \int_a^b K(s,t)\sigma_j(t)dt$$

$$= \sum_{j=1}^{\infty} u_j K(\sigma_j)$$

and putting this into (5) gives us

$$u_k = g_k + \sum_{j=1}^{\infty} u_j \langle K(\sigma_j), \sigma_k \rangle, \quad k = 1, 2, \ldots \qquad (7)$$

Observe that (7) is an infinite system of linear equations in the unknown constants $u_1, u_2, \ldots$. We may attempt to solve this system by the methods (or some generalization of them) of linear algebra. A question arises however. Does a solution to (7) give us a solution to (1) which is in $C[a,b]$? If not, then just what is it that we are finding here? Let us consider these questions a little. Since the function g was given to be in $C[a,b]$ and since $\{\sigma_j\}$ was given to be an orthonormal sequence, the sequence $\{g_k\}$, of numbers, must be in ℓ^2; by Bessel's inequality (Section 6, Theorem 2). If our solution u is to be in $C[a,b]$, then the sequence $\{u_j\}$ must also be in ℓ^2. Thus we must seek a solution $\{u_j\}$ to (7) for which

$$\sum_{j=1}^{\infty} |u_j|^2 < \infty$$

But now we come to the main point of the whole discussion. Even if we found a sequence $\{u_j\}$ with the properties just stated we must then find a function $u \in C[a,b]$ such that

$$\langle u, \sigma_j \rangle = u_j \text{ for every } j. \qquad (8)$$

Since $(C[a,b], \langle,\rangle)$ is not a Hilbert space this cannot be done in general (Exercises 6, problem 4).

This is the second time (the first was at the end of Section 7) that the lack of completeness of our inner product space has caused us trouble. The Lebesgue integral will enable us to define a Hilbert space, the "smallest" one, which contains $(C[a,b], \langle,\rangle)$ and has many useful properties. So let us turn now to a discussion of the integral and, after that, we shall show how it enables us to get around the difficulties we encountered in trying to solve (1) via (7), and those which arose at the end of Section 7.

We shall approach the integral through measure theory and so we must recall something about subsets of $\mathbf{R}$. For any a, b in $\mathbf{R}$, with $a < b$, $(a,b) = \{t \in \mathbf{R} | a < t < b\}$ is called an open interval. If $S \subseteq R$ and $y \in S$ we say that S is a neighborhood of y if for some a, b we have $y \in (a,b) \subseteq S$. The set S is called an open set if it is the empty set or it is a neighborhood of each of its points. An open interval is clearly an open set. Also:

Theorem 1

Every non-empty open subset of $\mathbf{R}$ is the union of a countable family of pairwise disjoint, open intervals.

The term "pairwise disjoint" when applied to a family of sets means that any two distinct sets in the family have empty intersection. We leave the proof of Theorem 1 to the exercises (see problem 1 for a series of hints). Now how does one assign a measure, or length, to a set of points? We start by assigning measure $b - a$ to (a,b) and measure zero to the empty set. Next, given any non-empty open set $\mathcal{O}$ of $\mathbf{R}$ we write

$$\mathcal{O} = \cup_{j=1}^{\infty} (a_j, b_j)$$

where the intervals $\{(a_j, b_j)\}_{j=1}^{\infty}$ are pairwise disjoint, and define the measure of $\mathcal{O}$, $m[\mathcal{O}]$, to be

$$\sum_{j=1}^{\infty} (b_j - a_j) \qquad (9)$$

Here we agree that $m[\mathcal{O}] = \infty$ if the series in (9) diverges. Now given any set $S \subseteq \mathbf{R}$ we define the outer measure of S, $m^*[S]$, as follows:

$$m^*[S] = \inf\{m[\mathcal{O}] | \mathcal{O} \text{ is open set containing } S\} \qquad (10)$$

This, too, can be infinite. Now there are really two problems that we are trying to solve here. What we want is a class of sets with some useful properties, and a function on this class which has useful properties also. Outer measure is defined on the class of all subsets of $\mathbf{R}$, we could not ask for more, but it does not have the properties we want a measure to have. What we do is this:

Definition 1

A set E is said to be a measurable set if for any set S, $m^*[S] = m^*[S \cap E] + m^*[S \cap E']$; here E' denotes the complement of E ($E' = \{t \in \mathbf{R} | t \notin E\}$). When E is a measurable set we define the measure of E, $m[E]$, to be $m^*[E]$.

This very strange definition requires some explanation and we shall give it below. First, we mention that all open sets and all closed sets (recall that a set is said to be a closed set if its complement is an open set) are measurable sets, the union of any countable family of measurable sets is a measurable set, and any set whose outer measure is zero is a measurable set. In fact, one has to work to construct a non-measurable set and to do so one needs the axiom of choice. The first example of a non-measurable set was constructed by Vitali in 1904 [10; p.63].

The class of measurable sets has some nice properties as we have just noted. Also, the measure defined on this class is well-behaved. In fact:

Theorem 2

If $\{E_j\}_{j=1}^\infty$ is a countable family of pairwise disjoint measurable sets, then their union is a measurable set and

$$m[\cup_{j=1}^\infty E_j] = \sum_{j=1}^\infty m[E_j] \qquad (11)$$

We say, to abbreviate Theorem 2, that Lebesgue measure, m, is a countably additive set function. This is not true of outer measure. The countable additivity of Lebesgue measure is what gives the Lebesgue integral so many useful properties.

Before we leave the discussion of measure we must explain where Definition 1 came from. Originally, one would choose, and fix, an interval (a, b) in $\mathbb{R}$; this could be $(-\pi, \pi)$ for example. Then any open subset of (a, b) would have finite measure since the series (9) would satisfy

$$\sum_{j=1}^n (b_j - a_j) \le \sum_{j=1}^{n+1} (b_j - a_j) \le b - a$$

for every n. The next thing one would do is define the measure of a closed set F in (a, b) as follows:

$$m[F] = (b - a) - m[F']$$

Here F' denotes the complement of F in (a, b), $F' = \{t \in (a, b) | t \notin F\}$, and this is open in (a, b) so it has a measure. The outer measure of any $S \subseteq (a, b)$ was defined as above, (10), but now we may also define the inner measure, $m_*[S]$, as follows

$$m_*[S] = \sup\{m[F] | F \text{ is a closed set contained in } S\} \qquad (12)$$

Finally, a set was said to be measurable if its inner measure and outer measure were equal. In this case the measure of the set was taken to be its outer measure as we have done. An early theorem stated that a set E is measurable, as just defined, if and only if

$$m^*[S] = m^*[S \cap E] + m^*[S \cap E']$$

for every set S. Now Caratheodory noticed that by taking this as the definition of a measurable set one could avoid all mention of inner measure, and one need not work inside a bounded interval. So there are advantages to the modern approach to measure but the price we pay for these advantages is a definition of measurable set which is highly non-intuitive.

We finish our discussion of measure by "proving" one result.

Theorem 3

Any countable subset of $\mathbb{R}$ is a measurable set and its measure is zero.

Proof. Let $S = \{s_j\}_{j=1}^{\infty}$ be any countable subset of $\mathbb{R}$. We mentioned above that if $m^*[S] = 0$, then S is a measurable set (and of course, since $m^*[S] = m[S]$ when S is measurable, $m[S] = 0$). So let us show that the outer measure of S is zero. Given $\varepsilon > 0$ we shall work with the open set $\mathcal{O}$ defined by

$$\mathcal{O} = \cup_{j=1}^{\infty} (s_j - \frac{\varepsilon}{2^{j+1}}, s_j + \frac{\varepsilon}{2^{j+1}})$$

Clearly $\mathcal{O}$ contains S and so

$$m^*[S] \le m[\mathcal{O}] \le \sum_{j=1}^{\infty} [(s_j + \frac{\varepsilon}{2^{j+1}}) - (s_j - \frac{\varepsilon}{2^{j+1}})] = \sum_{j=1}^{\infty} \frac{\varepsilon}{2^j} = \varepsilon$$

Since $\varepsilon > 0$ is arbitrary, this shows that $m^*[S] = 0$.

We should mention that a set of measure zero need not be countable and can be quite complicated.

Let us turn now to a discussion of the Lebesgue integral. A real-valued function f on $\mathbb{R}$, or on some interval, is said to be a measurable function if for every $r \in \mathbb{R}$ the set $\{t | f(t) \ge r\}$ is measurable. Since there are nonmeasurable sets, there are non-measurable functions. The function which is one on some non-measurable set and is zero everywhere else is an example. However, just about any function which arises in analysis or its applications is a measurable function. Now suppose that f is a bounded function on $[a, b]$; this means that the range of f, $\{f(t) | t \in \mathbb{R}\}$, is contained in some open interval. We may apply the definition of Riemann to f obtaining a set of numbers of the form

$$\sum_{j=0}^{n-1} f(x_j^*)(x_{j+1} - x_j)$$

where $a = x_0 < x_1 < \ldots < x_n = b$ and $x_j^* \in [x_j, x_{j+1}]$ for every j; we assume that this process is familiar from elementary calculus. If these sums have a limit as our subdivision of $[a, b]$ becomes finer and finer, then we say that f is Riemann integrable and we call the limit

$$\int_a^b f(t)dt$$

So some functions on $[a, b]$ are Riemann integrable and some are not. By contrast, the Lebesgue integration process, which we shall not describe here, applied to any bounded, measurable function g will always give us a real number as a limit and we denote this number by

$$\int_a^b g(t)dt$$

just as above. There is no need for another notation because every Riemann integrable function is bounded (by definition) and measurable (this requires proof) and its Riemann and Lebesgue integrals coincide. But it is important to realize that there are many bounded measurable functions, all of which have Lebesgue integrals, which are not integrable in the sense of Riemann.

We need to define the concept of an integrable function in the sense of Lebesgue. If we apply the integration process to a non-negative (not necessarily bounded) measurable function $g(t)$, we will get either a real-number or $+\infty$. Thus if $h(t)$ is any measurable function then $|h(t)|$ is also measurable and non-negative and we say that h is Lebesgue integrable if

$$\int_a^b |h(t)|dt < \infty \qquad (13)$$

If $h(t) = 0$ except perhaps on a set of measure zero, then (13) will be zero and the converse is also true. This means that if two integrable function f, g are equal, except perhaps on a set of measure zero, they will have the same integral; for

$$0 \leq |\int_a^b f(t)dt - \int_a^b g(t)dt| = |\int_a^b \{f(t) - g(t)\}dt| \leq \int_a^b |f(t) - g(t)|dt = 0$$

where we have used the fact that, for integrable functions,

$$|\int_a^b h(t)dt| \leq \int_a^b |h(t)|dt \qquad (14)$$

This leads us to a very important concept.

Definition 2

A condition or an equation (examples will follow) which holds at all points except perhaps those in a set of measure zero is said to hold almost everywhere; abbreviated "a.e."

THE SPACE $L^2[a,b]$

Consider now the set of all measurable functions on $[a,b]$ and let us call two such functions equal if they are equal almost everywhere; i.e., $f \equiv g$ iff $f(t) = g(t)$ a.e. This is an equivalence relation on our set of functions and it is customary to "abuse language" and call the equivalence classes functions.

Definition 3

The set (we will soon see that it is a vector space) $L^2[a,b]$ consists of all (equivalence classes of measurable) functions f such that

$$\int_a^b |f(t)|^2 dt < \infty \tag{15}$$

Since $[a,b]$ has finite measure one can show that (15) implies that f, and hence any function equivalent to f, is Lebesgue integrable, and any two functions in the same class have the same integral. An L^2-function can contain exactly one continuous function and it is customary to simply say that $C[a,b]$ is a linear manifold in $L^2[a,b]$.

Lemma 1

The set $L^2[a,b]$ is a vector space. Furthermore, for any $f, g \in L^2[a,b]$, $f(t)g(t)$ is Lebesgue integrable.

Proof. Suppose that $f, g \in L^2[a,b]$. Then

$$0 \le (|f(t)| - |g(t)|)^2 = |f(t)|^2 - 2|f(t)g(t)| + |g(t)|^2$$

and so

$$\left| \int_a^b f(t)g(t) dt \right| \le \int_a^b |f(t)g(t)| dt \le \frac{1}{2} \left(\int_a^b |f(t)|^2 dt + \int_a^b |g(t)|^2 dt \right)$$

This shows that fg is a Lebesgue integrable function. Also,

$$\int_a^b |f(t) + g(t)|^2 dt \le \int_a^b |f(t)|^2 dt + 2 \int_a^b |f(t)g(t)| dt + \int_a^b |g(t)|^2 dt$$

shows that $f + g$ is in $L^2[a,b]$.

We may now define an inner product on $L^2[a,b]$ as follows:

$$\langle f, g \rangle = \int_a^b f(t)g(t) dt$$

for complex-valued functions the $g(t)$ would be conjugated, we will discuss these functions below. The norm associated with the inner product will be denoted by $\|\cdot\|_2$. We have:

Theorem 4

The space $(L^2[a,b], \langle,\rangle)$ is a Hilbert space. The space of continuous functions on $[a,b]$, $C[a,b]$, is a linear manifold in this Hilbert space and, for any fixed $f \in L^2[a,b]$ and any $\varepsilon > 0$ we can find a function $\sigma \in C[a,b]$ such that $\|f - \sigma\|_2 < \varepsilon$.

We shall not prove this theorem. The first part of it can be obtained from the Lebesgue dominated convergence theorem (see below). Right now let us use the second part of Theorem 4 to prove:

Theorem 5

The sequence $\{\frac{e^{int}}{\sqrt{2\pi}} | n = 0, \pm 1, \pm 2, \ldots\}$ is an orthonormal basis for the Hilbert space $(L^2[-\pi, \pi], \langle,\rangle)$.

Proof. It is easy to see that the given sequence is an orthonormal set. To show that it is a basis for $L^2[a,b]$ it suffices to prove that if f satisfies

$$\langle f, e^{int} \rangle = 0 \text{ for } n = 0, \pm 1, \pm 2, \ldots \qquad (16)$$

then f is the zero vector in $L^2[a,b]$ (Section 6, Definition 4 and Theorem 3). So let us suppose that f satisfies (16) and is not the zero vector. Then we may assume that

$$\int_{-\pi}^{\pi} |f(t)| dt \neq 0$$

For if this were zero then we would have $|f(t)| = 0$ a.e. and hence $|f(t)|^2 = 0$ a.e. and hence

$$\|f\|_2 = \int_{-\pi}^{\pi} |f(t)|^2 dt = 0$$

showing that f is the zero vector in $L^2[a,b]$. With this in mind let us define a function F as follows

$$F(x) = \int_{-\pi}^{x} f(t) dt \qquad (17)$$

Now for any $k \neq 0$ and a constant C as yet undetermined we may write

$$\int_{-\pi}^{\pi} \{F(t) - C\} e^{-ikt} dt = \frac{\{F(t) - C\}e^{-ikt}}{-ik}\Big|_{-\pi}^{\pi} + \frac{1}{ik} \int_{-\pi}^{\pi} f(t) e^{-ikt} dt \qquad (18)$$

Our assumption on f, (16) above, shows that the second term on the right-hand side of (18) is zero. The first term is
$$\frac{-\{F(\pi)-C\}+\{F(-\pi)-C\}}{ik}$$
From (17) we see that $F(-\pi)=0$ and that
$$F(\pi)=\int_{-\pi}^{\pi}f(t)dt=\int_{-\pi}^{\pi}f(t)e^{-i0t}dt$$
and combining this with (16) we see that $F(\pi)=0$. Thus (18) shows that for all non-zero k we have
$$\int_{-\pi}^{\pi}\{F(t)-C\}e^{-ikt}dt=0 \tag{19}$$

At this point we choose the constant C so that (19) holds when $k=0$ as well.

Let us set $\sigma(t)=F(t)-C$ so that σ is a continuous, 2π-periodic function. By one of the corollaries to Fejér's theorem (Section 7, Corollary 2) we can, for any given $\varepsilon>0$, find a finite set, $2n+1$ say, of numbers A_k, $|k|\le n$ such that
$$\|\sigma(t)-\sum_{-n}^{n}A_k e^{ikt}\|_\infty < \varepsilon$$
Now (19) tells us that
$$\int_{-\pi}^{\pi}\sigma(t)\left[\sum_{-n}^{n}A_k e^{ikt}\right]dt=0$$
and so
$$\|\sigma\|_2^2 = \int_{-\pi}^{\pi}|\sigma(t)|^2 dt = \int_{-\pi}^{\pi}\sigma(t)\{\sigma(t)-\sum_{-n}^{n}A_k e^{ikt}\}dt$$
$$\le \int_{-\pi}^{\pi}\varepsilon\sigma(t)dt \le \left(\int_{-\pi}^{\pi}\varepsilon^2 dt\right)^{1/2}\left(\int_{-\pi}^{\pi}|\sigma(t)|^2 dt\right)^{1/2}$$
by the C.S.B. inequality (Section 4, Theorem 1). Hence
$$\|\sigma\|_2^2 \le (\varepsilon\sqrt{2\pi})\|\sigma\|_2$$
and if $\|\sigma\|_2 \ne 0$ this would give us
$$\|\sigma\|_2 \le \varepsilon\sqrt{2\pi}$$
for any $\varepsilon>0$. Thus we must conclude that $\sigma\equiv 0$ and hence that $F(t)\equiv C$. But now compare this with (17). We see that $f(t)=0$ a.e. as was to be proved.

Corollary 1

The space $L^2[-\pi, \pi]$ is isometrically isomorphic to the space ℓ^2.

Proof. This follows immediately from Theorem 5 and Theorem 4 of Section 6.

Corollary 2

For any $f \in L^2[-\pi, \pi]$ the series

$$\sum_{-\infty}^{\infty} \langle f, \frac{e^{int}}{\sqrt{2\pi}} \rangle \frac{e^{int}}{\sqrt{2\pi}}$$

converges to f for the L^2-norm. This is true, in particular, for any continuous, 2π-periodic function (Section 7, see the discussion just after the proof Corollary 3).

Proof. See Theorem 3 of Section 6.

We shall see in the next section that $L^2[a, b]$ also has a countable orthonormal basis. Thus if $\{\sigma_j\}$ is one of these the series

$$\sum_{j=1}^{\infty} u_j \sigma_j$$

will converge for the L^2-norm to a function in $L^2[a, b]$ whenever the series of constants $\sum |u_j|^2$ converges. So the process sketched above for solving the integral equation (1) will yield solutions in $L^2[a, b]$.

Let us finish our discussion of the Lebesgue integral by stating two important theorems. We shall have occasion to refer to these later on.

Theorem 6

(Lebesgue dominated convergence theorem) Let $\{f_n\}$ be a sequence of measurable functions on $[a, b]$ which converges pointwise to $f(t)$ at almost every point of this interval. Let $g(t)$ be an integrable function on $[a, b]$ such that $|f_n(t)| \leq g(t)$ a.e. for each n. Then each f_n is integrable over $[a, b]$, f is integrable over $[a, b]$ and

$$\lim \int_a^b f_n(t)dt = \int_a^b f(t)dt$$

Theorem 6 could be used to prove the completeness of the space $L^2[a,b]$. Our next result concerns double integrals. For $c,d \in \mathbb{R}$, $c < d$, let us take R to be the set $[a,b] \times [c,d] = \{(x,y) \in \mathbb{R}^2 | a \le x \le b \text{ and } c \le y \le d\}$. One can define measurable subsets of and measurable functions on R, and the discussion of these and of the Lebesgue integral for functions on R parallels that given above. One can also define the space $L^2(R)$.

Theorem 7

(Fubini) For any function $K(s,t) \in L^2(R)$ we have $K(s,t_0) \in L^2[a,b]$ for almost every $t_0 \in [c,d]$, and $K(s_0,t) \in L^2[c,d]$ for almost every $s_0 \in [a,b]$. Furthermore,

$$\int_R\int |K(s,t)|^2 ds\, dt = \int_a^b ds(\int_c^d |K(s,t)|^2 dt) = \int_c^d dt(\int_a^b |K(s,t)|^2 ds)$$

Finally, one can treat complex-valued functions by simply working with their real and imaginary parts separately or, where appropriate, working with their absolute value. Thus we may define "complex" $L^2[a,b]$ to be the vector space over $\mathbb{C}$ of all (equivalence classes of measurable) complex-valued functions f such that

$$\int_a^b |f(t)|^2 dt < \infty$$

For any f,g in this space we define

$$\langle f,g \rangle = \int_a^b f(t)\overline{g(t)} dt$$

and we find that $(L^2[a,b], \langle,\rangle)$ is a Hilbert space.

Exercises 8

1. Let $\mathcal{O}$ be any non-empty open subset of $\mathbb{R}$. Define a relation "$\sim$" on the points of $\mathcal{O}$ as follows: For x,y in $\mathcal{O}$, $x \sim y$ if, and only if, there are numbers a,b such that
   $$x,y \in (a,b) \subseteq \mathcal{O}$$

   a. Show that "$\sim$" is an equivalence relation on $\mathcal{O}$.

   b. Show that every equivalence class is an open interval contained in $\mathcal{O}$. Thus $\mathcal{O}$ is now the union of a family of pairwise disjoint open intervals.

c. Show that any family of pairwise disjoint open intervals of $\mathbf{R}$ must be countable. Hint: The set of rational numbers as countable.

2. Show that the union of any countable family of sets, each of which has measure zero, is a set of measure zero. Hint: There are several ways to do this problem, but try to use Theorem 2. So if $\{E_j\}$ is our family of sets we must replace this with a family of pairwise disjoint sets having the same union.

3. Let us work with the interval $[0,1]$. Suppose that $\{r_j\}_{j=1}^{\infty}$ is an enumeration of the rational numbers in this interval and define a sequence of functions as follows: For each fixed n, $f_n(r_j) = 1$ for $j \leq n$, and $f_n(t) = 0$ for all other $t \in [0,1]$. So $f_n(t)$ is zero except at n points. Finally, define $f_0(r_j) = 1$ for every $j = 1, 2, \ldots$ and $f_0(t) = 0$ for all other $t \in [0,1]$.

   a. Show that each f_n is Riemann integrable but f_0 is not. Hint: We must look at sums of the form

   $$\sum_{j=0}^{n-1} f(x_j^*)(x_{j+1} - x_j)$$

   where $0 = x_0 < x_1 < \ldots < x_n = 1$ and $x_j^* \in [x_j, x_{j+1}]$. This last interval contains both rational and irrational points.

   b. Let $g(t) = 1$ for all $t \in [0,1]$. Clearly g is both Riemann and Lebesgue integrable. Also, $|f_n(t)| \leq g(t)$ for each n and all t. Thus since $f_0(t) = \lim f_n(t)$ the function f_0 is Lebesgue integrable by Theorem 6.

   c. Give an example of a function h on $[0,1]$ which is not Riemann integrable but for which $|h|$ is Riemann integrable.

* 4. Suppose that $\{\sigma_n\}$ is an orthonormal basis for the space $L^2[a,b]$. We shall see an example of such a basis in Section 9. Let $S = [a,b] \times [a,b]$ and, for each pair of positive integers m, n let $\psi_{m,n}(s,t) = \sigma_m(s)\sigma_n(t)$. Show that $\{\psi_{m,n}|(m,n) \in \mathcal{N} \times \mathcal{N}\}$ is an orthonormal basis for $L^2(S)$. Hint: Suppose that $K(s,t) \in L^2(S)$ such that $\langle K(s,t), \psi_{m,n}(s,t)\rangle = 0$ for all m,n. Since the measure of S is finite we can conclude, from another version of Fubioni's theorem, that

$$\int_a^b \int_a^b [K(s,t)\sigma_m(s)\sigma_n(t)]dsdt = \int_a^b (\int_a^b K(s,t)\sigma_m(s)ds)\sigma_n(t)dt$$
$$= \int_a^b (\int_a^b K(s,t)\sigma_n(t)dt)\sigma_m(s)ds.$$

1.9 More About the Space $L^2[a,b]$

Our discussions above left us with some unanswered questions which we shall settle here. First, we would like to exhibit an orthonormal basis for $L^2[a,b]$; we found one for $L^2[-\pi, \pi]$ (Section 8, Theorem 5). Second, we would like to characterize those Hilbert spaces which have a countable orthonormal basis. Finally, we would like to say a little more about the relationship between the spaces $C[a,b]$ and $L^2[a,b]$. We begin with some terminology.

Definition 1

Let $(E, \|\cdot\|)$ be a normed space and let S be a subset of E. We shall say that S is dense in E if, given any $x \in E$ and any $\varepsilon > 0$ there is an $s \in S$ such that $\|x - s\| < \varepsilon$. We shall say that $(E, \|\cdot\|)$ is a separable normed space or, simply, that $(E, \|\cdot\|)$ is separable, if it has a subset which is both countable and dense in E. Finally, an inner product space $(V, \langle,\rangle)$ is said to be separable if the space $(V, \|\cdot\|)$, where $\|\cdot\|$ is the norm induced by $\langle,\rangle$, is separable.

By Theorem 4 of Section 8 the set $C[a,b]$ is dense in $(L^2[a,b], \|\cdot\|_2)$. Also, by Weierstrass' theorem (Section 7, Corollary 3 to Theorem 1), the set of polynomial functions is dense in $(C[a,b], \|\cdot\|_\infty)$. The latter space is separable (Exercises 7, problem 2(b) and 2(c)). We might also note that the "trigonometric polynomials" are dense in $C_p[-\pi, \pi]$ (Section 7, Corollary 2 to Theorem 1).

Lemma 1

Let $(H, \langle,\rangle)$ be a Hilbert space and let $\{v_j\}$ be an orthonormal sequence in H. Then $\{v_j\}$ is an orthonormal basis for H if, and only if, $lin\{v_j\}$ is dense in H.

Proof. We recall that $lin\{v_j\}$ is the set of all (finite) linear combinations of vectors in $\{v_j\}$ (Section 1, Definition 2).

Suppose that $\{v_j\}$ is an orthonormal basis for H and let $v \in H$, $\varepsilon > 0$ be given. Then we can write

$$v = \sum_{j=1}^{\infty} \langle v, v_j \rangle v_j$$

by Theorem 3 of Section 6. Here the series converges to v for the norm. Thus we may choose an integer n such that

$$\|v - \sum_{j=1}^{n} \langle v, v_j \rangle v_j\| < \varepsilon$$

But clearly
$$\sum_{j=1}^{n} \langle v, v_j \rangle v_j \in lin\{v_j\}$$

Now suppose that $lin\{v_j\}$ is dense in H and that for some $v \in H$ we have $\langle v, v_j \rangle = 0$ for every j. We must prove (Theorem 3, Section 6) that v is the zero vector. Given $\varepsilon > 0$ we find $w \in lin\{v_j\}$ such that $\|v - w\| < \varepsilon$. Now clearly $\langle v, w \rangle = 0$ because of our assumption that $\langle v, v_j \rangle = 0$ for all j. Thus

$$\|v\|^2 = \langle v, v \rangle = \langle v - w, v \rangle \le \|v - w\|\|v\| \le \varepsilon \|v\|$$

by the C.S.B. inequality (Theorem 1, Section 4). If v is not the zero vector then we could divide this last inequality by $\|v\|$ obtaining

$$\|v\| \le \varepsilon$$

for all $\varepsilon > 0$.

Let $(V, \langle, \rangle)$ be an inner product space and let $\{w_j\}$ be a countably infinite, linearly independent subset of V. Then no w_j can be the zero vector, and we can never write w_n as a linear combination of the vectors $w_1, \ldots, w_{n-1}$. Let us apply the Gram-Schmidt process to this sequence. Recall (Lemma 1, Section 6) that we first set $v_1 = \frac{w_1}{\|w_1\|}$, next let

$$w_2' = w_2 - \langle w_2, v_1 \rangle v_1$$

and set $v_2 = \frac{w_2'}{\|w_2'\|}$. It is easy to see, by direct calculation that $\{v_1, v_2\}$ is an orthonormal set, and that

$$lin\{v_1, v_2\} = lin\{w_1, w_2\}$$

Continue constructing vectors $v_1, \ldots$ in this way. After $v_1, v_2, \ldots, v_{k-1}$ have been constructed we let

$$w_k' = w_k - \sum_{j=1}^{k-1} \langle w_k, v_j \rangle v_j$$

and we set $w_k = \frac{w_k'}{\|w_k'\|}$. Then $v_1, \ldots, v_k$ is an orthonormal set and

$$lin\{v_1, \ldots, v_k\} = lin\{w_1, \ldots, w_k\}$$

Since this is true for every k our process yields a sequence $v_1, \ldots, v_n, \ldots$ which is orthonormal and satisfies:

$$lin\{v_j\}_{j=1}^{\infty} = lin\{w_j\}_{j=1}^{\infty}$$

Under certain conditions this process will give us an orthonormal basis.

Lemma 2

Let $(H, \langle,\rangle)$ be a Hilbert space and let $\{w_j\}$ be a linearly independent sequence in H whose linear span is dense in H. Then the orthonormal sequence obtained from $\{w_j\}$ by the Gram-Schmidt process is an orthonormal basis for H.

Proof. The Gram-Schmidt process will yield an orthonormal sequence $\{v_j\}$ such that
$$lin\{v_j\} = lin\{w_j\}$$
Since the latter is dense in H so is the former. Thus $\{v_j\}$ is an orthonormal basis for H by Lemma 1.

We can use these results to obtain an orthonormal basis for $L^2[a,b]$. We work with the sequence $1, t, t^2, t^3, \ldots$. Let us first show that this sequence is linearly independent.

Suppose that for some m we have scalars $a_0, \ldots, a_m$ such that
$$\|\sum_{j=0}^{m} a_j t^j\|_2 = 0$$
then
$$\int_a^b |\sum_{j=0}^{m} a_j t^j|^2 dt = 0$$
which tells us that (see Section 8 just after (13)).
$$\sum_{j=0}^{m} a_j t^j = 0 \quad \text{a.e.}$$

But this sum is a polynomial and, unless all of its coefficients are zero, it can be zero for at most m values of t. Since the equation is given to hold a.e. we must conclude that $a_0 = a_1 = \ldots = a_m = 0$.

Next we show that $lin\{1, t, t^2, \ldots\}$ is dense in $L^2[a,b]$. Given $f \in L^2[a,b]$ and $\varepsilon > 0$ we choose a function $h \in C[a,b]$ such that

$$\|h - f\|_2 < \frac{\varepsilon}{2} \tag{1}$$

This is possible by Section 8, Theorem 4. Now we use Weierstrass' theorem (Section 7, Corollary 3 to Theorem 1) to find a polynomial $p(t)$ (i.e., an element of $lin\{1, t, t^2, \ldots\}$) such that

$$\|p - h\|_\infty < \frac{\varepsilon}{2(b-a)^{1/2}} \tag{2}$$

Combining (1) and (2) we have

$$\|p - f\|_2 \leq \|p - h\|_2 + \|h - f\|_2 \leq \frac{\varepsilon}{2(b-a)^{1/2}}(\int_a^b dt)^{1/2} + \frac{\varepsilon}{2} < \varepsilon$$

Finally, by applying the Gram-Schmidt process to the set $1, t, t^2, \ldots$ we obtain, as Lemma 2 tells us, an orthonormal basis for $L^2[a,b]$. Note that the functions in this basis are polynomials. They may be conveniently described as follows:

Lemma 3

The sequence of polynomials $\{p_n(t)\}_{n=0}^\infty$ defined by

$$p_n(t) = \frac{1}{\gamma_n} \frac{d^{(n)}}{dt^n}[(t-a)(t-b)]^n$$

where for each fixed n, γ_n is a real constant chosen so that $\|p_n\|_2 = 1$, is an orthonormal basis for $L^2[a,b]$. When $a = -1$, $b = 1$ these are the Legendre polynomials.

Proof. Since $[(t-a)(t-b)]^n$ is a polynomial of degree $2n$ its nth derivative must have degree n. Thus $p_n(t) \in lin\{1, t, \ldots, t^n\}$. Now suppose, we shall prove it below, that $p_0(t), p_1(t), \ldots$ is an orthonormal set. Then, in particular, it is a linearly independent set. Let us use this observation to prove that $lin\{p_0(t), \ldots\}$ is dense in $L^2[a,b]$. By Lemma 1 this will show that $p_0(t), \ldots$ is an orthonormal basis for our space.

Since $p_0(t) \in lin\{1\}$ we see that 1 is a scalar multiple of $p_0(t)$. Next, $p_0(t)$ and $p_1(t)$ are in $lin\{1, t\}$ and they are independent. So each of them is a linear combination of 1 and t and we can solve these equations obtaining 1 as a linear combination of $p_0(t)$, $p_1(t)$ and t as a linear combination of these polynomials also. In other words

$$lin\{p_0(t), p_1(t)\} = lin\{1, t\}$$

But clearly this holds not just for $n = 1$ but for any fixed n; i.e.,

$$lin\{p_0(t), \ldots, p_n(t)\} = lin\{1, t, \ldots, t^n\}$$

We conclude that

$$lin\{p_0(t), \ldots, p_n(t), \ldots\} = lin\{1, t, \ldots, t^n, \ldots\}$$

and since the latter is dense in $L^2[a,b]$ so is the former. Thus to complete the proof of this lemma we need only show that our polynomials are orthonormal.

Recall that γ_n has been chosen so that $\|p_n(t)\|_2 = 1$ for each n. Thus we need only show that

$$\langle p_m(t), p_n(t) \rangle = 0 \text{ whenever } 0 \leq m < n$$

In order to do this it clearly suffices to prove that $\langle t^m, \gamma_n p_n(t) \rangle = 0$ whenever $0 \leq m < n$. We shall write out this last inner product, when $m \neq 0$, and integrate (repeatedly) by parts.

$$\begin{aligned}
\gamma_n \int_a^b t^m \frac{d^{(n)}}{dt^n}[(t-a)(t-b)]^n dt &= \gamma_n t^m \frac{d^{(n-1)}}{dt^{n-1}}[(t-a)(t-b)]^n \Big|_a^b \\
&\quad - \gamma_n m \int_a^b t^{m-1} \frac{d^{(n-1)}}{dt^{n-1}}[(t-a)(t-b)]^n dt \\
&= \cdots \\
&= (-1)^m m! \gamma_n \frac{d^{(n-m-1)}}{d^{n-m-1}}[(t-a)(t-b)]^n \Big|_a^b \\
&= 0
\end{aligned}$$

The case $m = 0$ requires only a trivial modification of this argument.

We have called a space separable if it has a countable dense subset. The separability of $L^2[a,b]$ can be proved in various ways. In particular, it follows from Lemma 3 and our next result.

Theorem 1

A Hilbert space has a countable orthonormal basis if, and only if, it is separable.

Proof. Let $(H, \langle,\rangle)$ be a Hilbert space that has a countable orthonormal basis $\{v_j\}$. We shall consider only the case of a countably infinite basis and the case of real scalars. The other possibilities we leave to the reader. Given any $v \in H$ and any $\varepsilon > 0$ we first write

$$v = \sum_{j=1}^{\infty} a_j v_j \tag{3}$$

where, for each j, $a_j = \langle v, v_j \rangle$. This series converges to v for the norm of H so we can choose an integer n such that

$$\|v - \sum_{j=1}^{n} a_j v_j\| < \frac{\varepsilon}{2} \tag{4}$$

Next we choose rational numbers $b_1, b_2, \ldots, b_n$ so that

$$|a_j - b_j| < \frac{\varepsilon}{2n}, \ j = 1, 2, \ldots, n \tag{5}$$

Combining (4) and (5) we have

$$\|v - \sum_{j=1}^{n} b_j v_j\| \leq \|v - \sum_{j=1}^{n} a_j v_j\| + \|\sum_{j=1}^{n} a_j v_j - \sum_{j=1}^{n} b_j v_j\|$$

$$\leq \frac{\varepsilon}{2} + \sum_{j=1}^{n} |a_j - b_j| < \varepsilon$$

Thus the set of all (finite) linear combinations of the vectors $v_1, v_2, \ldots$ with rational coefficients is dense in H. But this is a countable set, so H is separable.

Now suppose that we have a separable Hilbert space $(H, \langle,\rangle)$. Then we can find a countable set $\{x_j\}_{j=1}^{\infty} \subseteq H$ which is dense in H and we can choose a Hamel basis (Section 1, Definition 4) $\mathcal{H}$ for H. So for each $v \in H$ there is a finite set $h_1, \ldots, h_k$ in $\mathcal{H}$ and scalars $\alpha_1, \alpha_2, \ldots, \alpha_k$ such that

$$v = \sum_{j=1}^{k} \alpha_j h_j$$

Let $\mathcal{S}(v)$ be the finite set $\{h_1, \ldots, h_k\}$. For $j = 1, 2, \ldots$ consider the sets $\mathcal{S}(x_1)$, $\mathcal{S}(x_2), \ldots$. Each of these is a finite set but they may not be pairwise disjoint. So we replace them by the finite sets $\mathcal{S}(x_1)$, $\mathcal{S}(x_2) \setminus \mathcal{S}(x_1)$, $\mathcal{S}(x_3) \setminus \mathcal{S}(x_1) \cup \mathcal{S}(x_2)$, $\ldots$, $\mathcal{S}(x_n) \setminus \cup_{j=1}^{n-1} \mathcal{S}(x_j), \ldots$. These are all vectors in H and the set containing all of them, call it $\mathcal{B}$, is a countable subset of $\mathcal{H}$. So $\mathcal{B}$ is linearly independent; because $\mathcal{H}$ is linearly independent. Also, $lin\mathcal{B} \supseteq lin\{x_j\}$ showing that $lin\mathcal{B}$ is dense in H. We can now prove that H has a countable orthonormal basis by applying the Gram-Schmidt process to $\mathcal{B}$ (Lemma 2).

Corollary 1

Any separable, infinite dimensional Hilbert space over $\mathbb{R}$ (resp. $\mathbb{C}$) is isometrically isomorphic to real (resp. complex) ℓ^2. Any two separable, infinite dimensional Hilbert spaces over the same field are isometrically isomorphic.

Proof. The first statement follows from the theorem and Section 6, Theorem 4. See also Definition 2. The second statement we leave to the exercises (see problem 4).

As a special case of Corollary 1, we see that $L^2[a, b]$ is isometrically isomorphic to ℓ^2 and also to $L^2[-\pi, \pi]$.

We have said (Section 8 just after equation (8)) that $L^2[a, b]$ is the "smallest" Hilbert space which contains $C[a, b]$. Let us explain what we mean by this. The discussion requires some care.

Let $(V, \langle,\rangle)$ be an inner product space and let $(H, \langle,\rangle')$ be a Hilbert space (both over the same field $\mathcal{K}$). We shall say that H is a completion of V if H has a dense, linear manifold (Section 1, Definition 1) V' which is isometrically isomorphic (Section 6, Definition 2) to V.

We have:

Theorem 2

Any inner product space over $\mathcal{K}$ has a completion and any two completions of the same inner product space are isometrically isomorphic.

The proof is straightforward but very, very tedious and since we shall rarely, if ever, need the result we shall omit it. A proof can be found in [9; p. 331]. It is customary to speak of "the" completion of an inner product space, and to say that an inner product space is a linear manifold in its completion. Theorem 4 of Section 8 tells us that $L^2[a, b]$ is the completion of $C[a, b]$. We mention one final result.

Theorem 3

Let $(V, \langle,\rangle)$ be an inner product space. Then any Hilbert space H which contains (a linear manifold which is isometrically isomorphic to) V also contains (a linear manifold which is isometrically isomorphic to) the completion of V.

We cannot end our discussion of L^2 without mentioning a very remarkable result due to L. Carleson.

Theorem: Let $f \in L^2[-\pi, \pi]$. Then the Fourier series of f converges to this function pointwise almost everywhere.

We shall not need this result. A proof can be found in [2]. Note that, in particular, the Fourier series of a continuous function on $[-\pi, \pi]$ must converge to the function pointwise almost everywhere; so the set at which such a series diverges (or does not converge to the function) must have measure zero.

Exercises 9

1. Show that for any fixed n the spaces $(\mathbb{R}^n, \langle,\rangle)$ and $(C^n, \langle,\rangle)$ are separable.

* 2. Let $(E, \|\cdot\|)$ be a normed space.

a. Let $S \subseteq T \subseteq E$ and suppose that S is dense in T (we mean that given any $t \in T$ and any $\varepsilon > 0$ there is an $s \in S$ with $\|s - t\| < \varepsilon$) and that T is dense in E. Prove that S is dense in E.

b. A subset G of E is said to be a total set if $lin G$ is dense in E. Show that $(E, \|\cdot\|)$ is separable if, and only if, it has a countable total subset.

3. Let $(E, \|\cdot\|)$ be a normed space and let S be a subset of E. Show that S is dense in E if, and only if, every point of E is the limit of a sequence of points in S.

* 4. Let V_1, V_2 be two vector spaces over the same field $\mathcal{K}$ and let T be a map from V_1 to V_2 which is both one-to-one and onto. Define $T^{-1}: V_2 \to V_1$ as follows: $T^{-1}(v_2) = v_1$ if, and only if, $T(v_1) = v_2$. If T is linear, show that T^{-1} is linear (Section 5, Definition 1). If T is an isomorphism show that T^{-1} is one also.

a. Now suppose that $(V_1, \langle, \rangle)$ and $(V_2, \langle, \rangle)$ are two inner product spaces over the same field $\mathcal{K}$ and that T is an isometric isomorphism from V_1 onto V_2 (Section 6, Definition 2). Show that T^{-1} is an isometric isomorphism from V_2 to V_1.

b. Show that if two inner product spaces are both isometrically isomorphic to the same inner product space, then they are isometrically isomorphic to each other.

5. Show that the Hilbert space $L^2(S)$, where $S = [a, b] \times [a, b]$, is separable. Hint: See Exercises 8, problem 4.

2

LINEAR MAPPINGS AND OPERATORS

One of the classic books on functional analysis, "Operations Linears," was written by Stephan Banach in 1932. The English translation of the title is "Linear Operations." Banach explains, in the preface, that the theory of operations was created by V. Volterra and was concerned with the study of functions defined on infinite dimensional spaces. As the title indicates this chapter is concerned with linear functions of this kind. We mentioned earlier that there is little we can say about an infinite dimensional space unless it has some additional structure, like a norm or inner product for example. Similarly, there is little we can say about linear operators unless they are defined on a space with some additional structure and they are somehow tied in to that additional structure. We will be able to be more specific later on. Our purpose in this chapter is to define and study the basic concepts relating to linear mappings between, or linear operators upon, normed spaces. It is usually wise to define these concepts in some generality. This enhances their applicability but, unfortunately, it makes many of our results seem overly technical and sometimes awkward to state.

2.1 Bounded Linear Maps

Let us look once again at the integral equation

$$u(s) = g(s) + \int_a^b K(s,t)u(t)dt \qquad (1)$$

already considered in 1.8. Recall that the function K is a fixed, known function defined and continuous on $[a,b] \times [a,b]$, and that we seek a solution u to (1) for each given $g \in C[a,b]$. For any $f \in C[a,b]$ we defined a continuous function $K(f)$ as follows

$$K(f)(s) = \int_a^b K(s,t) f(t) dt$$

Let us look now at the operator, call it T, which takes each $f \in C[a,b]$ to the function $K(f) \in C[a,b]$. So

$$T[f] = K(f)$$

and

$$\begin{aligned} T[\alpha f + \beta g](s) &= K(\alpha f + \beta g)(s) \\ &= \int_a^b K(s,t)\{\alpha f(t) + \beta g(t)\} dt \\ &= \alpha \int_a^b K(s,t) f(t) dt + \beta \int_a^b K(s,t) g(t) dt \\ &= \alpha K(f)(s) + \beta K(g)(s) = \alpha T[f](s) + \beta T[g](s) \end{aligned}$$

giving us

$$T[\alpha f + \beta g] = \alpha T[f] + \beta T[g]$$

Thus T is a linear operator on $C[a,b]$. Furthermore, we may write (1) as

$$(I - T)[u] = g \tag{2}$$

where I is the identity operator (Exercises 1.5, problem 2b). Clearly (1) has a solution in $C[a,b]$ for the given $g \in C[a,b]$ if, and only if, this function is in the range of $I - T$ and, when this is the case, the solution will be unique if, and only if, the null space of $I - T$ contains only the zero vector (see 1.1 just after Definition 5 for this terminology). So we have shown how questions about the existence and uniqueness of solutions to (1) can be translated into questions about the range and null space of a certain linear operator. We shall come back to this many times below.

Let us notice one more thing about the operator T. For any $f \in C[a,b]$ we have

$$T[f](s) = \int_a^b K(s,t) f(t) dt$$

and so

$$\begin{aligned} \|T[f]\|_\infty &= \sup_{a \le s \le b} |\int_a^b K(s,t) f(t) dt| \le \sup \int_a^b |K(s,t)||f(t)| dt \\ &\le \int_a^b \{\sup |K(s,t)| \sup |f(t)|\} dt = M \|f\|_\infty \end{aligned}$$

where M is the number
$$(b-a)\sup |K(s,t)|$$
the supremum being taken over $[a,b] \times [a,b]$. Now what does this inequality say about the operator T? Recall that in elementary calculus we study continuity of a function at a fixed point, continuity at each point of some interval, or uniform continuity over an interval, and these are different things. However, as we shall soon see, for linear maps between normed spaces these three types of continuity are equivalent and each of them is equivalent to the fact that the map one is working with satisfies an inequality like the one we have just derived for T. Let us prove this.

Lemma 1

Let $(E_1, \|\cdot\|_1)$ and $(E_2, \|\cdot\|_2)$ be two normed spaces over the same field, and let T be a linear map from E_1 into E_2. If T has any one of the five following properties, it has all five of them:

a. (continuity at a point) For some fixed $x_0 \in E_1$ we have: Given $\varepsilon > 0$ there is a $\delta > 0$ such that $\|Tx - Tx_0\|_2 < \varepsilon$ whenever $x \in E_1$ and $\|x - x_0\|_1 < \delta$;

b. (continuity at zero) For any $\varepsilon > 0$ there is a $\delta > 0$ such that $\|Tx\|_2 < \varepsilon$ whenever $x \in E_1$ and $\|x\|_1 < \delta$;

c. (continuity at every point of E_1) For any $x \in E_1$ we have: Given $\varepsilon > 0$ there is a $\delta > 0$ such that $\|Tx - Ty\|_2 < \varepsilon$ whenever $y \in E_1$ and $\|x - y\|_1 < \delta$;

d. (uniform continuity on E_1) Given $\varepsilon > 0$ there is a $\delta > 0$ such that $\|Tx - Ty\|_1 < \varepsilon$ whenever $x, y \in E_1$ and $\|x - y\|_1 < \delta$;

e. (sequential continuity) Given any sequence $\{x_n\} \subseteq E_1$ which is convergent to a point $x_0 \in E$, the sequence $\{Tx_n\} \subseteq E_2$ is convergent to the point $Tx_0 \in E_2$.

Proof. The proof is easy but it does contain a few useful tricks. First assume that T has property (a). So for some $x_0 \in E_1$ and any $\varepsilon > 0$ we can choose $\delta > 0$ such that $\|Tx - Tx_0\|_2 < \varepsilon$ whenever $\|x - x_0\|_1 < \delta$. Then for any $w \in E_1$ with $\|w\|_1 < \delta$ we have $\|T(w + x_0) - Tx_0\|_2 < \varepsilon$ because $\|(w + x_0) - x_0\|_1 = \|w\|_1 < \delta$. But since T is linear this says that $\|Tw\|_2 < \varepsilon$ whenever $\|w\|_1 < \delta$ and we have shown that (a) implies (b).

Now suppose that T has property (b), let $\varepsilon > 0$ be given and choose $\delta > 0$ such that $\|Tw\|_2 < \varepsilon$ whenever $\|w\|_1 < \delta$. Then for any $x \in E_1$ we have $\|T(x - y)\|_2 < \varepsilon$ whenever $\|x - y\|_1 < \delta$; just use $x - y$ in place of w. Again recalling that T is linear we see that (b) implies (c). However (c), which is continuity at every point of E_1 clearly implies (a) which is continuity at some one fixed point of E_1. Thus (a), (b), and (c) are equivalent.

Let us show that (b) implies (d). Given $\varepsilon > 0$ we may choose $\delta > 0$ so that $\|w\|_1 < \delta$ implies $\|Tw\|_2 < \varepsilon$. Now if $x, y \in E_1$ and $\|x - y\|_1 < \delta$ then $\|T(x - y)\|_2 < \varepsilon$; again just use $x - y$ for w. Since T is linear (b) implies (d) and clearly (d) implies (b). Thus (a) through (d) are equivalent.

We will complete the proof by showing that (b) and (e) are equivalent. First assume that T has property (b) and let $\varepsilon > 0$ be given. Then we choose δ so that $\|Tw\|_2 < \varepsilon$ whenever $\|w\|_1 < \delta$. Now suppose that $\{x_n\} \subseteq E_1$ converges to x_0. Then we can find an integer N such that $\|x_0 - x_n\|_1 < \delta$ whenever $n \geq N$. Thus $\|T(x_0 - x_n)\|_2 < \varepsilon$ whenever $n \geq N$. Clearly this says that $\{Tx_n\}$ converges to Tx_0.

Now assume (e) and the negation of (b). So we are supposing that there is an $\varepsilon > 0$ such that for any $\delta > 0$ we can find $w \in E_1$ with $\|w\|_1 < \delta$ and $\|Tw\|_2 \geq \varepsilon$. Thus for this ε we can find w_1 such that $\|w_1\|_1 < 1$ and $\|Tw_1\|_2 \geq \varepsilon$; otherwise $\delta = 1$ will satisfy (b). Next we find w_2 with $\|w_2\|_1 < 1/2$ and $\|Tw_2\|_2 \geq \varepsilon$; again because $\delta = 1/2$ cannot satisfy the requirements of (b). Continue in this way. After $w_1, w_2, \ldots, w_n$ have been chosen so that $\|w_j\|_1 < 1/j$ and $\|Tw_j\|_2 \geq \varepsilon$, $1 \leq j \leq n$, we note that $\delta = \frac{1}{n+1}$ cannot satisfy (b) so we must be able to find w_{n+1} such that $\|w_{n+1}\|_2 < \frac{1}{n+1}$ and $\|Tw_{n+1}\|_2 \geq \varepsilon$. In this way we obtain a sequence $\{w_n\} \subseteq E_1$ which clearly converges to the zero vector. By (e) then, $\{Tw_n\}$ must converge to $T(0)$, the zero vector of E_2 (because T is linear so $Tx = T(x + 0) = Tx + T(0)$ giving $T(0) = 0$). But then we would have $\lim_{n \to \infty} \|Tw_n\|_2 = \|0\|_2 = 0$ (Exercises 1.3, problem 1) and this is impossible because $\|Tw_n\|_2 \geq \varepsilon$ for every n.

We should point out that for nonlinear mapping conditions (a)-(e) are not equivalent. The most practical means of showing that a linear map is continuous is contained in the next result.

Theorem 1

Let $(E_1, \|\cdot\|_1)$, $(E_2, \|\cdot\|_2)$ be two normed spaces over the same field and let T be a linear map from E_1 to E_2. If T has any of the three following properties it has all three of them.

1. (Boundedness) There is a real number M such that
$$\|Tx\|_2 \leq M\|x\|_1$$
for all $x \in E_1$;

2. (Bounded of the unit ball of E_1 (Exercises 1.2, problem 2c)) There is a real number M such that
$$\|Tx\|_2 \leq M$$

whenever $x \in E_1$ and $\|x\|_1 \leq 1$;

3. (Sequential continuity, (e)) The sequence $\{Tx_n\} \subseteq E_2$ converges to Tx_0 whenever $\{x_n\} \subseteq E_1$ converges to x_0.

Proof. Clearly (1) implies (2). Suppose T has property (2). Then for any $x \in E_1$, $x \neq 0$, $\frac{x}{\|x\|_1}$ has norm one; i.e., $\|\frac{x}{\|x\|_1}\|_1 = 1$. Hence by (2) we have

$$\|T\left(\frac{x}{\|x\|_1}\right)\|_2 \leq M$$

or, by the linearity of T,

$$\|Tx\|_2 \leq M\|x\|_1$$

for all non-zero elements of E_1. But clearly this holds for the zero vector as well and so (2) implies (1). Thus (1) and (2) are equivalent.

Let us show that (1) implies (3). Given a sequence $\{x_n\} \subseteq E_1$ that converges to $x_0 \in E_1$ we have, by (1),

$$\|T(x_0 - x_n)\|_2 \leq M\|x_0 - x_n\|_1$$

and since $\|x_0 - x_n\|_1$ tends to zero as n tends to infinity we see that $\{Tx_n\}$ converges to Tx_0.

Finally, let us show that (3) implies (2). We shall argue by contradiction. So assume (3) and the negation of (2). Then for any M we can find $x \in E_1$ with $\|x\|_1 \leq 1$ and $\|Tx\|_2 > M$. Take $M = 1^2$ and find x_1 in E_1 with $\|x_1\|_1 \leq 1$ and $\|Tx_1\|_2 > 1^2$. Next take $M = 2^2$ and find $x_2 \in E_1$ with $\|x_2\|_1 \leq 1$ and $\|Tx_2\|_2 > 2^2$. Continue in this way. After $x_1, \ldots, x_n$ have been chosen so that $\|x_j\|_1 \leq 1$ and $\|Tx_j\|_2 > j^2$, $1 \leq j \leq n$, we set $M = (n+1)^2$ and find x_{n+1} with $\|x_{n+1}\|_1 \leq 1$ and $\|Tx_{n+1}\|_2 > (n+1)^2$. In this way we generate a sequence $\{x_n\}$ of points of E_1. Now note that

$$\|x_n/n\|_1 \leq \frac{1}{n}$$

for every n and so $\{\frac{x_n}{x}\} \subseteq E_1$ converges to the zero vector. By linearity and (3) then, $\{T(\frac{x_n}{n})\}$ must converge to the zero vector of E_2. But this is impossible because

$$\|T(\frac{x_n}{x})\|_2 > \frac{n^2}{n} = n$$

for every n.

Definition 1

A linear mapping from E_1 into E_2 which has any, and hence all, of the properties listed in Theorem 1 will be called a bounded (or, sometimes, a continuous) linear map. A bounded, linear mapping from E_1 into itself will be called a bounded, linear operator on E_1.

CHAPTER 2   LINEAR MAPPINGS AND OPERATORS

Note that a bounded, linear mapping has all of the properties listed in Lemma 1. Suppose now that T is such a mapping and let M be a real number such that
$$\|Tx\|_2 \leq M\|x\|_1$$
for all x in E_1. Now any number larger than M will also satisfy this inequality. We want the smallest number which will satisfy it. This number is, dividing by $\|x\|_1$,
$$\sup\left\{\frac{\|Tx\|_2}{\|x\|_1} \mid x \neq 0\right\}$$
because this is as "large" as our lefthand side can be. It is easy to see that
$$\{\frac{\|Tx\|_2}{\|x\|_1}|x \neq 0\} = \{\|T(\frac{x}{\|x\|_1})\|_2 \mid x \neq 0\}$$
$$\subseteq \{\|Tx\|_2 \mid \|x\|_1 = 1\} \subseteq \{\|Tx\|_2 \mid \|x\|_1 \leq 1\}$$
and so
$$M_1 = \sup\{\frac{\|Tx\|_2}{\|x\|_1}|x \neq 0\} \leq M_2$$
$$= \sup\{\|Tx\|_2 \mid \|x\|_1 = 1\} \leq M_3 = \sup\{\|Tx\|_2 \mid \|x\|_1 \leq 1\}$$
But now, for all $x \neq 0$,
$$\frac{\|Tx\|_2}{\|x\|_1} \leq M_1$$
and this implies that
$$\|Tx\|_2 \leq M_1\|x\|_2$$
for all x which, in turn, gives
$$\|Tx\|_2 \leq M_1$$
for all $x \in E_1$ such that $\|x\|_1 \leq 1$. It follows that $M_3 \leq M_1$ and hence $M_1 = M_2 = M_3$.

Now what is the point of these calculations? To answer that we must raise the level of abstraction just a little bit. We want to consider the collection of all bounded, linear maps from E_1 into E_2. It is easy to see that this is a vector space under the usual definitions of addition and scalar multiplication of functions (Section 1 just after Definition 6); we shall actually prove this in the course of proving Theorem 2 below. We shall denote this vector space by

(*) $\mathcal{L}(E_1, E_2)$

or, when $E_1 = E_2$, by

(**) $\mathcal{L}(E_1)$

It may be helpful to see just what the zero vector in $\mathcal{L}(E_1, E_2)$ is. We have 0_1, the zero vector in E_1, 0_2, the zero vector in E_2, and 0, the zero vector of $\mathcal{L}(E_1, E_2)$. This last zero is the bounded, linear map from E_1 into E_2 such that $0(x) = 0_2$ for each $x \in E_1$.

The calculations done above lead us to define a function on $\mathcal{L}(E_1, E_2)$. We shall see that it is a norm and we shall call it the operator norm on $\mathcal{L}(E_1, E_2)$. This is a little misleading since our mappings are operators only when $E_1 = E_2$, however, it is the standard terminology.

Definition 2

For any $T \in \mathcal{L}(E_1, E_2)$ the number

$$\sup\{\frac{\|Tx\|_2}{\|x\|_1} | x \neq 0\} = \sup\{\|Tx\|_2 | \|x\|_1 = 1\} = \sup\{\|Tx\|_2 | \|x\|_1 \leq 1\}$$

is called the operator norm of T and is denoted by $\|\cdot\|$. Whenever we speak of $\mathcal{L}(E_1, E_2)$, or of $\mathcal{L}(E_1)$, as a normed space this is the norm we shall mean.

Let us now justify this terminology.

Theorem 2

Let E_1, E_2 be two normed spaces over the same field and let $\mathcal{L}(E_1, E_2)$ be the space of all bounded, linear mappings from E_1 into E_2. Then the function defined in Definition 2 is a norm on this space.

Proof. The properties of a norm are listed in Section 1.2, Definition 1. Given $S, T \in \mathcal{L}(E_1, E_2)$ we have

$$\begin{aligned} \|S+T\| &= \sup\{\|(S+T)(x)\|_2 | \|x\|_1 = 1\} \\ &= \sup\{\|Sx + Tx\|_2 | \|x\|_1 = 1\} \\ &\leq \sup\{\|Sx\|_2 + \|Tx\|_2 | \|x\|_1 = 1\} \end{aligned}$$

by the triangle inequality (Section 1.2, Definition 1 part (a)) for $\|\cdot\|_2$. Clearly then

$$\begin{aligned} \|S+T\| &\leq \sup\{\|Sx\|_2 + \|Tx\|_2 | \|x\|_1 = 1\} \leq \sup\{\|Sx\|_2 | \|x\|_1 = 1\} \\ &+ \sup\{\|Tx\|_2 | \|x\|_1 = 1\} \\ &= \|S\| + \|T\|. \end{aligned}$$

Next for any scalar α we have

$$\begin{aligned} \|\alpha T\| &= \sup\{\|\alpha T(x)\|_2 | \|x\|_1 = 1\} = \sup\{|\alpha| \|Tx\|_2 | \|x\|_1 = 1\} \\ &= |\alpha| \sup\{\|Tx\|_2 | \|x\|_1 = 1\} = |\alpha| \|T\|. \end{aligned}$$

Finally, suppose that $||T|| = 0$. Then

$$\sup\{||Tx||_2 | ||x||_1 = 1\} = 0$$

which says that Tx is the zero vector whenever $x \in E_1$ and $||x||_1 = 1$; because of property (c) of the norm $||\cdot||_2$. But for any $x \in E_1$ except the zero vector we have that $\frac{x}{||x||_1}$ has norm 1 for $||\cdot||_1$ and so T of this must be the zero vector of E_2. Clearly that says that $Tx = 0$ for every non-zero $x \in E_1$. But this is always true for the zero vector so $Tx = 0$ for every x in E_1 and this says that T is the "zero" linear map.

Exercises 1

1. Let $(E_1, ||\cdot||_1)$ and $(E_2, ||\cdot||_2)$ be normed spaces over the same field $\mathcal{K}$.

    a. If E_1 is finite dimensional (Section 1.1, Definition 4) show that every linear mapping from E_1 into E_2 is a bounded, linear map.

    b. Explain how Definition 2 gives us a norm on the space of all $m \times n$ matrices.

    c. A subset S of E_1 is said to be a bounded set if there is a number M such that $||x|| \leq M$ for every $x \in S$. Show that a linear map T from E_1 into E_2 is a continuous, linear map if, and only if, $T(S) = \{Tx | x \in S\}$ is a bounded subset of $(E_2, ||\cdot||_2)$ whenever S is a bounded subset of $(E_1, ||\cdot||_1)$.

* 2. Let $(E_1, ||\cdot||_1)$ be a normed space and let $\{x_n\} \subseteq E$ be a Cauchy sequence (Section 1.3, Definition 2) in E. Show that $\{x_n\}$ is a bounded subset of E as defined in problem 1(c).

* 3. Let S, T be two bounded, linear operators on the normed space $(E_1, ||\cdot||_1)$.

    a. Show $S \cdot T$ and $T \cdot S$ (Exercises 1.5, problem 2(b)) are bounded, linear operators on E.

    b. Show that $||S \cdot T|| \leq ||S||||T||$ and these need not be equal. Hint: Find two nonzero square matrices whose product is zero.

    c. Define $T^0 = I$, the identity operator (Exercises 1.5, problem 2(b)), $T^1 = T$ and for any integer $n > 1$ let $T^n = T \cdot (T^{n-1})$. Show that $||T^n|| \leq ||T||^n$ for any fixed n.

* 4. Show that an isometry (Section 1.6, Definition 2) must have norm one. In particular, an isometric isomorphism has norm one.

* 5. We have defined two operators on the space ℓ^2 (see Exercises 1.5, problem 2(i) and 2(ii)). Compute the norm of each of these operators. Hint: First compute the norm of A_r. Then use the fact that $A_\ell A_r = I$ to get an upper bound on $\|A_\ell\|$.

2.2 Comparing Norms

We have worked with the sup norm and also the L^2-norm on the space $C[a,b]$, and we have seen several norms on $\mathbb{R}^2$ (Exercises 1.2, problem 1). One can compare two norms and there is a very reasonable notion of equivalence between two norms. The practical importance of this is that in certain problems one can replace a norm that is difficult to work with by an equivalent one which is much easier to use. Equivalence of norms is usually expressed in terms of an inequality which we shall present soon (see also Exercises 1.3, problem 2). Before doing that, however, we shall try to present the ideas behind the definition. We must introduce some terminology which will be familiar to many readers.

Definition 1

Let $(E, \|\cdot\|)$ be a normed space, let x_0 be an arbitrary, but fixed, point of E, and let $\varepsilon > 0$ be given. Then $\mathcal{B}_\varepsilon(x_0) = \{x \in E | \|x - x_0\| \leq \varepsilon\}$ is called the closed ball of radius ε centered at x_0, and $\mathcal{O}_\varepsilon(x_0) = \{x \in E | \|x - x_0\| < \varepsilon\}$ is called the open ball of radius ε centered at x_0. A subset S of E is said to be a neighborhood of x_0 if it contains an open ball centered at x_0; i.e., if for some $\varepsilon > 0$ we have $\mathcal{O}_\varepsilon(x_0) \subseteq S$.

Observe that the unit ball of E is just $\mathcal{B}_1(0)$. Furthermore,

$$\mathcal{B}_\varepsilon(0) = \varepsilon \mathcal{B}_1(0) = \varepsilon\{x \in E | \|x\| \leq 1\} \equiv \{\varepsilon x | \|x\| \leq 1\}$$

and

$$\mathcal{B}_\varepsilon(x_0) = x_0 + \varepsilon \mathcal{B}_1(0) = \{y \in E | y = x_0 + \varepsilon x \text{ for } x \in E \text{ with } \|x\| \leq 1\}$$

because

$$x_0 + \varepsilon \mathcal{B}_1(0) = \{x_0 + \varepsilon x | \|x\| \leq 1\}$$

and anything in here is within ε of x_0 hence in $\mathcal{B}_\varepsilon(x_0)$. Similarly,

$$\mathcal{O}_\varepsilon(x_0) = x_0 + \varepsilon \mathcal{O}_1(0)$$

These observations show that the neighborhoods of any point of E are just translates of the neighborhoods of zero. Thus the "local properties" of E

are the same at every point because "near" any point x_0 the space behaves as it does "near" zero. This imprecise but intuitively useful comment will become clear as we go. Note however that we have already seen that continuity at any point, for linear maps, is the same as continuity at zero (Section 1, Lemma 1(a), (b)).

Let us note that the definition of a convergent sequence (Section 1.3, Definition 1) can be conveniently formulated in terms of open balls. A sequence $\{x_n\}$ of points of E converges to the point $x_0 \in E$ if, and only if, given any open ball centered at x_0, say $\mathcal{O}_\varepsilon(x_0)$, there is an integer N such that

$$x_n \in \mathcal{O}_\varepsilon(x_0)$$

whenever $n \geq N$.

Definition 2

Let $(E, \|\cdot\|)$ be a normed space. A subset of E is said to be an open set if it is empty, or it is non-empty and is a neighborhood of each of its points. A subset of E is said to be a closed set if its (set-theoretic) complement in E is an open set.

We leave it to the reader (see the Exercises) to show that an open ball is an open set and a closed ball is a closed set. Of course, many sets are neither open nor closed, and the sets $\emptyset, E$ are both open and closed (they are the only two sets for which this is true). The union of any family of open sets is an open set; because any x_0 in the union must be in some set, say $\mathcal{O}$, of the family and $\mathcal{O}$ (being open) is a neighborhood of x_0, hence the union (which contains $\mathcal{O}$) is a neighborhood of x_0. Similarly, one can show that the intersection of any finite family of open sets is an open set.

The family of all open subsets of $(E, \|\cdot\|)$ is the topology defined on E by the given norm. It is this, the topology, that determines which functions on E are continuous, as the following result shows:

Theorem 1

Let $(E, \|\cdot\|)$ and $(F, \|\|\cdot\|\|)$ be two normed spaces over the same field and let T be a map from E into F. Then T is continuous on E if, and only if, the set

$$T^{-1}(\mathcal{O}) = \{x \in E | Tx \in \mathcal{O}\}$$

is open in $(E, \|\cdot\|)$ whenever the set $\mathcal{O}$ is open in $(F, \|\|\cdot\|\|)$.

Remark. There are several things to notice here before we begin the proof. First, we are not saying that T has an inverse even though it may seem like it. The notation $T^{-1}(S)$ is used to mean $\{x \in E | Tx \in S\}$, $S \subseteq F$. So $T^{-1}(S)$

is a set and it could be empty even if S is not; for S may not intersect the range of T. Second, the theorem is true even if T is nonlinear. In such a case continuity on E means: For any $\varepsilon > 0$ and any $x \in E$ there is a $\delta > 0$ such that $|||Tx - Ty||| < \varepsilon$ whenever $||x - y|| < \delta$.

Proof. Suppose first that T is continuous on E and that $\mathcal{O}$ is a non-empty, open subset of F. Choose $x_0 \in T^{-1}(\mathcal{O})$. Since $\mathcal{O}$ is open there is an $\varepsilon > 0$ such that $\mathcal{O}_\varepsilon(Tx_0) \subseteq \mathcal{O}$. By the continuity of T there is a $\delta > 0$, corresponding to the ε just found, such that

$$|||Tx - Tx_0||| < \varepsilon \text{ whenever } ||x - x_0|| < \delta$$

to put this another way

$$Tx \in \mathcal{O}_\varepsilon(Tx_0) \text{ whenever } x \in \mathcal{O}_\delta(x_0)$$

Thus $\mathcal{O}_\delta(x_0) \subseteq T^{-1}(\mathcal{O})$ showing that this last set is a neighborhood of x_0.

Now suppose that T satisfies our condition, and let x_0 be any point of E. Choose and fix $\varepsilon > 0$ and consider the set $\mathcal{O}_\varepsilon(Tx_0)$. Since this is open so also is

$$T^{-1}[\mathcal{O}_\varepsilon(Tx_0)]$$

and the latter must then be a neighborhood of x_0; for it clearly contains this point. Hence there is a $\delta > 0$ such that

$$\mathcal{O}_\delta(x_0) \subseteq T^{-1}[\mathcal{O}_\varepsilon(Tx_0)]$$

or

$$T[\mathcal{O}_\delta(x_0)] \subseteq \mathcal{O}_\varepsilon(Tx_0)$$

This last inclusion says that $|||Tx - Tx_0||| < \varepsilon$ whenever $x \in \mathcal{O}_\delta(x_0)$; i.e., whenever $||x - x_0|| < \delta$.

We have just seen that a norm determines a topology and a topology determines the continuous functions on the space. So we shall compare norms by comparing the topologies they determine: The stronger norm gives us the stronger topology (meaning more open sets) and hence more continuous functions. Now how can we compare topologies? Well, a topology is just a family of open sets, an open set is one which is a neighborhood of each of its points, and a neighborhood of any point is just a translate of a neighborhood of zero, and a neighborhood of zero is just a multiple of the unit ball. So we can compare the topologies that two norms give us by comparing the unit balls in each of these norms.

Definition 3

Let E be a vector space over $\mathcal{K}$ and let $||\cdot||_1$ and $||\cdot||_2$ be two norms on this space. We shall say that $||\cdot||_1$ is weaker than $||\cdot||_2$, and we shall write

$\|\cdot\|_1 \leq \|\cdot\|_2$, if there is a positive number λ such that $\lambda \mathcal{B}_1(\|\cdot\|_2) \subseteq \mathcal{B}_1(\|\cdot\|_1)$. We shall say that $\|\cdot\|_1$ and $\|\cdot\|_2$ are equivalent, and we shall write $\|\cdot\|_1 \equiv \|\cdot\|_2$, if we have both $\|\cdot\|_1 \leq \|\cdot\|_2$ and $\|\cdot\|_2 \leq \|\cdot\|_1$.

Suppose that $\|\cdot\|_1 \leq \|\cdot\|_2$ and that S is any non-empty subset of E which is open for $\|\cdot\|_1$. Then given any $x_0 \in S$ we must have

$$\mathcal{O}_{\varepsilon 1}(x_0) = \{x \in E \mid \|x - x_0\|_1 < \varepsilon\} \subseteq S$$

for some $\varepsilon > 0$, where the subscript 1 refers to $\|\cdot\|_1$. Thus

$$\mathcal{O}_{\varepsilon 1}(x_0) = x_0 + \varepsilon \mathcal{O}_{11}(0) \subseteq S$$

where $\mathcal{O}_{11}(0) = \{x \in E \mid \|x\|_1 < 1\}$. Now $\lambda \mathcal{B}_1(\|\cdot\|_2) \subseteq \mathcal{B}_1(\|\cdot\|_1)$ because $\|\cdot\|_1 \leq \|\cdot\|_2$ and so

$$\frac{\lambda}{2}\mathcal{O}_{12}(0) = \frac{\lambda}{2}\{x \in E \mid \|x\|_2 < 1\} \subseteq \mathcal{O}_{11}(0)$$

From this we see that

$$\frac{\varepsilon\lambda}{2}\mathcal{O}_{12}(0) \subseteq \varepsilon\mathcal{O}_{11}(0)$$

and hence that

$$x_0 + \frac{\varepsilon\lambda}{2}\mathcal{O}_{12}(0) \subseteq x_0 + \varepsilon\mathcal{O}_{11}(0) = \mathcal{O}_{\varepsilon 1}(x_0)$$

However

$$x_0 + \frac{\varepsilon\lambda}{2}\mathcal{O}_{12}(0) = \mathcal{O}_{\frac{\varepsilon\lambda}{2} 2}(x_0)$$

and we see that S is a neighborhood of x_0 for $\|\cdot\|_2$. Thus any subset of E which is open for $\|\cdot\|_1$ is also open for $\|\cdot\|_2$; so the stronger norm does in fact give us the stronger topology (i.e., a topology with more open sets).

Once again, let $\|\cdot\|_1, \|\cdot\|_2$ be two norms on E with $\|\cdot\|_1 \leq \|\cdot\|_2$ and let λ be a positive number for which

$$\lambda \mathcal{B}_1(\|\cdot\|_2) \leq \mathcal{B}_1(\|\cdot\|_1)$$

For any nonzero vector x in E we have $\frac{x}{\|x\|_2} \in \mathcal{B}_1(\|\cdot\|_2)$ and so λ times this must be in $\mathcal{B}_1(\|\cdot\|_1)$. In other words

$$\|\lambda \frac{x}{\|x\|_2}\|_1 \leq 1$$

hence

$$\lambda \|x\|_1 \leq \|x\|_2$$

for all nonzero elements of E, and clearly this holds for the zero vector also. If we also have $\|\cdot\|_2 \le \|\cdot\|$ then

$$\mu\|x\|_2 \le \|x\|_1$$

for all $x \in E$ and some fixed, positive number μ. Combining these inequalities we have: Two norms $\|\cdot\|_1$ and $\|\cdot\|_2$ a vector space E are equivalent if, and only if, there are positive constants λ and μ such that

$$\lambda\|x\|_1 \le \|x\|_2 \le \mu\|x\|_2$$

for all $x \in E$; note that we have changed our notation slightly; μ is now $1/\mu$.

Theorem 2

Any two norms on a finite dimensional vector space are equivalent.

Proof. Let F be a finite dimensional space and let $\|\cdot\|_1$ and $\|\cdot\|_2$ be two norms on F. Choose a Hamel basis $x_1, x_2, \ldots, x_n$ for F and define a third norm, $|\cdot|$, as follows: For each x in F there is a unique set of scalars $\alpha_1, \alpha_2, \ldots, \alpha_n$ such that $x = \sum_{j=1}^n \alpha_j x_j$. Take $|x|$ to be the maximum of the numbers $|\alpha_1|, |\alpha_2|, \ldots, |\alpha_n|$. Suppose that each of the norms $\|\cdot\|_1$ and $\|\cdot\|_2$ is equivalent to $|\cdot|$. Then there are positive scalars m_1, M_1 and m_2, M_2 such that

$$m_1|x| \le \|x\|_1 \le M_1|x| \text{ and } m_2|x| \le \|x\|_2 \le M_2|x|$$

for each x in F. It follows that

$$(\frac{m_2}{M_1})\|x\|_1 \le m_2|x| \le \|x\|_2 \le M_2|x| \le (\frac{M_2}{m_1})\|x\|_1$$

for each x in F, and hence $\|\cdot\|_1$ and $\|\cdot\|_2$ are equivalent.

Now let $\|\cdot\|$ denote either $\|\cdot\|_1$ or $\|\cdot\|_2$. We shall show that $\|\cdot\|$ is equivalent to $|\cdot|$. If $x = \sum_{j=1}^n \alpha_j x_j$ then

$$\|x\| \le \sum |\alpha_j| \|x_j\| \le |x|(\sum \|x_j\|)$$

Since $(\sum \|x_j\|)$ is a constant, we see that $\|\cdot\|$ is weaker than $|\cdot|$ on F. Let $S = \{x \text{ in } F | |x| = 1\}$ and choose a sequence $\{y_n\}$ of points of S such that $\lim \|y_k\| = \inf\{\|x\| | x \in S\}$. For each y_k there are scalars $\alpha_{k1}, \alpha_{k2}, \ldots, \alpha_{kn}$ such that $y_k = \sum \alpha_{kj} x_j$. Since $y_k \in S$, $|\alpha_{kj}| \le 1$ for $j = 1, 2, \ldots, n$ and every k.[†] These inequalities imply that there is a subsequence of $\{y_k\}$ (call it

---

[†] $\{\alpha_{k1}\}$ has a convergent subsequence since it is a bounded sequence of complex numbers. Thus, by choosing a subsequence of $\{y_k\}$, we may assume that $\{\alpha_{k1}\}$ converges. But, by the same argument, we may take another subsequence of $\{y_k\}$ and get $\{\alpha_{k2}\}$ to converge. Of course $\{\alpha_{k1}\}$ will still converge. By repeating this argument n times we can get a sequence, which we may as well call $\{y_k\}$ also, such that $\{\alpha_{kj}\}$ converges for $j = 1, 2, \ldots, n$.

$\{y_k\}$ also) such that $\lim \alpha_{kj}$ exists and equals, say, α_j for $j = 1, 2, \ldots, n$. Let $y = \sum \alpha_j x_j$ and note that $\{y_k\}$ converges to y for $|\cdot|$; i.e., $\lim |y - y_k| = 0$. It follows that $|y| = 1$ (Exercises 1.3 problem 1) and hence $y \neq 0$.

At this point we make an observation: Let λ be the maximum of the numbers $\|x_j\|$, $j = 1, 2, \ldots, n$. If a positive ε is given then the elements $x = \sum \beta_j x_j$ and $z = \sum \gamma_j x_j$ of F will satisfy the inequality $\|x - z\| < \varepsilon$ if $|\beta_j - \gamma_j| < \frac{\varepsilon}{\lambda n}$ for each j; i.e., if $|x - z| < \frac{\varepsilon}{\lambda n}$. This fact, together with the remarks contained in the last paragraph, implies that $\lim \|y_k\| = \|y\|$. So $\|y\| = \inf\{\|x\| | x \in S\}$ and since $y \neq 0$ this infinum is positive. Now if x is any nonzero element of F, then $\frac{x}{|x|} \in S$ and hence $\|x\| \geq \|y\||x|$. It follows that $|\cdot|$ is weaker than $\|\cdot\|$ on F and hence that these norms are equivalent.

This theorem has some remarkable corollaries. Before we can state them we need:

Definition 4

Let E and F be two normed spaces over the same field and let T be an isomorphism from E onto F. We shall say that T is a topological isomorphism if both T and T^{-1} (Exercises 1.9, problem 4) are continuous mappings. Two normed spaces are said to be topologically isomorphic if there is a topological isomorphism from one of these spaces onto the other.

The term isomorphism is defined in Section 1.1, Definition 5. A special case of the next result is the fact that any n-dimensional normed space over $\mathbb{R}$ (resp. $\mathbb{C}$) is topologically isomorphic to $\mathbb{R}^n$ (resp. $\mathbb{C}^n$). Earlier we proved a similar result for inner product spaces (Section 1.6, Theorem 1).

Corollary 1

If two finite dimensional normed spaces over the same field have the same dimension, then they are topologically isomorphic.

Proof. It suffices to show that if a normed space $(F, \|\cdot\|)$ over $\mathbb{R}$ (resp. $\mathbb{C}$) has (finite) dimension n then it is topologically isomorphic to the space $\mathbb{R}^n$ (resp. $\mathbb{C}^n$) with the Euclidean norm. Let us work over $\mathbb{R}$. There is an isomorphism u from F onto $\mathbb{R}^n$. We can use this map to define a new norm on F as follows: For each x in F let $\|\|x\|\|$ be the norm of $u(x)$ in $\mathbb{R}^n$ (Exercises 1.2, problem 5). When F is given this new norm the map u becomes a topological isomorphism. However $\|\|\cdot\|\|$ is equivalent to $\|\cdot\|$ on F because F is finite dimensional.

Corollary 2

Any finite dimensional normed space over $\mathcal{K}$ is a Banach space (Section 1.3, Definition 2).

Proof. Let $(F, \|\cdot\|_0)$ be an n-dimensional normed space over $\mathbb{R}$. By Corollary 1 we have a topological isomorphism T from this space onto $\mathbb{R}^n$ with the Euclidean or any other norm $\|\cdot\|$. Now let $\{x_n\} \subseteq F$ be a Cauchy sequence in this space. We have two constants M_1 and M_2 such that

(∗) $\quad \|Tx\| \leq M_1 \|x\|_0$ and $\|T^{-1}y\|_0 \leq M_2 \|y\|$

for all $x \in F$ and all $y \in \mathbb{R}^n$ respectively. Given $\varepsilon > 0$ we choose N so that

$$\|x_n - x_m\|_0 < \frac{\varepsilon}{M_1}$$

whenever $m, n \geq N$. Then for the sequence $\{Tx_n\}$ we have

$$\|Tx_n - Tx_m\| \leq M_1 \|x_n - x_m\|_0 < M_1 \left(\frac{\varepsilon}{M_1}\right)$$

whenever $m, n \geq N$. Thus $\{Tx_n\}$ is a Cauchy sequence in $\mathbb{R}^n$. Since $\mathbb{R}^n$ is complete (Section 1.3 just after Definition 2), $\{Tx_n\}$ converges to some $y \in \mathbb{R}^n$, and since T is onto, $y = Tx_0$ for some $x_0 \in F$.

Finally, since

$$\|x_0 - x_n\|_0 = \|T^{-1}(Tx_0) - T^{-1}(Tx_n)\|_0 \leq M_2 \|Tx_0 - Tx_n\|$$

we see that $\{x_n\}$ converges to x_0.

Recall that a finite dimensional inner product space is always a Hilbert space (Section 1.6, Corollary 1 to Theorem 1). This result is a special case of Corollary 2 above. We might remind the reader that some simple properties of equivalent norms were given in Exercises 1.3, problem 2.

Exercises 2

1. Let E be a vector space over $\mathcal{K}$ and let $\|\cdot\|_1, \|\cdot\|_2$ be two norms on E. Suppose that $\|\cdot\|_1 \equiv \|\cdot\|_2$ and that $(E, \|\cdot\|_1)$ is a Banach space (Section 1.3, Definition 2). Show that $(E, \|\cdot\|_1)$ is also a Banach space. Hint: Follow the proof of Corollary 2 above.

* 2. Let E, F be two normed spaces over the same field and let T be a topological isomorphism from E onto F. Denote the norm on E by $\|\cdot\|_E$ and the norm on F by $\|\cdot\|_F$.

    a. For each x in E define $\||x\||$ to be $\|Tx\|_F$. Show that $\|\|\cdot\|\|$ is a norm on E (Exercises 1.2, problem 5) and that it is equivalent to $\|\cdot\|_E$.

b. For each $y \in F$ there is a unique $x \in E$ such that $Tx = y$. Define $|||y||| = ||x||_E$. Show that $|||\cdot|||$ is a norm on F and that it is equivalent to $||\cdot||_F$.

c. Suppose now that E is just a vector space, F is a normed space and T is an isomorphism from E onto F. We know (Exercises 1.2, problem 5) that the function defined in (a) is still a norm on E. Show that when E is given this norm, T becomes an isometric isomorphism from E onto F. If E is a normed space and F is just a vector space, similar remarks can be made about the function defined in (b).

3. Let $(E, ||\cdot||)$ be a normed space.

   a. Show that any open ball in E is an open subset of E and that any closed ball in E is a closed subset of E.

   b. Show that the intersection of any family of closed subsets of E is a closed subset of E.

   c. Show that the union of any finite family of closed subsets of E is a closed subset of E, and the intersection of any finite family of open subsets of E is an open subset of E. Show that these statements can be false for infinite families.

* 4. Let E, F be two normed spaces over the same field and let T be a map from E into F. Show that T is continuous on E if, and only if, $T^{-1}(C) = \{x \in E | Tx \in C\}$ is a closed subset of E whenever C is a closed subset of F.

2.3 The Spectrum of a Linear Operator

The basic ideas of spectral theory first arise in linear algebra. We recall that if T is a linear operators on $\mathbb{R}^n$ and if $v_1, \ldots, v_n$ is a basis for this space then

$$Tv_j = \sum_{i=1}^{n} \alpha_{ij} v_i, \quad j = 1, 2, \ldots, n$$

and T is then represented by the matrix

$$(\alpha_{ij}), \ 1 \leq i \leq n, \ 1 \leq j \leq n$$

This representation of T is extremely useful for doing calculations and these become especially simple if the matrix is diagonal; i.e., if $\alpha_{ij} = 0$ for all $i \neq j$. Thus we seek a basis $w_1, \ldots, w_n$ of $\mathbb{R}^n$ such that

$$Tw_j = \lambda_j w_j, \ i \leq j \leq n$$

The scalars obtained here are called the eigenvalues of T. In general, λ is an eigenvalue for T if there is a nonzero vector v such that $Tv = \lambda v$ and the set of all eigenvalues of T is called the spectrum of T.

It seems to have been Fredholm who first observed that certain integral equations have eigenvalues and this fact was exploited by Hilbert and his students. The spectral theory for linear operators on a Hilbert space was an outgrowth of this work. In this setting, the eigenvalues are a subset of the spectrum and only rarely do they give us the entire spectrum. Once again, we shall work in a general setting so as to maximize the applicability of our results. Our first few theorems contain technical facts about spaces of operators. Let us stress that such a space is always assumed to have the operator norm (Section 1, Definition 2).

Theorem 1

Let E_1, E_2 be two normed spaces over the same field. If E_2 is a Banach space, then so also is $\mathcal{L}(E_1, E_2)$ (see Section 1, Theorem 2 for this notation).

Proof. Let $\{T_n | n = 1, 2, \ldots\} \subseteq \mathcal{L}(E_1, E_2)$ be a Cauchy sequence. So for every $\varepsilon > 0$ we have an integer N such that

$$\|T_n - T_m\| < \varepsilon$$

whenever $m, n \geq N$. Then for each fixed x in E_1 the sequence $\{T_n x\}$ is a Cauchy sequence in E_2; for $\|T_n x - T_m x\|_2 \leq \|T_n - T_m\| \|x\|_1$. Since E_2 is a Banach space $\{T_n x\}$ must converge to a point of E_2 which we shall denote by Tx. In this way we obtain a map T from E_1 into E_2; for each $x \in E_1$, Tx is the limit in E_2 of $\{T_n x\}$. It is a straightforward application of the definitions to show that T is a linear map. We shall show first that it is a bounded map, and hence belongs to $\mathcal{L}(E_1, E_2)$, and then that $\{T_n\}$ converges to T for the operator norm.

Since $\{T_n\}$ is a Cauchy sequence,

$$M = \sup\{\|T_n\| | n = 1, 2, \ldots\} < \infty$$

by Exercises 1, problem 2. Now for any x in E_1 we have

$$\|T_n x - T_m x\|_2 \leq \|T_n - T_m\| \|x\|_1 \leq \varepsilon \|x\|_1 \tag{1}$$

whenever $m, n \geq N$; here we mean, of course, that given $\varepsilon > 0$ we can choose N so that (1) holds for every $x \in E_1$. Letting n tend to infinity, which we may do because $\{T_n x\}$ is convergent, we obtain

$$\|Tx - T_m x\|_2 \leq \varepsilon \|x\|_1 \tag{2}$$

for each x in E_1. But then

$$\|Tx\|_2 \leq \|Tx - T_m x\|_2 + \|T_m x\|_2 \leq \varepsilon\|x\|_1 + M\|x\|_1 \leq (M + \varepsilon)\|x\|_1 \quad (3)$$

for every $x \in E_1$. This shows that T is a bounded, linear map (Section 1, Definition 1).

Now that we know that $T \in \mathcal{L}(E_1, E_2)$ we can talk about the norm of $T - T_m$. Using (2) we have

$$\|T - T_m\| = \sup\{\|(T - T_m)x\|_2 \mid \|x\|_1 \leq 1\} \leq \varepsilon$$

for all $m \geq N$. But since $\varepsilon > 0$ is arbitrary, we have shown that $\{T_n\}$ converges to T for the operator norm.

As a special case of Theorem 1, we see that the space of all bounded, linear operators on a Banach space (or on a Hilbert space) is a Banach space. Spaces of operators do not generally have an inner product, so they are not Hilbert spaces.

We recall that the composition ST of two bounded, linear operators S and T on a normed space E, is a bounded, linear operator on E, and that $\|ST\| \leq \|S\|\|T\|$ (Exercises 1, problem 3).

Lemma 1

Let E be a normed space, and let $S, T \in \mathcal{L}(E)$ and let $\{T_n\} \subseteq \mathcal{L}(E)$ converge to T. Then $\{ST_n\}$ converges to ST, and $\{T_n S\}$ converges to TS.

Proof. Recall that $\mathcal{L}(E)$ is the space of all bounded, linear operators on E. We have

$$\|ST - ST_n\| \leq \|S\|\|T - T_n\|$$

and our first statement follows from this. The second statement is proved similarly.

Lemma 2

Let E be a normed space, let I be the identity operator on E, and let $S \in \mathcal{L}(E)$. Then:

    a. If there is an operator $T \in \mathcal{L}(E)$ such that $ST = TS = I$, then there is only one;

    b. If there is an operator T as in part (a), then the operator S is both one-to-one and onto.

Proof. Suppose that $T_1, T_2 \in \mathcal{L}(E)$ and that $T_1 S = ST_1 = I$ and $T_2 S = ST_2 = I$. Then

$$T_1 = (I)T_1 = (T_2 S)T_1 = T_2(ST_1) = T_2(I) = T_2$$

This proves part (a).

For part (b) we are assuming that we have an operator T for which $ST = TS = I$. Then the equation $Sx = Sy$ is possible if, and only if, $x = y$, because we may apply T to both sides to get $x = y$. This shows that S must be a one-to-one mapping. It is also onto, for if $y \in E$ then $S(Ty) = y$.

Definition 1

Let E be a normed space, let $S \in \mathcal{L}(E)$ and let I be the identity operator on E. We say that S is an invertible operator if there is an element $T \in \mathcal{L}(E)$ such that $ST = TS = I$. When this is the case the operator T, which is unique by Lemma 2(a), is called the inverse of S and is denoted by S^{-1}.

We notice that if we have two invertible operators $S, T \in \mathcal{L}(E)$ then $S \cdot T$ is an invertible operator on E and

$$(S \cdot T)^{-1} = T^{-1} \cdot S^{-1}$$

This follows from Lemma 2(a) and the equations:

$$(S \cdot T)(T^{-1} \cdot S^{-1}) = S \cdot (T \cdot T^{-1}) \cdot S^{-1} = S \cdot I \cdot S^{-1} = S \cdot S^{-1} = I$$
$$(T^{-1} \cdot S^{-1}) \cdot (S \cdot T) = T^{-1} \cdot (S^{-1} \cdot S) \cdot T = T^{-1} \cdot I \cdot T = T^{-1} \cdot T = I$$

The next theorem has several important corollaries. Also, when suitably interpreted, it tells us something about integral equations (see Exercises 4, problem 6 below).

Theorem 2

Let $\mathcal{B}$ be a Banach space and let $T \in \mathcal{L}(\mathcal{B})$. If $\|T\| < 1$, then $I - T$ is an invertible operator. Furthermore,

$$(I - T)^{-1} = \sum_{n=0}^{\infty} T^n$$

Proof. Since $\mathcal{B}$ is a Banach space so also is $\mathcal{L}(\mathcal{B})$ by Theorem 1. Thus in order to prove the convergence of the series

$$\sum_{n=0}^{\infty} T^n \qquad (4)$$

it is sufficient to prove the convergence of the series

$$\sum_{n=0}^{\infty} \|T^n\| \tag{5}$$

(Section 1.3, Theorem 1). Now, for any integer n, we have

$$0 \le \|T^n\| \le \|T\|^n \tag{6}$$

(Exercises 1, problem 3(c)). Hence by the comparison theorem, familiar from calculus, (5) will converge if

$$\sum_{n=0}^{\infty} \|T\|^n \tag{7}$$

does (remember that (5) and (6) are series of non-negative numbers, (4) is a series of operators). Note however that (6) is a geometric series and since $\|T\| < 1$, it is convergent. Thus

$$\sum_{n=0}^{\infty} T^n \in \mathcal{L}(\mathcal{B})$$

Now, by Lemma 1,

$$(I - T)\sum_{0}^{\infty} T^n = (I - T)(\lim_{p \to \infty} \sum_{0}^{p} T^n) = \lim_{p \to \infty}(I - T)\sum_{0}^{p} T^n$$

$$= \lim_{p \to \infty}(\sum_{0}^{p} T^n - \sum_{0}^{p} T^{n+1}) = \lim_{p \to \infty}(T^0 - T^{p+1}) = I$$

because $\|T\| < 1$ implies that

$$\lim_{p \to \infty} T^{p+1} = 0$$

A similar argument shows that

$$(\sum_{0}^{\infty} T^n)(I - T) = I$$

and so, by Lemma 2(a), the proof is complete.

Corollary 1

Let $S, T \in \mathcal{L}(\mathcal{B})$, suppose that T is an invertible operator and that

$$\|T - S\| < \frac{1}{\|T^{-1}\|}$$

Then S is an invertible operator. Thus the set of invertible operators is an open subset (Section 2, Definition 2) of $\mathcal{L}(B)$.

Proof. We have

$$\|I - T^{-1}S\| = \|T^{-1}(T - S)\| \leq \|T^{-1}\|\|T - S\| < 1$$

by hypothesis. Thus, by our theorem,

$$I - (I - T^{-1}S) = T^{-1}S$$

is an invertible operator. Call this invertible operator D. Then

$$S = TD$$

and so

$$D^{-1}T^{-1}$$

is the inverse of S showing that S is an invertible operator.

Corollary 2

Let $T \in \mathcal{L}(B)$ and let λ be a complex number such that $\|T\| < |\lambda|$. Then $T - \lambda I$ is an invertible operator and its inverse is

$$(-\frac{1}{\lambda}) \sum_{0}^{\infty} (\frac{T}{\lambda})^n$$

Proof. Since $\frac{T}{\lambda}$ has norm less than one our theorem tells us that $I - \frac{T}{\lambda}$ is invertible and that its inverse is

$$\sum_{0}^{\infty} (\frac{T}{\lambda})^n$$

Clearly then $(I - \frac{T}{\lambda})(-\lambda I)$ is an invertible operator because it is the product of the two invertible operators $(-\lambda I)$ and $(I - \frac{T}{\lambda})$. Furthermore, the inverse of this product is

$$(-\lambda I)^{-1}(I - \frac{T}{\lambda})^{-1} = (-\frac{1}{\lambda}) \sum_{0}^{\infty} (\frac{T}{\lambda})^n$$

This proves the theorem because

$$(I - \frac{T}{\lambda})(-\lambda I) = (T - \lambda I)$$

We come now to one of the most important concepts in all of operator theory.

Definition 2

Let $T \in \mathcal{L}(B)$. A complex number λ is called a regular point of T if the operator $T - \lambda I$ is invertible. The set of all regular points of T is called the resolvent set of T and will be denoted by $\rho(T)$. The complement, in $\mathbb{C}$, of $\rho(T)$ is called the spectrum of T and will be denoted by $\sigma(T)$.

Corollary 2 tells us that the spectrum of any bounded, linear operator T on a Banach space is a bounded set (Exercises 1, problem 1(c)) because

$$\sigma(T) \subseteq \{\lambda \in \mathbb{C} | |\lambda| \leq \|T\|\}$$

Clearly then $\rho(T) \neq 0$.

Corollary 3

For any $T \in \mathcal{L}(B)$ the set $\rho(T)$ is an open subset of $\mathbb{C}$ and hence $\sigma(T)$ is a closed, bounded subset of $\mathbb{C}$.

Proof. We have already observed that $\sigma(T)$ must be a bounded set, and since $\sigma(T) = \mathbb{C} \setminus \rho(T) = \{\lambda \in \mathbb{C} | \lambda \notin \rho(T)\}$, we need only show that $\rho(T)$ is an open set (Section 2, Definition 2). Let λ be any point of $\rho(T)$, so that $(T - \lambda I)^{-1} \in \mathcal{L}(B)$. Next choose μ such that

$$|\mu - \lambda| < \|(T - \lambda I)^{-1}\|^{-1}$$

and let $T' = T - \lambda I$, $T'' = T - \mu I$. Now T' is invertible and

$$\|T' - T''\| = \|(T - \lambda I) - (T - \mu I)\| = \|(\mu - \lambda)I\| = |\mu - \lambda| < \|(T')^{-1}\|^{-1}$$

It follows from Corollary 1 that T'' is an invertible operator. Thus $\mu \in \rho(T)$, and we have shown that any $\lambda \in \rho(T)$ has a neighborhood which is contained in $\rho(T)$; specifically

$$\{\mu \in \mathbb{C} | |\mu - \lambda| < \|(T')^{-1}\|^{-1}\} \subseteq \rho(T).$$

It follows that $\rho(T)$ is open and hence its complement, $\sigma(T)$, is closed.

We have already mentioned the eigenvalues of an operator on $\mathbb{R}^n$. Let us define these points in some generality.

Definition 3

Let B be a Banach space and let $T \in \mathcal{L}(B)$. A complex number λ will be called an eigenvalue of T if there is a nonzero vector x in B such that

$$(T - \lambda I)x = 0$$

We call x an eigenvector corresponding to λ. Finally, when λ is an eigenvalue of T then
$$\{x \in B | (T - \lambda I)x = 0\}$$
is called the eigenspace of λ.

Observe that the zero vector of B is never an eigenvector. However, every eigenspace contains this vector; it is clear that the eigenspace is just the null space of $T - \lambda I$ and so it certainly is a vector space (Section 1.1, just after Definition 5).

A complex number λ is an eigenvalue of T if, and only if, $T - \lambda I$ has a non-trivial null space (i.e., this null space contains more than just the zero vector). Thus $T - \lambda I$ could not be one-to-one and hence this operator could not be invertible (Lemma 2(b)) when λ is an eigenvalue. So every eigenvalue of T is in $\sigma(T)$.

We should mention that there is an important difference between what a mathematician means by the "eigenvalues of an operator" and what a physicist means by this phrase. This difference affects the question of whether or not the eigenvalues exhaust the spectrum of the operator. We shall use the mathematical terminology, but we shall explain more fully the terminology used by physicists in connection with some specific operators and wherever else it is appropriate to do so.

Exercises 3

1. Suppose that T is an arbitrary linear operator on $\mathbb{R}^n$ (or $\mathbb{C}^n$).

    a. Show that T is a bounded, linear operator.

    b. Let $R(T)$ denote the range of T and let $N(T)$ denote the null space of this operator (Section 1.1, just after Definition 5). Show that dim $R(T)$+dim $N(T) = n$ (note: dim $R(T)$ denotes the dimension of $R(T)$, see Section 1.1, Definition 4).

    c. Use (b) to conclude that T is invertible if, and only if, it is one-to-one. In fact, T is onto if, and only if, it is one-to-one.

    d. Use (c) to show that $\sigma(T)$ coincides with the set of all eigenvalues of T.

    e. Show that λ is an eigenvalue for T if, and only if, it satisfies a certain polynomial equation. Conclude that $\sigma(T) \neq \emptyset$.

    f. Let $\lambda_1, \lambda_2, \ldots, \lambda_n$ be n distinct eigenvalues of T. For each j let v_j be an eigenvector of T corresponding to λ_j. Show that the set $\{v_j | j = 1, \ldots, n\}$ is linearly independent. Use induction on n.

* 2. Let B be a Banach space, and let $S \in \mathcal{L}(B)$, and let λ, μ be two regular values of S. Show that

   i. $(S - \lambda I)^{-1}(S - \mu I)^{-1} = (S - \mu I)^{-1}(S - \lambda I)^{-1}$
      Hint: $(S-\lambda I)^{-1} = (S-\lambda I)^{-1}(S-\mu I)(S-\mu I)^{-1}$ and $(S-\mu I)^{-1} = (S - \lambda I)^{-1}(S - \lambda I)(S - \mu I)^{-1}$.

   ii. $(S - \lambda I)^{-1} - (S - \mu I)^{-1} = (\lambda - \mu)(S - \lambda I)^{-1}(S - \mu I)^{-1}$
      Hint: $(S - \lambda I)^{-1} - (S - \mu I)^{-1} = (S - \lambda I)^{-1}[(S - \mu I) - (S - \lambda I)](S - \mu I)^{-1}$.

* 3. Let $(H, \langle,\rangle)$ be a Hilbert space and let A be an isometry on H (Section 1.6, Definition 2). Then $A \in \mathcal{L}(H)$ and $\|A\| = 1$, Exercises 1, problem 4. Show that

   a. Any eigenvalue of A (it may not have any, as we shall see) is a complex number having absolute value one.

   b. Eigenvectors corresponding to distinct eigenvalues of A are orthogonal.

2.4 Some Operators and their Spectra

Here we shall look at some specific operators and calculate their spectra. Then we shall discuss a class of operators on $L^2[a, b]$.

(a) Recall that ℓ^2 is the space of all sequences $\{z_j\}$ such that

$$\sum_{j=1}^{\infty} |z_j|^2 < \infty$$

and that for $\vec{z} = \{z_j\}$, $\vec{w} = \{w_j\}$ in this space

$$\langle \vec{z}, \vec{w} \rangle = \sum_{j=1}^{\infty} z_j \overline{w}_j$$

(Section 1.5, Definition 1). The space $(\ell^2, \langle,\rangle)$ is a Hilbert space (Section 1.5) and the sequence $\{\vec{e}_n\}_{n=1}^{\infty}$, where each $\vec{e}_n$ is the sequence with nth term one and all other terms zero (Exercises 5, problem 1) is an orthonormal basis for this space (Section 1.6 just before Theorem 4). We may define an operator A_r on ℓ^2 as follows:

$$A_r(\vec{z}) = A_r(\{z_1, z_2, \ldots\}) = \{0, z_1, z_2, \ldots\}$$

(Exercises 1.5, problem 2). This is called the right-shift operator on ℓ^2. It is clearly linear and it is an isometry. Since

$$\|A_r \vec{z}\|_2 = \|\vec{z}\|_2$$

we see that $\|A_r\| = 1$ (Section 1, Definition 2) and so $\sigma(A_r) \subseteq \{\lambda \in \mathbb{C} | |\lambda| \leq 1\}$ by the remark made in Section 3 just after Definition 2.

Let us show that every point of the disk (Recall that $\{\lambda \in \mathbb{C} | |\lambda| \leq 1$ is called the closed unit disk in $\mathbb{C}$. In our terminology it is the closed unit ball in the normed space $(\mathbb{C}, |\cdot|)$.) is in $\sigma(A_r)$. First suppose that λ is a fixed complex number with

$$0 < |\lambda| \leq 1$$

We shall show that λ is in the spectrum of A_r by showing that $(A_r - \lambda I)$ is not onto. By Section 3, Lemma 2(b) this will show that $A_r - \lambda I$ has no inverse. In fact, consider the equation

$$(*) \quad (A_r - \lambda I)\vec{z} = \vec{e}_1$$

We shall show that $(*)$ has no solution in ℓ^2. We argue by contradiction. Suppose that $\vec{z} \in \ell^2$ satisfies $(*)$. Then

$$\vec{z} = \sum_{k=1}^{\infty} a_k \vec{e}_k$$

and so, if this is to satisfy $(*)$,

$$(A_r - \lambda I)\vec{z} = \sum_{k=1}^{\infty} a_k A_r(\vec{e}_k) - \sum_{k=1}^{\infty} \lambda a_k I(\vec{e}_k) = \sum_{k=1}^{\infty} (a_{k-1} - \lambda a_k)\vec{e}_k = \vec{e}_1$$

where $a_0 \equiv 0$. We take the inner product of both sides of this last equation first with $\vec{e}_1$, then with $\vec{e}_2$, and so on to obtain

$$a_0 - \lambda a_1 = 1, \; a_{k-1} - \lambda a_k = 0 \text{ for all } k \geq 2$$

Now $a_0 \equiv 0$ (from the way the shift operator is defined) and so

$$a_1 = \frac{-1}{\lambda}, \; a_k = \frac{-1}{\lambda^k} \text{ for all } k \geq 2$$

This gives us

$$\vec{z} = \sum_{k=1}^{\infty} a_k \vec{e}_k = -\sum_{k=1}^{\infty} \frac{1}{\lambda^k} \vec{e}_k$$

and this is our desired contradiction because $\vec{z} \in \ell^2$ implies

$$\sum |a_k|^2 < \infty$$

while for $0 < |\lambda| < 1$

$$\sum \frac{1}{|\lambda|^{2k}}$$

is divergent since

$$\lim_{k \to \infty} \frac{1}{|\lambda|^{2k}} \neq 0$$

Our calculation shows that

$$\{\lambda \in \mathbb{C} | 0 < |\lambda| \leq 1\} \subseteq \sigma(A_r)$$

Now we already have that

$$\sigma(A_r) \subseteq \{\lambda \in \mathbb{C} | |\lambda| \leq 1\}$$

and since $\sigma(A)$ is a closed set (Section 3, Corollary 3 to Theorem 2) our next result shows that these last two sets are equal.

Theorem 1

Let $(E, \|\cdot\|)$ be a normed space and let C be a non-empty subset of E. Then C is a closed set if, and only if, every sequence of points of C which is convergent, converges to a point of C.

Remark: Since $(\mathbb{C}, |\cdot|)$ is a normed space and $\sigma(A_r)$ is a closed subset of this space, the inclusions

$$\{\lambda \in \mathbb{C} | 0 < |\lambda| \leq 1\} \subseteq \sigma(A_r) \subseteq \{\lambda \in \mathbb{C} | |\lambda| \leq 1\}$$

show that $\sigma(A_r)$ coincides with the disk.

Proof of Theorem 1. Suppose first that C is a closed set and that $\{x_n\} \subseteq C$ is convergent to the point $y \in E$. If $y \notin C$ then y is in the complement of C which is an open set; i.e., $y \notin C$ implies y is in the open set $E \setminus C$. Then for some $\varepsilon > 0$ we must have

$$\mathcal{O}_\varepsilon(y) \subseteq E \setminus C$$

(Section 2, Definition 2). But since $\{x_n\}$ converges to y we can find, for this particular ε, an integer N such that

$$\|y - x_n\| < \varepsilon$$

whenever $n \geq N$. This says that $x_n \in \mathcal{O}_\varepsilon(y)$ whenever $n \geq N$ which is a contradiction.

Now suppose that C is not a closed set. Then $E \setminus C$ is not an open set so it could not be empty. Furthermore, there must be a point, say y, in this set such that

$$\mathcal{O}_\varepsilon(y) \cap C \neq \emptyset$$

for every $\varepsilon > 0$; otherwise, $E \setminus C$ would be a neighborhood of each of its points. We shall use this now to obtain a sequence $\{x_n\} \subseteq C$ which converges to $y \in E \setminus C$. Simply take

$$x_n \in \mathcal{O}_{1/n}(y) \cap C$$

for each $n = 1, 2, 3, \ldots$.

Theorem 2

The right-shift operator on ℓ^2 has no eigenvalues and its spectrum is the set $\{\lambda \in \mathbb{C} | |\lambda| \leq 1\}$.

Proof. The only thing left to show is that A_r has no eigenvalues. We argue by contradiction. Suppose that $\lambda \in \mathbb{C}$ is an eigenvalue of this operator. Any eigenvector $\vec{z} \in \ell^2$ for λ gives

$$\langle \vec{z}, \vec{z} \rangle = \langle A_r \vec{z}, A_r \vec{z} \rangle = \langle \lambda \vec{z}, \lambda \vec{z} \rangle = |\lambda|^2 \langle \vec{z}, \vec{z} \rangle$$

and so $|\lambda| = 1$. Let us write

$$\vec{z} = \{z_j\}_{j=1}^{\infty}$$

and suppose that z_ℓ is the first non-zero term of this sequence. Then

$$0 = \langle A_r \vec{z}, \vec{e}_\ell \rangle = \langle \lambda \vec{z}, \vec{e}_\ell \rangle = \lambda z_\ell$$

because $A_r \vec{z}$ has its first non-zero term in the $(\ell + 1)$st position. But $|\lambda| = 1$ so $\lambda \neq 0$ and our last equation gives $z_\ell = 0$ contradicting the choice of this number.

(b) We shall now study an operator defined on $L^2[a, b]$. Recall that a, b are real numbers with $a < b$, we consider all measurable functions on $[a, b]$ and agree that two such functions are equivalent if they are equal almost everywhere; i.e., the set of points at which they are different has measure zero. Next we single our those (equivalence classes of) functions f such that

$$\int_a^b |f(t)|^2 dt < \infty$$

The set of all these is $L^2[a, b]$. Given $f, g \in L^2[a, b]$ we define

$$\langle f, g \rangle = \int_a^b f(t) \overline{g(t)} dt$$

and we find that $(L^2[a, b], \langle, \rangle)$ is a Hilbert space (Section 1.8, Theorem 4). We recall one more fact, already used in our definition of the inner product:

If $f, g \in L^2[a,b]$) then $f(t)g(t)$ is a Lebesque integrable function (Section 1.8, Lemma 1). Thus:

$$\int_a^b |f(t)g(t)|dt < \infty$$

(Section 1.8, equation (13)).

Let us define an operator P on $L^2[a,b]$ as follows:

$$P(f) = tf(t)$$

for every $f \in L^2[a,b]$. Since $d(t) = t$ is certainly an L^2-function and f is given to be in this space, $tf(t)$ is a Lebesque integrable function. Furthermore,

$$\int_a^b |tf(t)|^2 dt \le M^2 \|f\|_2^2$$

where $M = \max\{|a|, |b|\}$, showing $P(f) \in L^2[a,b]$. The linearity of P is easily checked:

$$P(\alpha f + \beta g) = t(\alpha f(t) + \beta g(t)) = \alpha t f(t) + \beta t g(t) = \alpha P(f) + \beta P(g)$$

and so P is a bounded, linear operator on $L^2[a,b]$. We shall compute the spectrum of P.

Observe first that P has no eigenvalues. To see this note that, for $\lambda \in \mathbb{C}$,

$$\|(P - \lambda I)f\|_2^2 = \|(t - \lambda)f(t)\|_2^2 = \int_a^b |t - \lambda|^2 |f(t)|^2 dt$$

for any f. Now suppose λ is an eigenvalue of P and f is a corresponding eigenvector; so $f \ne 0$. Then we must have

$$(P - \lambda I)f = 0$$

and coupling this with our observation we have

$$\int_a^b |t - \lambda|^2 |f(t)|^2 dt = 0$$

It follows from this that

$$|t - \lambda|^2 |f(t)|^2 = 0 \text{ a.e.}$$

Now $|t - \lambda| = 0$ only at the single point $t = \lambda$. Thus we must have

$$|f(t)| = 0 \text{ a.e.}$$

and this contradicts the fact that f is an eigenvector since it says that $f = 0$.

Next let $\lambda \in \mathbb{C}$, $\lambda \notin [a,b]$. Then the operator $P - \lambda I$ is certainly one-to-one because, otherwise, λ would be an eigenvalue of P and it does not have any. This operator, $P - \lambda I$, is also onto as we now show. Let $g \in L^2[a,b]$ be arbitrary and note that

$$(P - \lambda I)\left[\frac{g(t)}{t - \lambda}\right] = g(t)$$

So we will have shown that $P - \lambda I$ is onto once we show that

$$\frac{g(t)}{t - \lambda}$$

is in L^2. To do this first let

$$d = \inf\{|t - \lambda| \mid t \in [a,b]\}$$

and note that $d > 0$ since $\lambda \notin [a,b]$ and the latter set is closed. Hence

$$\int_a^b |\frac{g(t)}{t - \lambda}|^2 dt \leq \int_a^b \frac{|g(t)|^2}{d^2} = \frac{\|g\|_2^2}{d^2}$$

To sum up: For $\lambda \notin [a,b]$ and fixed the operator $P - \lambda I$ is a bounded, linear mapping from $L^2[a,b]$ to $L^2[a,b]$ which is both one-to-one and onto.

It is easy to see that $P - \lambda I$ must have an inverse and this inverse must be a linear map; but is it a bounded, linear map? As a matter of fact, the answer is "yes." This is not obvious and requires that we use a very powerful theorem.

Theorem 3

(Open-Mapping Theorem). Let T be a bounded, linear map from one Banach space onto a second Banach space. Then T maps any open subset of its domain onto an open subset of its range.

We omit the rather sophisticated proof (see [3; p. 30] for a proof). Note however the important consequence of this:

Corollary 1

A bounded, linear map from one Banach space onto a second Banach space which is one-to-one, has a bounded, linear inverse (compare with Section 3, Lemma 2(b)).

Proof. Let B_1, B_2 be Banach spaces and let $T : B_1 \to B_2$ be bounded, linear mapping from B_1 onto B_2 which is one-to-one. Then T^{-1} is a linear

CHAPTER 2 LINEAR MAPPINGS AND OPERATORS

map from B_2 onto B_1 which is one-to-one. Let $\mathcal{O}$ be any open subset of B_1. In order to show that $T^{-1} : B_2 \to B_1$ is continuous, and hence bounded, we must show that

$$(T^{-1})^{-1}(\mathcal{O})$$

is open in B_2 (Section 2, Theorem 1). But clearly,

$$(T^{-1})^{-1}(\mathcal{O}) = T(\mathcal{O})$$

and this is open in B_2 by the theorem.

Returning now to the space $L^2[a,b]$, which is a Banach space, and the multiplication operator P we see from our summary and this last corollary that: For any $\lambda \notin [a,b]$ the map $(P - \lambda I)$ has a bounded, linear inverse. Thus no such λ is in $\sigma(P)$ (Section 3, Definition 2). We conclude

$$\sigma(P) \subseteq [a,b]$$

Let us now show that these last two sets are in fact equal. Choose, and fix, $\lambda \in [a,b]$ and define, for each integer n, a function $C_n(t)$ as follows:

$$C_n(t) = 1 \text{ for } t \in [\lambda - \frac{1}{n}, \lambda + \frac{1}{n}] \cap [a,b]$$

$$C_n(t) = 0 \text{ for all other values of } t \in [a,b]$$

The functions thus defined are nonzero and belong to $L^2[a,b]$. Let us set, for each n,

$$f_n(t) = \frac{C_n(t)}{\|C_n(t)\|_2}$$

thereby obtaining a sequence of L^2-functions with unit norm. Observe that

$$\|(P - \lambda I)f_n\|_2^2 = \int_a^b |t - \lambda|^2 |f_n(t)|^2 dt \leq \frac{1}{n^2} \int_a^b |f_n(t)|^2 dt = \frac{1}{n^2}$$

because $f_n(t)$ is nonzero only on the set

$$[\lambda - \frac{1}{n}, \lambda + \frac{1}{n}] \cap [a,b]$$

and for all t in this set

$$|t - \lambda|^2 \leq \frac{1}{n^2}$$

Thus we have

$$\lim_{n \to \infty} \|(P - \lambda I)f_n\|_2 = 0$$

Now if λ is not in $\sigma(P)$, then $P - \lambda I$ has a bounded, linear inverse $(P - \lambda I)^{-1}$ and we can write

$$f_n = (P - \lambda I)^{-1}[(P - \lambda I)f_n]$$

However, we could then also write

$$\lim_{n\to\infty} f_n = \lim(P - \lambda I)^{-1}[(P - \lambda I)f_n] = (P - \lambda I)^{-1}[\lim(P - \lambda I)f_n] = 0$$

by Theorem 1 of Section 1 and the boundedness of $(P - \lambda I)^{-1}$. But this last equation is impossible because $\|f_n\|_2 = 1$ for all n (Exercises 1.3, problem 1). Thus we must have $\lambda \in \sigma(P)$.

Theorem 4

The multiplication operator on $L^2[a, b]$ has no eigenvalues and its spectrum coincides with the set $[a, b]$.

The discussion given above illustrates the different way that physicists use the term "eigenvalue." We have $\lambda \in [a, b]$ and a sequence $\{C_n(t)\} \subseteq L^2[a, b]$, $C_n(t)$ is zero except when t is in

$$[\lambda - \frac{1}{n}, \lambda + \frac{1}{n}] \cap [a, b]$$

and at such points this function is one. Hence

$$\int_a^b |C_n(t)|^2 dt = \frac{2}{n}$$

This is true when $\lambda \in (a, b)$. For $\lambda = a$ or $\lambda = b$ the integral is $1/n$, but this is a minor matter. The functions $f_n(t)$ were defined to be

$$\frac{C_n(t)}{\|C_n(t)\|_2}$$

and so

$$f_n(t) = \sqrt{\frac{n}{2}} \text{ for } t \in [\lambda - \frac{1}{n}, \lambda + \frac{1}{n}] \cap [a, b]$$
$$f_n(t) = 0 \text{ for all other points of } [a, b].$$

Now we saw above that
$$\lim(P - \lambda I)f_n = 0$$
and since $P - \lambda I$ is a bounded, linear map a physicist would take the limit inside $(P - \lambda I)$ to write
$$(P - \lambda I)\delta_\lambda(t) = 0$$
where $\delta_\lambda(t)$ is the "limit" of the sequence $\{f_n(t)\}$; the limit being taken pointwise. Notice that as $n \to \infty$ the set

$$[\lambda - \frac{1}{n}, \lambda + \frac{1}{n}] \cap [a, b]$$

shrinks down to $\{\lambda\}$ and $f_n(t) = \sqrt{\frac{n}{2}}$ tends to infinity. So the "function" $\delta_\lambda(t)$ is infinite when $t = \lambda$ and is zero for all other values of t. We call this the Dirac δ-function at λ. A physicist would say that λ is a "continuous eigenvalue" of P with eigenfunction $\delta_\lambda(t)$. Every $\lambda \in [a, b]$ is an eigenvalue of P in this sense.

Mathematicians also recognize the special nature of these points. Their terminology is this:

Definition 1

Let B be a Banach space and let $T \in \mathcal{L}(B)$. A complex number λ is said to be a generalized eigenvalue of T if there is a sequence $\{x_n\}$ of unit vectors in B (i.e., vectors whose length is one) such that

$$\lim(T - \lambda I)x_n = 0$$

Our discussion shows: Every point in the spectrum of the multiplication operator on $L^2[a, b]$ is a generalized eigenvalue of this operator.

(c) For a function $K(s, t)$ which is continuous on $[a, b] \times [a, b]$ we have defined

$$K(f)(s) = \int_a^b K(s, t) f(t) dt$$

for each $f \in C[a, b]$ and we saw that $K(f) \in C[a, b]$ (Section 1). Furthermore, the map which takes each f to $K(f)$ is a bounded, linear operator on $C[a, b]$ (Section 1, equation (3)). Suppose now that $K(s, t) \in L^2(R)$ where $R = [a, b] \times [a, b]$ and, for $f \in L^2[a, b]$, consider

$$K(f)(s) = \int_a^b K(s, t) f(t) dt$$

We shall show that this function is in $L^2[a, b]$. Recall that

$$K(s, t) \in L^2[a, b]$$

(as a function of t) for almost every fixed $s \in [a, b]$ and so

$$\int_a^b |\int_a^b K(s,t)f(t)dt|^2 \, ds \leq \int_a^b \left(\int_a^b |K(s,t)|^2 dt \int_a^b |f(t)|^2 dt \right) ds = M^2 \|f\|_2^2$$

by the C.S.B. inequality (Section 4, Theorem 1), and by Fubini's theorem (Section 1.8, Theorem 7); Fubini's theorem tell us that

$$M^2 = \int_a^b (\int_a^b |K(s,t)|^2 dt) ds = \int_a^b \int_a^b |K(s,t)|^2 dt ds = \|K\|_2^2$$

We conclude that
$$K(f) \in L^2[a,b]$$
whenever f is in this space and that the map, call it T, which takes each f in $L^2[a,b]$ to the function $K(f)$ is a linear operator on this space which satisfies
$$\|Tf\|_2 = \|K(f)\|_2 \leq \|K\|_2 \|f\|_2$$
and so (Section 1, Definition 2)
$$\|T\| \leq \|K\|_2$$
We call T a Hilbert-Schmidt operator on $L^2[a,b]$ with kernel $K(s,t)$; keep in mind that $K \in L^2(R)$. The spectrum of this kind of operator will be investigated later.

Exercises 4

1. Calculate the spectrum of the left shift operator A_ℓ on ℓ^2 (Exercises 1.5, problem 2). Hint: It is easy to see that $\|A_\ell\| \leq 1$. Also since $I = A_\ell A_r$, $1 = \|I\| = \|A_\ell A_r\| \leq \|A_\ell\| \|A_r\| = \|A_\ell\|$ because $\|A_r\| = 1$. Thus $\sigma(A_\ell) \subseteq \{\lambda \in \mathbb{C} | |\lambda| \leq 1\}$.

* 2. a. Let E be a normed space, let $C \subseteq E$ and suppose that every Cauchy sequence (Section 1.3, Definition 2) of points of C converges to a point of C. Show that C is a closed set.

    b. Show that any finite dimensional, linear manifold in a normed space is a closed set (See Section 2, Theorem 2 and its corollaries).

    c. Show that the null space (Section 1.1, just after Definition 5) of any bounded, linear operator on E is a closed set. In fact, the null space of any bounded, linear map from E into another normed space F is a closed set.

    d. Let B be a Banach space and let $C \subseteq B$ be a closed set. Show that any Cauchy sequence of points of C is convergent to a point of C.

* 3. Let B be a Banach space and let $T \in \mathcal{L}(B)$. Show that for any fixed complex number λ the following are equivalent:

    a. For any $\varepsilon > 0$ there is a unit vector x_ε in B such that
    $$\|Tx_\varepsilon - \lambda x_\varepsilon\| < \varepsilon;$$

    b. For any $\varepsilon > 0$ there is a nonzero vector x_ε in B such that $\|Tx_\varepsilon - \lambda x_\varepsilon\| < \varepsilon \|x_\varepsilon\|$;

c. λ is a generalized eigenvalue of T (Definition 1).

* 4. Show that the generalized eigenvalues of an operator are in the spectrum of that operator. Hint: This was done for the multiplication operator above.

* 5. Consider again the right-shift operator, A_r, on ℓ^2. We have seen that $\sigma(A_r) = \{\lambda \in \mathbb{C} | |\lambda| \leq 1\}$. Which points of this set are generalized eigenvalues of A_r and which are not?

6. Let $K(s,t) \in L^2(R)$ where $R = [a,b] \times [a,b]$ and consider

$$u(s) = g(s) + \int_a^b K(s,t)u(t)dt$$

where g is an arbitrary, but fixed, element of $L^2[a,b]$. Show that if $\|K\|_2 < 1$, then this equation has a unique solution in $L^2[a,b]$. Hint: See Section 3, Theorem 2.

3

COMPACT OPERATORS

There is still more to be learned from the simple integral equation considered in Section 1. Recall

$$u(s) = g(s) + \int_a^b K(s,t)u(t)dt \tag{1}$$

where a, b are real numbers with $a < b$, $g \in C[a,b]$ is assumed known and $K(s,t)$ is a continuous function on $[a,b] \times [a,b] = R$. For any fixed $f \in C[a,b]$ we defined

$$K(f)(s) = \int_a^b K(s,t)f(t)dt$$

and observed that $K(f) \in C[a,b]$ also. Finally, we defined a linear operator T on $C[a,b]$ by

$$T[f] = K(f)$$

for all f in this space. The operator T is bounded, in fact:

$$\|T[f]\|_\infty \leq M\|f\|_\infty$$

where $M = \max\{|K(s,t)| | a \leq s \leq b, a \leq t \leq b\}(b-a)$. Now (1) may be written

$$(I - T)u = g \tag{2}$$

and clearly this has a solution in $C[a,b]$ if, and only if, g is in the range of $I - T$. Solutions to (2) will be unique if, and only if, the null space of $I - T$ contains only the zero vector of $C[a,b]$.

Let us now examine the operator T a little more closely. Suppose that $\{f_n\}$ is a sequence in $C[a,b]$ which is bounded; i.e. there is a number ℓ such that
$$\|f_n\|_\infty \leq \ell \text{ for } n = 1, 2, 3, \ldots$$
Then
$$|T[f_n](s) - T[f_n](t)| \leq \ell(b-a) \max\{|K(s,u) - K(t,u)| | a \leq u \leq b\} \quad (3)$$
because
$$|T[f_n](s) - T[f_n](t)| \leq \int_a^b |K(s,u) - K(t,u)| |f_n(u)| du$$

Now $K(s,t)$ is continuous on R and since this is a closed, bounded subset of $\mathbb{R}^2$ the function K is uniformly continuous over R [10; p. 46]. Hence given $\varepsilon > 0$ there is a $\delta > 0$ such that
$$|K(s,u) - K(t,u)| < \frac{\varepsilon}{\ell(b-a)} \quad (4)$$
whenever $|s - t| < \delta$. It follows from (3) and (4) that
$$|T[f_n](s) - T[f_n](t)| < \varepsilon \text{ for every } n \quad (5)$$
whenever $|s - t| < \delta$.

Now what does (5) tell us? Recall that a family $\mathcal{F}$ of functions on $[a,b]$ is said to be equicontinuous if given $\varepsilon > 0$ there is a $\delta > 0$ such that
$$|f(s) - f(t)| < \varepsilon \text{ for all } t \in \mathcal{F}$$
whenever $|s - t| < \delta$. This says, in particular, that each $f \in \mathcal{F}$ is uniformly continuous on $[a,b]$ but it says more than that because for the given ε the same δ works for every $f \in \mathcal{F}$. There is a well-known theorem due to Ascoli and Arzela which says that if $\mathcal{F}$ is both bounded (i.e., $\|f\|_\infty \leq \ell$ for all $f \in \mathcal{F}$, ℓ independent of f) and equicontinuous, then any sequence of functions in $\mathcal{F}$ has a subsequence which converges uniformly (i.e., for $\|\cdot\|_\infty$) over $[a,b]$ [10; p. 177]. Returning now to the operator T we see that, since T is a bounded, linear operator and $\{f_n\}$ is a bounded sequence $\{T[f_n]\}$ which, by (5), is also equicontinuous. Thus: For any bounded sequence $\{f_n\} \subseteq C[a,b]$ the sequence $\{T[f_n]\} \subseteq C[a,b]$ has a subsequence which converges for $\|\cdot\|_\infty$.

We can use this property of T to learn something about the behavior of solutions to (1). Operators with this property arise frequently in applications.

3.1 Compact Sets

Definition 1

A bounded, linear operator S on a normed space E is said to be a compact operator if for any bounded sequence $\{x_n\} \subseteq E$ the sequence $\{Sx_n\} \subseteq E$ has a convergent subsequence. A subset of E is said to be a compact set if it is empty or it is non-empty and every sequence of points of the set has a subsequence which converges to a point of the set.

We shall explore these concepts and their interrelationships below. For now let us return to equation (1) in the form (2). Using our new terminology we have:

$$(I - T)u = g \qquad (1)$$

where $g \in C[a,b]$ is known, I is the identity operator and T is a compact operator on this space. Observe that the null space of $I-T$, let us call it $N(I-T)$, is a closed linear manifold (Exercises 2.4, problem 2(c)) in $(C[a,b], \|\cdot\|_\infty)$. Furthermore, the unit ball in this manifold, call it $\mathcal{B}_1$, is a compact set: If $\{x_n\} \subseteq \mathcal{B}_1$, then $\{x_n\} \subseteq N(I-T)$ so $x_n - Tx_n = 0$ for every n. But T is a compact operator so $\{Tx_n\}$ has a convergent subsequence, $\{T x_{n_k}\}$ say. Clearly then $\{x_{n_k}\}$ is a subsequence of $\{x_n\}$ which converges to a point of $\mathcal{B}_1$ (Section 2.4, Theorem 1).

We have just observed that, when T is a compact operator, $N(I - T)$ is a normed space whose unit ball is a compact set. There is a nice way to characterize such spaces.

Theorem 1

The unit ball of a normed space is a compact set if, and only if, the space is finite dimensional.

Proof. If $(F, \|\cdot\|)$ is a finite dimensional (Section 1.1, Definition 4) normed space, then it is a topologically isomorphic to either $\mathbb{R}^n$ or $\mathbb{C}^n$, where n is the dimension of F (Section 2.2, Definition 4 and the discussion just after it). Since any closed, bounded, subset of $\mathbb{R}^n$ (or $\mathbb{C}^n$) is a compact set [10; p. 42] it is easy to see that the unit ball of F must also be a compact set (see problem 1 for a series of hints).

Now suppose that $(F, \|\cdot\|)$ is a normed space and that the unit ball of F is a compact set. We shall argue by contradiction. If F is not finite dimensional then we can find an infinite sequence $\{x_n\} \subseteq F$ such that any finite subset of $\{x_n\}$ is linearly independent. For each k let F_k be the linear manifold in F spanned by (Section 1.1, Definition 2) $\{x_1, x_2, \ldots, x_k\}$ and note that

F_k is a closed set (Section 2.2, Corollary 2 to Theorem 2 and Exercises 2.4, problem 2(a)). In fact we have

$$F_k \underset{\neq}{\subset} F_{k+1}$$

and F_k is a closed, linear manifold in F_{k+1} for each k. Suppose, we shall prove it in a moment, that we can choose, for every k, a point $y_k \in F_{k+1}$ such that $\|y_k\| = 1$ and

$$\|x - y_k\| \geq 1/2$$

for all $x \in F_k$. Then the sequence $\{y_k\}$ is in the unit ball of F, but since $y_k \in F_{k+1}$, $y_{k+1} \in F_{k+2}$ we have

$$\|y_k - y_{k+1}\| \geq 1/2$$

for every k. Clearly then no subsequence of $\{y_k\}$ can be convergent. This however contradicts the fact that the unit ball of F is a compact set. Hence we must conclude that we cannot choose an infinite linearly independent set in F; i.e., F must be finite dimensional.

Finally, let us show that we can find the vectors y_k as described above. We shall show that for any closed, linear manifold $G \underset{\neq}{\subset} F$ and any δ, $0 \leq \delta < 1$, we can find $y_0 \in F$ such that $\|y_0\| = 1$ and

$$\|x - y_0\| \geq \delta$$

for all x in G. First choose $y \in F$, $y \notin G$ and let

$$d = \inf\{\|y - x\| | x \in G\}$$

Since G is closed $d > 0$. Next notice that for any fixed $x_0 \in G$

$$\|x - (y - x_0)\|y - x_0\|^{-1}\| = \|\{x\|y - x_0\| - (y - x_0)\}\|y - x_0\|^{-1}\|$$
$$= \|\{(x\|y - x_0\| + x_0) - y\}\|y - x_0\|^{-1}\|$$

and since $x\|y - x_0\| + x_0$ is in G while $y \in F$ this last term is

$$\geq d\|y - x_0\|^{-1}$$

for every $x \in G$. Hence y_0 can be found as follows: First take $x_0 \in G$ such that $\|y - x_0\| < d\delta^{-1}$ and then take y_0 to be $(y - x_0)\|y - x_0\|^{-1}$.

Theorem 1 tells us that the general solution to the integral equation (1) contains only a finite number of arbitrary constants. We see this as follows: First we write (1) in the form

$$(I - T)u = g \qquad (2)$$

where T is a compact operator. By our theorem $N(I-T)$ is a finite dimensional linear manifold in $C[a,b]$ so it has a (finite) Hamel basis $u_1(t), u_2(t), \ldots, u_k(t)$ say. Given $g \in C[a,b]$ suppose that $u_g(t)$ is any solution to (1) for this g. Then any other solution u(t) of this equation must satisfy the condition $u(t) - u_g(t) \in N(I-T)$. Hence

$$u(t) = u_g(t) + \sum_{j=1}^{k} \alpha_j u_j(t)$$

where $\alpha_1, \ldots, \alpha_k$ are arbitrary constants.

The result just derived for the equation (1) will also be true for the equation

$$u(s) = g(s) + \int_a^b K(s,t)u(t)dt$$

where g is an arbitrary, but fixed, element of $L^2[a,b]$, and $K \in L^2(R)$, $R = [a,b] \times [a,b]$, provided we can show that the map which takes each $f \in L^2[a,b]$ to

$$K(f)(s) = \int_a^b K(s,t)f(t)dt$$

is a compact operator on this space; i.e., if we can show that any Hilbert-Schmidt (Section 2.4, (c)) operator on $L^2[a,b]$ is a compact operator. We shall be able to do that after we study compact sets and operators in more detail.

Definition 2

Let S be a subset of the normed space E. A point x_0 of E is said to be an adherent point of S if every neighborhood of x_0 contains infinitely many points of S. The set of all adherent points of S will be denoted by S' and the set $S \cup S'$, which we shall call the closure of S, will be denoted by $\bar{S}$.

We leave it to the reader to show that $\bar{S}$ is a closed set and that it is the "smallest" closed set containing S (see problem 4(c)). Incidentally, a linear manifold in E which is also a closed set will be called a *linear subspace* of E. This terminology seems to be used only in books about Hilbert spaces and we shall use it here. Is the closure of a linear manifold a linear subspace? This, too, we leave to the exercises.

We introduced the notion of closure here in order to give a characterization of compact operators which is often convenient to work with. The result we have in mind, which is a corollary of the next theorem, will help us prove that any Hilbert-Schmidt operator is a compact operator. First we need one piece of terminology: A subset of a normed space is said to be a relatively compact set if its closure is a compact set.

Theorem 2

A compact subset of a normed space E is both closed and bounded. A subset S of E is relatively compact if, and only if, every sequence of points of S has a subsequence which converges to a point of E.

Proof. Let C be a compact subset of E. To prove that C is closed we shall use Theorem 1 of Section 4. Suppose that $\{x_n\}$ is a sequence of points of C which converges to the point of $y \in E$. We must show that $y \in C$. But since C is a compact set, $\{x_n\}$ has a subsequence $\{x_{n_k}\}$ say, which converges to a point x_0 of C. We shall complete the proof by showing that $y = x_0$, and hence $y \in C$. We do this by contradiction. Suppose that $\|x_0 - y\| = \varepsilon > 0$. Then, for this fixed number, we can choose two integers N_1, N_2 such that

$$\|y - x_n\| < \frac{\varepsilon}{2} \text{ and } \|x_0 - x_{n_k}\| < \frac{\varepsilon}{2}$$

whenever $n \geq N_1$ and $k \geq N_2$. However, since $\{x_{n_k}\}$ is a subsequence of $\{x_n\}$ we can find an integer N_3 such that $N_1 \leq n_k$ whenever $N_3 \leq k$; because $n_1 < n_2 < n_3 < \ldots$ by the definition of a subsequence. Hence if we take N to be the maximum of the three integers N_1, N_2, N_3 then, for all $k \geq N$, we have both

$$\|x_0 - x_{n_k}\| < \frac{\varepsilon}{2} \text{ and } N_1 \leq n_k$$

But then we must have both

$$\|x_0 - x_{n_k}\| < \frac{\varepsilon}{2} \text{ and } \|y - x_{n_k}\| < \frac{\varepsilon}{2}$$

whenever $k \geq N$. This gives us the desired contradiction because

$$\|x_0 - y\| \leq \|x_0 - x_{n_k}\| + \|x_{n_k} - y\| < \varepsilon$$

for any $k \geq N$; just recall how ε was chosen.

Next let us show that C must be a bounded set. Again we argue by contradiction. If C is not bounded then, for each n, we can choose $x_n \in C$ such that $n < \|x_n\|$. The sequence obtained in this way is contained in C and hence must have a subsequence $\{x_{n_k}\}$ which converges to $x_0 \in C$. But then

$$\lim \|x_{n_k}\| = \|x_0\|$$

by Exercises 1.3, problem 1. This, of course, is our contradiction since $\|x_0\| < \infty$, $n_k < \|x_{n_k}\|$ for every k, and $\lim n_k = \infty$.

Now suppose that $S \subseteq E$ is a relatively compact set. Then $\bar{S}$ is a compact set and so any sequence of points of $\bar{S}$, in particular any sequence of points of S, must have a subsequence which converges to a point of $\bar{S} \subseteq E$. Thus S satisfies our condition. Conversely, suppose that a subset S of E has the property that any sequence of points of S has a subsequence which converges to a point of E. We must show that $\bar{S}$ is a compact set. Let $\{y_n\}$ be a sequence in $\bar{S}$. If $\{y_n\} \subseteq S$, then by our condition it has a convergent subsequence. But $\bar{S}$ is a closed set so the limit of this subsequence will be in $\bar{S}$ (Section 4, theorem 1). So this case causes no problems. Suppose than that $\{y_n\} \subseteq \bar{S}\setminus S$; i.e., $\{y_n\} \subseteq S'$ (Definition 2). For each fixed n,

$$\{x \in E \mid \|x - y_n\| < \frac{1}{n}\}$$

is a neighborhood of y_n and so contains a point $s_n \in S$; again simply from Definition 2. Now $\{s_n\} \subseteq S$ and by our condition it has a subsequence $\{s_{n_k}\}$ which converges to a point $y_0 \in E$; clearly $y_0 \in S' \subseteq \bar{S}$. Now

$$\|s_{n_k} - y_{n_k}\| \leq \|y_0 - s_{n_k}\| + \|s_{n_k} - y_{n_k}\|$$

tells us that $\{y_{n_k}\}$ converges to y_0. This completes the proof.

Corollary 1

For any bounded, linear operator T on a normed space E, these are equivalent:

a. T is a compact operator;

b. T maps the unit ball of E onto a relatively compact subset of E;

c. T maps any bounded subset of E onto a relatively compact subset of E.

Proof. Let us first show that (a) implies (b). So suppose that T is a compact operator and let $\mathcal{B}_1$ be the unit ball of E. We must show that

$$T(\mathcal{B}_1) = \{Tx \mid x \in \mathcal{B}_1\}$$

is a relatively compact set. We use the theorem. Let $\{y_n\}$ be any sequence of points of $T(\mathcal{B}_1)$. Then, for each n, there is an $x_n \in \mathcal{B}_1$ such that $y_n = Tx_n$. Now the sequence $\{x_n\}$ is a bounded sequence because it is contained in $\mathcal{B}_1$, hence the sequence $\{Tx_n\}$ (i.e., the sequence $\{y_n\}$) has a convergent subsequence (Definition 1).

Next we suppose that T maps the unit ball $\mathcal{B}_1$ of E onto a relatively compact set, we choose any bounded set subset $\mathcal{S}$ of E and we consider

$$T(\mathcal{S}) = \{Tx \mid x \in \mathcal{S}\}$$

Now $\mathcal{S}$ is a bounded set so for some ℓ we have

$$\|x\| \leq \ell$$

for all $x \in \mathcal{S}$, and this tells us that $\mathcal{S} \subseteq \ell \mathcal{B}_1$ or

$$\frac{1}{\ell}\mathcal{S} \subseteq \mathcal{B}_1$$

Applying the linear operator T we have

$$\frac{1}{\ell}T(\mathcal{S}) \subseteq T(\mathcal{B}_1)$$

and the latter is a relatively compact set. Now if $\{y_n\}$ is any sequence in $T(\mathcal{S})$, then $\{\frac{y_n}{\ell}\}$ is a sequence of points of $T(\mathcal{B}_1)$ and hence it has a convergent subsequence. But since ℓ is a constant this means that $\{y_n\}$ has a convergent subsequence and so $T(\mathcal{S})$ is a relatively compact set. Thus (b) implies (c).

Finally, assume that T satisfies (c). To prove (a) we need only show that $\{Tx_n\}$ has a convergent subsequence whenever the sequence $\{x_n\}$ is bounded. But this clearly follows from (c).

Exercises 1

1. Let E, F be two normed spaces over the same field.

    a. If $T : E \to F$ is a bounded, linear map and $C \subseteq E$ is a compact set, show that $T(C)$ is a compact set. Hint: Let $\{y_n\} \subseteq T(C)$, then for each n there is an $x_n \in C$ such that $Tx_n = y_n$. Now consider $\{x_n\}$. Note that the linearity of T plays no role here; all we need is sequential continuity (Section 1, Lemma 1).

    b. Suppose that the unit ball of F is a compact set. Show that any closed, bounded subset of F must also be a compact set.

    c. Suppose that T is a topological isomorphism (Section 2.2, Definition 4) from E onto F and that F has a compact unit ball. Show that the unit ball of E must be compact. Hint: Let $\mathcal{B}_1$ be the unit ball of E. Then $T(\mathcal{B}_1)$ is a bounded subset of F. It is also a closed set because $T(\mathcal{B}_1) = (T^{-1})^{-1}(\mathcal{B}_1)$, $\mathcal{B}_1$ is a closed set and T^{-1} is continuous (Exercises 2, problem 4). Now use (b) and then (a).

* 2. Let E be a normed space, let T be a bounded, linear operator on E, and let S be a compact operator on E.

    a. Show that both TS and ST are compact operators.

    b. Show that, for any integer n, there is a compact operator S_n on E such that $(I - S)^n = I - S_n$. Hint: Use (a) and the binomial theorem.

* 3. Let E be a normed space and let $\{x_n\}$ be a sequence of points of E.

    a. Suppose that infinitely many points of $\{x_n\}$ are distinct and that y is an adherent point of this set of points. Show that some subsequence of $\{x_n\}$ converges to y. Concludes that a subset C of E is compact iff any infinite subset of C has an adherent point in C.

    b. Now suppose that $\{x_n\}$ is a Cauchy sequence and that it has a convergent subsequence. Show that $\{x_n\}$ must also converge and to the same limit. Conclude that a Cauchy sequence in a compact set always converges.

* 4. Let E be a normed space and let S be a non-empty subset of E.

    a. Show that the point $y \in E$ is an adherent point of S iff there is a sequence of distinct points of S that converges to y.

    b. Show that $(S')' \subseteq S'$; i.e., any adherent point of the set of adherent points of S is an adherent point of S. Conclude that S' is a closed set.

    c. Show that $\bar{S}$ is a closed set, and that any closed set which contains S must also contain $\bar{S}$. Thus $\bar{S} = \cap \{C \subseteq E | C$ is a closed set and $S \subseteq C\}$.

    d. Let H be a linear manifold in E. Show that $\bar{H}$ is a linear subspace of E. Hint: We need only prove that $\bar{H}$ is a linear manifold since, by (c), it is closed. So take x, y in $\bar{H}$ and any scalar λ and show that $x + y$ and λx are in $\bar{H}$.

    e. Show that S is dense in E iff $\bar{S} = E$ (Section 1.9, Definition 1).

* 5. Let E be a normed space and let T be a bounded, linear operator on E. We shall say that T has finite rank, or that T is an operator of finite rank, if the range of T is a finite dimensional linear manifold in E. Show that any operator of finite rank is a compact operator.

* 6. Let E, F be two normed spaces over the same field. A bounded, linear map S from E into F is called a compact mapping if for any bounded sequence $\{x_n\} \subseteq E$, the sequence $\{Sx_n\} \subseteq F$ has a convergent subsequence.

    a. Prove an analogue of Corollary 1 to Theorem 2 for compact maps.

    b. Show that the set of all compact maps from E onto F is a linear manifold in $\mathcal{L}(E, F)$ (Equation (*) just after Definition 1 of Section 1).

3.2 More about Compact Sets

In the last section we saw that the general solution to the integral equation

$$u(s) = g(s) + \int_a^b K(s,t)u(t)dt \tag{1}$$

where $g \in C[a,b]$ is fixed and K is continuous on $R = [a,b] \times [a,b]$, contains only a finite number of arbitrary constants. We proved this by first noting that the operator T which takes each $f \in C[a,b]$ to the function

$$K(f)(s) = \int_a^b K(s,t)f(t)dt \tag{2}$$

is a compact operator. Then we noted that the null space of $I - T$ is a normed space whose unit ball is a compact set, and such a space is finite dimensional (Section 1, Theorem 1). Returning to (1) let us suppose now that $g \in L^2[a,b]$ and $K \in L^2(R)$. We have seen (Section 2.4, part (c)) that for any $f \in L^2[a,b]$ their function $K(f)$, defined by equation (2), is also in $L^2[a,b]$, and that the map T which takes each such f to $K(f)$ is a bounded, linear operator on $L^2[a,b]$. We called this the Hilbert-Schmidt operator with kernel K and we saw that

$$\|T\| \leq \|K\|_2$$

If we can show that T is a compact operator then our result concerning the general solution to (1) will be valid for this, more general, type of equation — the same proof goes through word for word. So, the purpose of this section is to show that every Hilbert-Schmidt operator whose kernel $K \in L^2(R)$ is a compact operator. This will take some doing.

The first thing we do is choose, and fix, an orthonormal basis (Section 1.6, Definition 4) for $L^2[a,b]$ and recall that the set

$$\{\psi_{n,m}(s,t) = \sigma(s)\sigma_m(t) | m, n \text{ integers}\}$$

is an orthonormal basis for $L^2(R)$ (Exercises 1.8, problem 4). Thus we may write

$$K(s,t) = \sum C_{n,m}\psi_{n,m}(s,t) \tag{3}$$

where the sum is over all pairs of integers, and the series will converge to K for the norm of $L^2(R)$. Now we used K to define a function $K(f)$ for any $f \in L^2[a,b]$. For each fixed integer N we define a similar function $K_N(f)$ as follows:

$$K_N(f)(s) = \sum C_{nm} \int_a^b \psi_{n,m}(s,t)f(t)dt = \sum C_{nm}\left(\int_a^b \sigma_m(t)f(t)dt\right)\sigma_n(s) \tag{4}$$

MORE ABOUT COMPACT SETS

where the summation is over all pairs of integers (n, m) such that $1 \le m \le N$ and $1 \le n \le N$. The last sum in (4) is a finite sum of L^2-functions, hence $K_N(f)$ is certainly in this space. Furthermore, $K_N(f)$ is actually in the linear span in $L^2[a, b]$ of the finite set $\sigma_1, \ldots, \sigma_N$. Thus the map T_N which takes each f in $L^2[a, b]$ onto $K_N(f)$ is a linear map whose range is a finite dimensional, linear subspace of $L^2[a, b]$. In other words each T_N is an operator of finite rank, and such operators are known to be compact operators (Exercises 1, problem 5).

One last observation. For any f with $\|f\|_2 \le 1$ we can write:

$$\begin{aligned}
\|K(f) - K_N(f)\|_2^2 &= \langle \sum C_{nm}(\int_a^b \sigma_m(t)f(t)dt)\sigma_n(s), \\
&\qquad \sum C_{nm}(\int_a^b \sigma_m(t)f(t)dt)\sigma_n(s)\rangle \\
&= \sum |C_{nm}|^2 |\int_a^b \sigma_m(t)f(t)dt|^2 \|\sigma_n\|_2^2 \\
&= \sum |C_{nm}|^2 |\langle \sigma_m(t), f(t)\rangle|^2 \|\sigma_n\|_2^2 \\
&\le \sum |C_{nm}|^2 \|\sigma_m\|_2^2 \|\sigma_n\|_2^2 = \sum |C_{nm}|^2
\end{aligned} \quad (5)$$

where our summations are over all pairs of integers (n, m) with either n or m greater than N. However, by (3) and the Parseval relation (Section 1.6, Theorem 3) we have

$$\|K\|_2^2 = \sum |C_{nm}|^2 \qquad (6)$$

where the sum is over all pairs of integers. Thus the right-hand side of (5) tends to zero as N tends to infinity and we have shown that

$$\lim_{N \to 0} \|K(f) - K_N(f)\|_2 = 0 \qquad (7)$$

uniformly over $f \in L^2[a, b]$, $\|f\|_2 \le 1$; i.e., given $\varepsilon > 0$ we can choose N_1 such that

$$\|K(f) - K_N(f)\|_2 < \varepsilon$$

for all f with $\|f\|_2 \le 1$, whenever $N \ge N_1$. To put this another way

$$\|T[f] - T_N[f]\|_2 \le \varepsilon$$

for all f with $\|f\|_2 \le 1$ whenever $N \ge N_1$, or still another way (Section 2.1, Definition 2)

$$\lim_{N \to \infty} \|T - T_N\| = 0 \qquad (8)$$

What we have shown is this: Any Hilbert-Schmidt operator (with kernel in L^2) is the limit, for the operator norm, of a sequence of operators of finite rank. Thus to show that any such operator is compact it suffices to show that the limit of any sequence of finite rank (or, more generally, compact) operators is a compact operator. This is what we shall do.

Let us say that a subset C of a normed space E is a complete set (or, simply, complete) if it is empty or it is non-empty and every Cauchy sequence (Section 1.3, Definition 2) of points of C converges to a point of C. It is clear that any closed subset of a Banach space is a complete set (Section 1.3, Definition 2 and Section 4, Theorem 1). Also, any complete set is a closed set (Exercises 2.4, problem 2(a)).

Definition 1

A subset S of a normed space $(E, \|\cdot\|)$ is said to be a precompact (or totally bounded) set if for any $\varepsilon > 0$ there is a finite subset $s_1, s_2, \ldots, s_n$ of S such that any point of S is within ε of one of these points; i.e., for any $s \in S$ there is an s_j, $1 \leq j \leq n$, such that $\|s - s_j\| < \varepsilon$.

The finite set $s_1, \ldots, s_n$ corresponding to ε is called an ε-net in S. Using this term we can say that a set is precompact if, and only if, it has a (finite) ε-net corresponding to every $\varepsilon > 0$. One might also say, more informally, that a precompact set is one which can be guarded by a finite number of arbitrarily short sighted policemen.

It is easy to see that a precompact set S must be a bounded set: Choose a 1-net $s_1, \ldots, s_n$ in S and let $M = \max\{\|s_j\| | 1 \leq j \leq n\}$. Then for any $s \in S$ we have $\|s\| \leq \|s - s_j\| + \|s_j\| \leq M + 1$. The unit ball in ℓ^2 is a bounded set which is not precompact. We can see this as follows: Let $\{\vec{e}_n\}_{n=1}^{\infty}$ be the standard orthonormal basis for ℓ^2 (Exercises 1.5, problem 1) and note that $\|\vec{e}_k - \vec{e}_j\| = \sqrt{2}$ for $i \neq j$. Since each $\vec{e}_n$ is in the unit ball of ℓ^2. This unit ball could not contain an ε-net for any $\varepsilon < \frac{\sqrt{2}}{2}$.

Theorem 1

A subset of a normed space E is a compact set if, and only if, it is both precompact and complete.

Proof. Let $C \subseteq E$ be a compact set. We have already noted (Exercises 1, problem 3(b)) that any such set is complete. Let us show now that C must also be precompact. We will argue by contradiction. If C is not precompact, then for some $\varepsilon > 0$ no finite subset of C is an ε-net in C. Choose $x_1 \in C$ and, since $\{x_1\}$ is not an ε-net in C, find $x_2 \in C$ with $\|x_1 - x_2\| \geq \varepsilon$. Now $\{x_1, x_2\}$ is not an ε-net in C hence we can choose $x_3 \in C$ with $\|x_3 - x_j\| \geq \varepsilon$, $j = 1, 2$. Continue in this way. After $x_1, x_2, \ldots, x_k$ have been chosen so that $\|x_k - x_j\| \geq \varepsilon$ for $j = 1, 2, \ldots, k-1$, we observe that $\{x_1, \ldots, x_n\}$ is not an ε-net in C and so there must be a point x_{k+1} in this set such that $\|x_{k+1} - x_j\| \geq \varepsilon$,

$1 \leq j \leq k$. The sequence so generated is contained in C but it clearly has no convergent subsequence, and this contradicts the fact that C is a compact set.

Now suppose that we have a set $C \subseteq E$ that is both precompact and complete. In order to show that C is compact we must show that every sequence of points of C has a subsequence which converges to a point of C (Section 1, Definition 1). Since C is complete we need only show that every sequence of points of C has a subsequence which is a Cauchy sequence. So, let $\{x_n\}$ be any sequence of points of C and let $y_{11}, y_{12}, \ldots, y_{1k(1)}$ be a 1-net in C. Then every x_n is within one of one of these points and so we can find a subsequence $\{x_{1n}\}$ of $\{x_n\}$ such that
$$\|x_{1n} - x_{1m}\| < 2$$
for all m, n; there are only a finite number of y's so for one of them, say y_{1j}, infinitely many x_n's satisfy
$$\|x_n - y_{1j}\| < 1$$
and this gives us
$$\|x_n - x_m\| \leq \|x_n - y_{1j}\| + \|y_{1j} - x_m\| < 2$$
for all m, n in this infinite set. Next let $y_{21}, \ldots, y_{2k(2)}$ be a 1/2-net in C. Again since this set is finite while $\{x_{1n}\}$ is infinite we must have a subsequence $\{x_{2n}\}$ of $\{x_{1n}\}$ such that
$$\|x_{2n} - x_{2m}\| < 1$$
for all m, n. Now choose a 1/3-net $y_{31}, \ldots, y_{3k(3)}$ in C and use it to obtain a subsequence $\{x_{3n}\}$ of $\{x_{2n}\}$ such that
$$\|x_{3n} - x_{3m}\| < 2/3$$
for all n, m. Continue in this way. After a subsequence $\{x_{kn}\}$ of $\{x_{k-1\,n}\}$ has been chosen so that
$$\|x_{kn} - x_{km}\| < 2/k$$
for all m, n, we choose an $\frac{1}{k+1}$-net in C and use it to find a subsequence $\{x_{k+1\,n}\}$ of $\{x_{kn}\}$ such that
$$\|x_{k+1\,n} - x_{k+1\,m}\| < \frac{2}{k+1}$$
for all m, n. Consider now the sequence $\{x_{kk}\}_{k=1}^{\infty}$. Clearly $\{x_{kk}\}_{k=1}^{\infty}$ is a subsequence of $\{x_{1n}\}$ and so any two of its terms differ by less than 2. Also $\{x_{kk}\}_{k=2}^{\infty}$ (note we start at two here) is a subsequence of $\{x_{2n}\}$, so any two of its terms differ by less than 1. Again $\{x_{kk}\}_{k=3}^{\infty}$ is a subsequence of $\{x_{3n}\}$ hence any two of its terms differ by less than 2/3. In fact, $\{x_{kk}\}$ is a subsequence of $\{x_n\}$ which is a Cauchy sequence. This completes the proof.

Corollary 1

The unit ball of any infinite dimensional Banach space is a bounded set which is not precompact. Furthermore, a subset of $\mathbb{R}^n$, or $\mathbb{C}^n$, is precompact if, and only if, it is a bounded set.

Proof. The first statement follows from the theorem and Theorem 1 of Section 1. The second statement follows easily from the fact that any closed, bounded subset in $\mathbb{R}^n$, or $\mathbb{C}^n$, is a compact set.

Recall now that a bounded, linear map T from the normed space E into the normed space F (i.e., an element of $\mathcal{L}(E, F)$) is a compact mapping if the set $T(\mathcal{B}) \subseteq F$ is relatively compact, meaning its closure is a compact set, whenever the set $\mathcal{B} \subseteq E$ is bounded (Exercises 1, problem 6).

Theorem 2

Let E, F be two normed spaces over the same field, let $\{T_n\} \subseteq \mathcal{L}(E, F)$ be a sequence of compact mappings which converges to $T \in \mathcal{L}(E, F)$. If F is a Banach space, then T is a compact mapping.

Proof. Let $\mathcal{B}$ be any bounded subset of E with $\|x\| \leq k$, say, for all x in $\mathcal{B}$. Given $\varepsilon > 0$ choose N so that

$$\|T - T_N\| < \frac{\varepsilon}{2k}$$

and note that, for all x in $\mathcal{B}$, we then have

$$\|Tx - T_N x\| < \frac{\varepsilon}{2} \tag{9}$$

Now T_N is a compact mapping and so $\overline{T_N(\mathcal{B})}$ is a compact subset of F (Section 1, Corollary 1 to Theorem 2). Thus, by Theorem 1, we can choose a finite $\varepsilon/2$-net in this set. However, by (9), some member of this finite set is within ε of every point of $\overline{T(\mathcal{B})}$. Clearly then these is a 2ε-net in $\overline{T(\mathcal{B})}$. But $\varepsilon > 0$ was arbitrary, hence the latter set is precompact. Since it is obviously complete, since it is a closed subset of the Banach space F, it is a compact set.

Corollary 2

The set of all compact mappings from a normed space E into a Banach space F is a linear subspace (i.e., a closed, linear manifold) in $\mathcal{L}(E, F)$.

Proof. By Exercises 1, problem 6(b), this set is a linear manifold in $\mathcal{L}(E, F)$ and, by the theorem, it is also a closed set.

As a special case of Corollary 1 we see that the set of all compact operators on a Banach space B is a linear subspace of $\mathcal{L}(B)$.

Corollary 3

Every Hilbert-Schmidt operator on $L^2[a,b]$ (with kernel in $L^2(R)$) is a compact operator.

Proof. We saw at the beginning of this Section that any such operator is the limit of a sequence of operators which have finite rank, and hence are compact.

Exercises 2

1. Let S be a precompact subset of the normed space E.

    a. Show that any subset of S is a precompact set.

    b. For any fixed scalar λ let $\lambda S = \{\lambda s | s \in S\}$. Show that λS is precompact.

    * c. Show that there is a countable set $S_0 \subseteq S$ such that $S \subseteq \bar{S}_0$. Conclude that $\bar{S}_0 = \bar{S}$.

    * d. Let $T \in \mathcal{L}(E, F)$ where F is another normed space and suppose that T is a compact mapping. Show that the range of T, $R(T) = \{Tx | x \in E\}$, contains a countable, dense subset (Section 1.9, Definition 1).

* 2. Give the details of the proof, sketched above, that the unit ball in ℓ^2 is not precompact.

* 3. Define two operators on ℓ^2 as follows:

    $$A(\{x_n\}) = \{\frac{x_n}{n}\}, \quad B(\{x_n\}) = \{\frac{x_{n-1}}{n}\} \text{ with } x_0 \equiv 0$$

    a. Show that both A and B are bounded, linear operators on ℓ^2.

    b. Use Theorem 2 to show that both A and B are compact operators on ℓ^2; in fact, show that each of these operators is the limit of a sequence of operators each having finite rank.

* 4. Let $(B, \|\cdot\|)$ be a Banach space, let S be a subset of B and let $f : S \to B$ be uniformly continuous (Section 2.1, Lemma 1(d)). We do not assume that f is linear.

a. Show that there is a unique function $\bar{f} : \bar{S} \to B$ such that:

    i. $\bar{f}(s) = f(s)$ for all $s \in S$;

    ii. $\bar{f}$ is uniformly continuous on $\bar{S}$.

b. Suppose that $f(s) = 0 \in B$ for every $s \in S$. What can we say about $\bar{f}$?

c. The assumption that f be uniformly continuous on S, and not just continuous, is necessary for (a) to be true. To see this consider $f(x) = \frac{1}{x}$ on the set $(0,1)$.

3.3 The Spectrum of a Compact Operator

We can say quite a lot about the spectrum of a compact operator on a Banach space as we shall see. We shall present a set of results which are known, collectively, as Riesz theory. The development of this theory, and its applications to integral equations, was one of the first successes of functional analysis. Let us begin with some general remarks. Every linear operator on a finite dimensional space is a compact operator and its spectrum coincides with the set of eigenvalues of the operator (Exercises 2.3, problem 1(d)). So in this case the spectrum is a finite set.

Now, and throughout this section, let T denote a compact operator on an infinite dimensional Banach space B. We first observe that the number zero is in the spectrum of T. We see this as follows: If $0 \notin \sigma(T)$ then $(T - 0I)$ has a bounded, linear inverse; i.e., T has a bounded, linear inverse T^{-1}. Now $T^{-1}T$ is a compact operator because it is the composition of the bounded operator T^{-1} with the compact operator T (Exercises 1, problem 2(a)). But $T^{-1}T = I$ so we are saying that I is a compact operator on B. It follows that I must map the unit ball of B onto a relatively compact set (Section 1, Corollary 1 to Theorem 2). Clearly this says that the unit ball of B is a compact set which, in turn, says that B is finite dimensional (Section 1, Theorem 1). Thus we have arrived at a contradiction.

It can happen that the spectrum of a compact operator contains only the number zero (see Remark 3 below). This possibility must be borne in mind. The reader will recall that if one can find a basis of eigenvectors of an operator on a finite dimensional space, then the matrix of that operator for this special basis is just a diagonal matrix — we will discuss this in more detail in a later section. Of course, this is not always possible and so we try to find vectors which will enable us to put the matrix of our operator into some convenient form — the Jordan canonical form for example. In order to do this one makes

THE SPECTRUM OF A COMPACT OPERATOR

use of "generalized eigenvectors" of the operator. A vector v is a generalized eigenvector of T of order m if

$$(T - \lambda I)^{m-1} v \neq 0 \text{ but } (T - \lambda I)^m v = 0$$

Clearly the null space of $(T - \lambda I)^{m-1}$ is contained in the null space of $(T - \lambda I)^m$, but the latter is often a larger space. With this little bit of background, let us make some definitions. For any fixed $\lambda \neq 0$ and each $n = 0, 1, 2, \ldots$ let N_n be the null space of $(T - \lambda I)^n$; recall that $(T - \lambda I)^0 \equiv I$ and $(T - \lambda I)^n = (T - \lambda I)(T - \lambda I)^{n-1}$ for all $n \geq 1$. Clearly

$$\{0\} = N_0 \subseteq N_1 \subseteq N_2 \subseteq \ldots \subseteq N_n \subseteq N_{n+1} \subseteq \ldots$$

for if $v \in N_n$ then $(T-\lambda I)^{n+1}v = (T-\lambda I)[(T-\lambda I)^n v] = (T-\lambda I)0 = 0$, hence $v \in N_{n+1}$. Since our operators are bounded each of these linear manifolds is closed; i.e., they are subspaces of B. Moreover:

Theorem 1

For each n the space N_n is finite dimensional. Furthermore, there is an integer p such that $N_n \neq N_{n+1}$ for $n = 0, 1, 2, \ldots, p-1$ but $N_n = N_{n+1}$ for all $n \geq p$.

Proof. We have already proved that N_1 is finite dimensional (Section 1, Theorem 1 and the discussion just before it). However, since $(T - \lambda I)^n = T_n - \lambda^n I$ for some compact operator T_n (Exercises 1, problem 2(b)) we see that N_n is finite dimensional for every n.

Suppose now that $N_n \neq N_{n+1}$ for every $n = 0, 1, 2, \ldots$. For each fixed n choose $v_n \in N_{n+1}$ such that $\|v_n\| = 1$ and

$$\inf\{\|v_n - w\| \mid w \in N_n\} \geq 1/2$$

The reader may want to refer to the last paragraph in the proof of Theorem 1 in Section 1. In this way we obtain a sequence $\{v_n\}$ which is bounded, and we apply T to these vectors. If $s > r$

$$\|Tv_s - Tv_r\| = |\lambda| \|v_s + \{\frac{1}{\lambda}(T - \lambda I)v_s + v_r + \frac{1}{\lambda}(T - \lambda I)v_r\| \geq \frac{|\lambda|}{2}$$

But this contradicts the fact that T is a compact operator and so we must conclude that $N_n = N_{n+1}$ for some n. Let p be the first integer for which this is so. We shall now show that $N_n = N_{n+1}$ for all $n \geq p$. Suppose $n > p$ and that $v \in N_n$. Then

$$(T - \lambda I)^{p+1}[(T - \lambda I)^{n-p-1} v] = (T - \lambda I)^n v = 0$$

It follows from this that the element $(T - \lambda I)^{n-p-1}v$ is in N_{p+1}. But $N_{p+1} = N_p$ and so

$$(T - \lambda I)^{n-1}v = (T - \lambda I)^p[(T - \lambda I)^{n-p-1}v] = 0$$

showing that v, an arbitrary element of N_n, is in N_{n-1}.

Before we can exploit Theorem 1 we need to prove an analogous result for the ranges of the operators $(T - \lambda I)^n$, $n = 0, 1, \ldots$. For each n let $R_n = \{(T - \lambda I)^n v | v \in B\}$. In this way we get a sequence of linear manifolds such that

$$B = R_0 \supseteq R_1 \supseteq R_2 \supseteq \ldots \supseteq R_n \supseteq R_{n+1} \supseteq \ldots$$

(see problem 5 below). We shall prove that this sequence, just as for $N_0 \subseteq N_1 \subseteq \ldots$, is finite (i.e., the manifolds become equal from some point on). In proving this for the null spaces we used the fact that when $N_n \neq N_{n+1}$ we could choose a vector $v_n \in N_{n+1}$ such that $\|v_n\| = 1$ and

$$\inf\{\|v_n - w\| | w \in N_n\} \geq 1/2$$

This is possible because N_n, being finite dimensional, is a closed, linear manifold in N_{n+1} (Exercises 2.4, problem 2(b)) — again the reader may want to refer to the last paragraph in the proof of Theorem 1 in Section 1. We shall need this fact in order to prove that the ranges become equal from some point on and so we need first to prove that each R_n is closed; i.e., it is a linear subspace of B. We warn the reader that this fact depends heavily on the way R_n was defined and on the fact that T is a compact operator. The range of a general bounded, linear operator is not closed. To prove the fact that we need we require a technical result which will be used a number of times below. At first it is difficult to see what this result means. However, it plays a crucial role in our development and its meaning becomes clear once we have seen it used.

Lemma 1

Let B be a Banach space, T a compact operator on B and $\lambda \neq 0$ a fixed complex number. There is a number M with the following property: For each y in R_1 the range of $(T - \lambda I)$, there is an x in B such that $y = (T - \lambda I)x$ and $\|x\| \leq M\|y\|$.

Proof. Suppose that the lemma is false. Then for each positive integer n there is a point $y_n \in R_1$ such that whenever $x_n \in B$ satisfies the equation $y_n = (T - \lambda I)x_n$ it must also satisfy the inequality $\|x_n\| > n\|y_n\|$. Clearly $y_n \neq 0$ because the inequality is strict; i.e., if $y_n = 0$ then $x_n = 0$ certainly satisfies our equation but not the inequality. For $n = 1, 2, \ldots$ choose $w_n \in B$

such that $(T - \lambda I)w_n = y_n$. Now, as we just observed, no y_n is zero and so no w_n is in $N_1 = N(T - \lambda I)$, the null space of $T - \lambda I$, and the latter is a closed set (Exercises 2.4, problem 2(c)). This last remark shows that each of the numbers

$$d_n = \inf\{\|w_n - x\| \mid x \in N_1\}$$

is positive. Hence for each n we can choose v_n in N_1 such that

$$d_n \leq \|w_n - v_n\| < 2d_n$$

For $n = 1, 2, \ldots$ define

$$z_n = \frac{w_n - v_n}{\|w_n - v_n\|}$$

and note that $\{z_n\}$ is a bounded sequence. Thus, since T is a compact operator, we may assume that the sequence $\{Tz_n\}$ is convergent (Section 1, Definition 1). Now $(T - \lambda I)$ applied to $\|w_n - v_n\|z_n$ is just $(T - \lambda I)w_n$, because $v_n \in N_1$, and this is just y_n. Hence $\|w_n - v_n\|\|z_n\| > n\|y_n\|$. It follows that

$$\|(T - \lambda I)z_n\| = \|y_n\|\|w_n - v_n\|^{-1} < \frac{\|z_n\|}{n} = \frac{1}{n}$$

and so

$$\lim (T - \lambda I)z_n = 0$$

Now we can write

$$z_n = \frac{1}{\lambda}[Tz_n - (T - \lambda I)z_n]$$

and so it is clear that $\lim z_n$ exists, and we shall call it z. Then, since

$$(T - \lambda I)z = \lim(T - \lambda I)z_n = 0$$

we see that $z \in N_1$. However,

$$\begin{aligned}
\|z_n - z\| &= \|\|w_n - v_n\|^{-1}(w_n - v_n) - z\| \\
&= \|w_n - (v_n + \|w_n - v_n\|z)\|\|w_n - v_n\|^{-1} \\
&\geq d_n\|w_n - v_n\|^{-1} \geq 1/2
\end{aligned}$$

This contradicts the fact that $\lim z_n = z$.

Corollary 1

Let B be a Banach space, T a compact operator on B and $\lambda \neq 0$ a fixed complex number. Then R_1, the range of $T - \lambda I$, is a linear subspace of B.

Proof. Since R_1 is a linear manifold in B we need only show here that it is closed. Let $\{y_n\}$ be a sequence of points of R_1 which converges to $y \in B$. Let M be the number whose existence was proved in Lemma 1. Then for each integer n we can choose x_n in B such that $y_n = (T-\lambda I)x_n$ and $\|x_n\| \leq M\|y_n\|$. Since $\{y_n\}$ was chosen to be a convergent sequence it is bounded (Exercises 2.1, problem 2), and so $\{x_n\}$ is bounded. Hence, we may assume that $\{Tx_n\}$ is convergent. Now, for each n, we have

$$x_n = \lambda^{-1}(Tx_n - y_n)$$

just solve $y_n = (T - \lambda I)x_n$ for x_n. It follows from this that $\{x_n\}$ converges to, say, $x \in B$. But then

$$(T - \lambda I)x = \lim(T - \lambda I)x_n = \lim y_n = y$$

showing that y is in the range of $T - \lambda I$.

One thing we shall do is show that a nonzero complex number λ is in the spectrum of a compact operator T if, and only if, λ is an eigenvalue of T. We have seen (Section 2.3, just after Definition 3) that the eigenvalues of T are always in $\sigma(T)$. What we are trying to show here is that all of the nonzero points of $\sigma(T)$ are eigenvalues of T. This is certainly not true in general (see Section 2.4 for examples). Let us examine how we might go about proving this. Take $\lambda \in \mathbb{C}$, $\lambda \neq 0$, and suppose that λ is not an eigenvalue of T. We must show that $\lambda \notin \sigma(T)$; i.e., we just show that $(T - \lambda I)$ has a bounded, linear inverse on B. Now since λ is not an eigenvalue of T the null space of $T - \lambda I$, N_1, contains only the zero vector. Thus $T - \lambda I$ is a bounded, linear map which takes B onto R_1, the range of $T - \lambda I$, and it is a one-to-one map. By the open-mapping theorem (Section 2.4, Theorem 3) $(T - \lambda I)$ has a bounded, linear inverse which maps R_1 onto B; note that to apply this theorem we need the fact that R_1 is closed in the Banach space B and hence it is itself a Banach space. But this does not give us an inverse for $T - \lambda I$ defined on all of B. If we happened to know that $R_1 = B$, then we would have $\lambda \notin \sigma(T)$. There is another way to show that $T - \lambda I$ has an inverse on B in the case that $R_1 = B$ which avoids the open-mapping theorem. We can use Lemma 1. By that result there is an M such that, given any $y \in R_1$ there is an $x \in B$ with $y = (T - \lambda I)x$ and $\|x\| \leq M\|y\|$. Now we have observed that, since λ is not an eigenvalue of T, $T - \lambda I$ is one-to-one, and we are assuming that it is onto (we assumed $R_1 = B$), thus it has a linear inverse S defined on all of B. Now is S bounded? Well we have

$$y = (T - \lambda I)x$$

and $\|x\| \leq M\|y\|$ so if we apply S to our equation we get

$$Sy = x$$

and, putting this in our inequality we find that

$$\|Sy\| = \|x\| \leq M\|y\|$$

showing that S is bounded (Section 2.1, Definition 1).
We shall return to this very soon (Corollary 2 to Theorem 3).

Theorem 2

Each of the linear manifolds R_n is closed. Furthermore, there is an integer q such that $R_n \neq R_{n+1}$ for $n = 0, 1, \ldots, q-1$ but $R_n = R_{n+1}$ for all $n \geq q$.

Proof. The corollary to Lemma 1 tells us that R_1 is closed. However, since $(T - \lambda I)^n = T_n - \lambda^n I$ for some compact operator Tn (Exercises 1, problem 2(b)) we see that R_n is closed for every n.

Now that we know that each R_n is closed the argument used in the second paragraph of the proof of Theorem 1 can be used here to show that $R_n = R_{n+1}$ for some n. Let q be the first integer for which this is true. Then since $R_{q+1} = R_q$ and $(T - \lambda I)R_q = R_{q+1}$ we must have

$$R_{q+2} = (T - \lambda I)R_{q+1} = (T - \lambda I)R_q = R_{q+1}$$

Similarly, we can prove that $R_{q+k} = R_q$ for every integer $k \geq 1$.

Theorem 3

Let q be the integer whose existence was proved in Theorem 2. Then:

1. $N_q \cap R_q = \{0\}$;
2. $N_q + R_q = \{x + y | x \in N_q, y \in R_q\} = B$;
3. $T(N_q) \subseteq N_q$;
4. $T(R_q) \subseteq R_q$.

Furthermore, the restriction of $T - \lambda I$ to the space R_q has a bounded, linear inverse.

Proof. (1) We shall show that $N_m \cap R_q = \{0\}$ for any integer m. Suppose that z is in this intersection. For each $n \geq q$ we must have a point z_n in B such that $z = (T - \lambda I)^n z_n$ because $R_n = R_q$ for $n \geq q$. If we assume that $z \neq 0$ then $z_n \notin N_n$ for each n. But clearly z_n does belong to N_{m+n} because $z \in N_m$; remember that $N_n \subseteq N_{n+1} \subseteq \ldots$. However, if n is large enough, $N_{m+n} = N_n$ and we have reached a contradiction.

(2) If $z \in B$ there is a y in B such that $(T-\lambda I)^q z = (T-\lambda I)^{2q} y$; because $(T-\lambda I)^q z$ is in $R_q = R_{2q}$. Hence

$$(T-\lambda I)^q [z - (T-\lambda I)^q y] = 0$$

and we have shown that $[z - (T-\lambda I)^q y]$ is in N_q. But then z, which is equal to

$$[z - (T-\lambda I)^q y] + (T-\lambda I)^q y$$

is in $N_q + R_q$ and so this sum must be all of B.

(3) Since $N_0 = \{0\}$ we certainly have $T(N_0) \subseteq N_0$. Now

$$(T-\lambda I)N_k \subseteq N_{k-1}$$

for if $v \in N_k$ then $(T-\lambda I)^k v = 0 = (T-\lambda I)^{k-1}[(T-\lambda I)v]$. However, $N_n \subseteq N_{n+1}$ for all n so

$$(T-\lambda I)N_k \subseteq N_k$$

Hence

$$T(N_k) = [\lambda I + (T-\lambda I)](N_k) \subseteq N_k + N_k = N_k$$

(4) To prove that $T(R_q) \subseteq R_q$ we observe that

$$R_{q+1} = (T-\lambda I)R_q = R_q$$

and argue as in (3).

Finally, let $(T-\lambda I)|R_q$ denote the restriction of $T-\lambda I$ to the space R_q. By (4) this restriction is a linear operator on R_q (i.e., it maps R_q to R_q) and it is continuous because $T-\lambda I$ is continuous on all of B. The null space of the restriction is $N_1 \cap R_q = \{0\}$ by (1) and so $(T-\lambda I)|R_q$ is one-to-one. Thus this restriction has a linear inverse on R_q. The fact that this inverse is continuous follows from Lemma 1 or the open-mapping theorem as discussed above (just before Theorem 2).

Corollary 2

The integers p and q defined in Theorems 1 and 2 are equal.

Proof. Let x be any point in N_{q+1} and find $y \in N_q$, $z \in R_q$ such that $x = y + z$; we can do this by (2). Clearly

$$(T-\lambda I)^{q+1} z = (T-\lambda I)^{q+1}(x-y) = 0$$

But $z \in R_q$, $(T-\lambda I)|R_q$ has a bounded, linear inverse, and $(T-\lambda I)R_q \subseteq R_q$. It follows that $z = 0$ and hence that $x \in N_q$. So $N_{q+1} = N_q$ and we must conclude that $q \geq p$ because p is the first integer for which these spaces coincide. Now $(T - \lambda I)^p N_p$ is the zero subspace, hence

$$R_p = (T - \lambda I)^p B = (T - \lambda I)^p R_q + (T - \lambda I)^p N_q = (T - \lambda I)^p R_q = R_{q+p} = R_q$$

showing that $p \geq q$.

Corollary 3

Let λ be a nonzero complex number. If λ is not an eigenvalue of T, then $\lambda \notin \sigma(T)$. Equivalently: All nonzero points of the spectrum of T are eigenvalues of this operator.

Proof. If λ is not an eigenvalue for T then the null space of $T - \lambda I$, N_1, is the zero subspace. Clearly then $N_0 = N_1 = \ldots$ hence the number p defined in Theorem 1 is 1. Thus, by Corollary 1, $q = 1$ also and so, by (2) of Theorem 3, $R_1 + N_1 = B$; i.e., R_1, the range of $T - \lambda I$, is B. As discussed above, just before Theorem 2, the operator $T - \lambda I$ has a bounded, linear inverse showing that $\lambda \notin \sigma(T)$.

Theorem 4

The restriction of T to R_q, $T|R_q$, is a compact operator on R_q and $\sigma(T|R_q) = \sigma(T) \sim \{\lambda\}$.

Proof. By (4) of Theorem 3 we see that $T|R_q$, call this S, is a bounded, linear operator on R_q. Since T is a compact operator and R_q is closed, S is a compact operator. Now $0 \in \sigma(T)$ because B is infinite dimensional and since $B = N_q + R_q$ and N_q has finite dimension, R_q must be infinite dimensional. Thus $0 \in \sigma(S)$. Now S, T are both compact operators so their nonzero spectra consist entirely of eigenvalues of these operators. Clearly any eigenvalue of S is an eigenvalue of T. Suppose that $\mu \neq \lambda$ is an eigenvalue of T, $\mu \neq 0$, and let x be a corresponding eigenvector. Then $Tx = \mu x$ and $(T - \lambda I)x = (\mu - \lambda)x$, $(T - \lambda I)^q x = (\mu - \lambda)^q x$. Hence $x = (T - \lambda I)^q (\mu - \lambda)^{-q} x$ is in R_q, and this says μ is an eigenvalue of S. We have now shown that

$$\sigma(T) \sim \{\lambda\} \subseteq \sigma(S) \subseteq \sigma(T)$$

However, $(T - \lambda I)|R_q$ is invertible (Theorem 3) and so λ is not in $\sigma(S)$ showing that $\sigma(S) = \sigma(T) \sim \{\lambda\}$.

Corollary 4

Let T be a compact operator on an infinite dimensional Banach space B. If λ is an adherent point (Section 5, Definition 2) of $\sigma(T)$, then $\lambda = 0$.

Proof. We know that $\sigma(T)$ is a compact subset of $\mathbb{C}$ (Section 2.3, Corollary 3 to Theorem 2) hence every adherent point of $\sigma(T)$ is in this set; for a compact set is closed and a closed set contains all of its adherent points. Let $\lambda \in \sigma(T)$, $\lambda \neq 0$. For this constant we find the space N_q and R_q of Theorem 1 and 2 and we recall that, by Theorem 4, $S \equiv (T - \lambda I)|R_q$ is a compact operator on R_q whose spectrum is $\sigma(T) \sim \{\lambda\}$. But R_q is a Banach space and so $\sigma(S)$ is a closed subset of $\mathbb{C}$; in fact, it is a compact subset of $\mathbb{C}$. Thus λ, since it is not in this set, must have a neighborhood which does not contain any points of $\sigma(S)$; otherwise every neighborhood of λ would meet $\sigma(S)$ showing that $\lambda \in \sigma(S)$ and this is not the case. Thus λ is not an adherent point of $\sigma(T) \sim \{\lambda\}$ proving the corollary.

Theorem 5

Let T be a compact operator on an infinite dimensional Banach space B. Then $0 \in \sigma(T)$, each nonzero point of $\sigma(T)$ is an eigenvalue of T whose eigenspace (Section 2.3, Definition 3) is finite dimensional, $\sigma(T)$ is either a finite set or it is a sequence which converges to zero.

Proof. We have already seen that $0 \in \sigma(T)$ when B is infinite dimensional. The second statement was proved in Corollary 2 to Theorem 3. If $\lambda \neq 0$ is in $\sigma(T)$ then the null space of $T - \lambda I$ is finite dimensional by Theorem 1 (this is so even if $\lambda \notin \sigma(T)$). Let us now prove the last statement. For each integer n, let $\Lambda_n = \{\lambda \in \mathbb{C} | \frac{1}{n} \leq |\lambda|\}$. This is a closed set and so $\sigma(T) \cap \Lambda_n$ is a compact set. If this set were infinite it would have an adherent point (Exercises 1, problem 3(a)), say λ_0. But then $\lambda_0 \in \sigma(T) \cap \Lambda_0$ shows that λ_0 is not zero, because $\frac{1}{n} \leq |\lambda_0|$, and λ_0 is an eigenvalue of T. This would contradict Corollary 1 to Theorem 4. Hence $\sigma(T) \cap \Lambda_n$ is a finite set for each fixed n. Label the elements of $\sigma(T) \cap \Lambda_1$ in any way; i.e., write them as $\lambda_1, \lambda_2, \ldots, \lambda_k$. Next label the elements of $\sigma(T) \cap \Lambda_2$ which do not already have labels as $\lambda_{k+1}, \lambda_{k+2}, \ldots$. Continue in this way. If at any point we have $\sigma(T) \cap \Lambda_n = \emptyset$ then $\sigma(T)$ is a finite set. Otherwise our process shows that $\sigma(T)$ is a sequence, for it is contained in the union of the countably many finite sets (i.e., $\sigma(T) \subseteq \bigcup_{n=1}^\infty \sigma(T) \cap \Lambda_n$), and our construction shows that it converges to zero.

Remark 1. We can, at this point, conveniently introduce some useful terminology.

- If T is a bounded, linear operator on a Banach space B and if λ is an eigenvalue of T then the null space of $T - \lambda I$ is called the eigenspace of T and its dimension, which can be infinite, is called the multiplicity of λ. The nonzero eigenvalues of a compact operator all have finite multiplicity.

- Let T be a bounded, linear operator on a Banach space B. A linear subspace H of B is said to invariant under T if $T(H) = \{Tx | x \subseteq H\}$. When T is a compact operator and $\lambda \neq 0$ is an eigenvalue of T then each of the spaces N_q, R_q are invariant under T.

- Let X be a vector space and let Y, Z be two linear manifolds in X. We say that Y and Z are supplementary linear manifolds, we also say that each is a supplement for the other in X, if: (i) $Y \cap Z = \{0\}$ and (ii) $Y + Z = X$. Two linear subspaces G, H of a Banach space B are said to be complementary subspaces if they are supplementary as linear manifolds. We also say that each is a complement for the other in B. When T is a compact operator on B and $\lambda \neq 0$ is an eigenvalue for T, then N_q and R_q are complementary subspaces of B.

Remark 2. An integral equation in which the unknown function occurs under the integral sign and nowhere else is called an equation of the first kind. If an equation is not of the first kind, then we say that it is of the second kind. A typical equation of the second kind is this:

$$f(s) = g(s) + \int_a^b K(s,t)f(t)dt \tag{1}$$

where $g \in L^2[a,b]$ and $K \in L^2[R]$, $R = [a,b] \times [a,b]$, are known. Since the limits on the integral are constants this is called a Fredholm equation. Were we to replace b by s we would have a Volterra equation. This, perhaps superficial seeming, distinction turns out to be important. Now define

$$K(f)(s) = \int_a^b K(s,t)f(t)dt \tag{2}$$

and let T be the operator on $L^2[a,b]$ which takes each f to $K(f)$; i.e., T is the Hilbert-Schmidt operator with kernel $K(s,t)$ (Section 2.4, example (c) just before the Exercises). We may now write (1) as

$$(I - T)f = g \tag{3}$$

In the theory of integral equations one defines the characteristic values to be those scalars λ for which there is some nonzero $f \in L^2[a,b]$ with $\lambda T[f] = I[f]$; or

$$f(s) = \lambda \int_a^b K(s,t)f(t)dt$$

For such λ the equation does not have a unique solution. Note that λ could not be zero so we may write

$$f(s) = g(s) + \lambda \int_a^b K(s,t)f(t)dt \tag{4}$$

$$\lambda^{-1}f(s) = \lambda^{-1}g(s) + \int_a^b K(s,t)f(t)dt \tag{5}$$

$$(\lambda^{-1}I - T)f = \lambda^{-1}g \tag{6}$$

and this last formulation shows that λ is a characteristic value for (1) if, and only if, λ^{-1} is an eigenvalue of T.

Now T is a compact operator (Section 2, Corollary 2 to Theorem 2) and we have just learned that the eigenvalues of such an operator either form a finite set or a sequence which converges to zero. Thus the characteristic values of a Fredholm integral equation of the second kind form either a finite set or a sequence $\{\lambda_j\}$ of complex numbers such that $\lim |\lambda_j| = \infty$.

Remark 3. Let us show here that a Voltera integral equation of the second kind, with continuous kernel, has no characteristic values. This will give us an example of a compact operator on a Banach space whose spectrum is $\{0\}$.

Let $K(s,t)$ be continuous for $a \leq t \leq s \leq b$ and consider

$$f(s) = g(s) + \lambda \int_a^s K(s,t)f(t)dt \tag{7}$$

where $g \in C[a,b]$ is known. As we have done so often before, let us set

$$K(f)(s) = \int_a^s K(s,t)f(t)dt$$

and let $T(f) = K(f)$ for each $f \in C[a,b]$. We leave it to the reader to show that $K(f) \in C[a,b]$ and that T is a compact operator on $(C[a,b], \|\cdot\|_\infty)$; see Section 1. Let

$$M = \sup\{|K(s,t)| | a \leq t \leq s \leq b\}$$

and note that for any f and any λ we have

$$|\lambda K(f)(s)| = |\lambda \int_a^s K(s,t)f(t)dt| \leq |\lambda| M \|f\|_\infty (s-a)$$

We can consider iterates of λK; i.e.,

$$(\lambda K)^n(f) = (\lambda K)((\lambda K)^{n-1}(f))$$

Let us show that

$$|(\lambda K)^n(f)(s)| \leq |\lambda|^n M^n \|f\|_\infty \frac{(s-a)^n}{n!} \tag{8}$$

for every n. We have already shown that this is true when $n = 1$. Assume that it is true when $n = k$ and consider the case when $n = k+1$ (The reader familiar with mathematical induction will reorganize this argument. Others might note that since we have it for $n = 1$ our calculation will show it is true for $n = 2$ and that, in turn, will show it true for $n = 3$ and so on). We have

$$\begin{aligned}|(\lambda K)^{k+1}(f)(s)| &= |\lambda K((\lambda K)^k(f))(s)| = |\lambda \int_a^s K(s,t)(\lambda K)^k(f)(t)dt| \\ &\leq |\lambda| \int_a^s K(s,t)|\lambda|^k M^k \|f\|_\infty \frac{(s-a)^k}{k!} dt| \\ &\leq \frac{|\lambda|^{k+1} M^{k+1}(s-a)^{k+1}}{(k+1)!} \|f\|_\infty \end{aligned}$$

Return now to equation (7). In operator form it becomes

$$(I - \lambda T)f = g \tag{9}$$

and we have just seen that

$$\|(\lambda T)^n\| \leq \frac{|\lambda|^n M^n}{n!}(b-a)^n \tag{10}$$

Thus

$$\|\sum_{n=0}^\infty (\lambda T)^n\| \leq \sum_{n=0}^\infty \frac{[M|\lambda|(b-a)]^n}{n!} = e^{M|\lambda|(b-a)} \tag{11}$$

showing that

$$\sum_{n=0}^\infty (\lambda T)^n \tag{12}$$

converges in the Banach space $\mathcal{L}(C[a,b])$ (see Section 2.3, the proof of Theorem 2). But the series (12), since it converges to a bounded, linear operator on $C[a,b]$ is the inverse of $I - \lambda T$ as a direct calculation shows. Thus (2) always has a unique solution for any $\lambda \neq 0$. It follows that the compact operator T has no nonzero points in its spectrum.

Exercises 3

* 1. Recall the Hilbert space ℓ^2 and the operator

$$A(\{x_n\}) = \{\frac{x_n}{n}\}$$

which, we have seen (Exercises 2, problem 3(b)), is a compact operator.

a. Compute the spectrum of A. What are the eigenvalues of this operator?

b. Let $\lambda \neq 0$ be an eigenvalue of A. Find the null space and range of $A - \lambda I$. Show directly that there are complementary subspaces of ℓ^2.

c. Recall the operator
$$B(\{x_n\}) = \{\frac{x_{n-1}}{n}\}, \quad x_0 \equiv 0$$
This, too, we have seen to be compact. Answer questions (a) and (b) for B.

2. Referring to the operators A, B defined in problem 1 let ℓ be an integer, $\ell > 1$, and consider A^ℓ, B^ℓ. Compute the spectra of each of these operators.

3. Again we shall work with the operators A, B of problem 1.

  a. Let $p(t) = \sum_{j=0}^{n} \alpha_j t^j$ be a polynomial and suppose that $\alpha_0 = 0$. So $p(t) = \alpha_1 t + \alpha_2 t^2 + \ldots + \alpha_n t^n$ and we have no constant term. For any operator T, $p(T)$ can be taken to be $\alpha_1 T + \alpha_2 T^2 + \cdots + \alpha_n T^n$. Find the spectrum of $p(A)$ and $p(B)$ — note that since $p(t)$ has no constant term these operators are compact.

  b. We recall that $\sin t = \sum_{k=0}^{\infty} \frac{(-1)^k t^{2k+1}}{(2k+1)!}$ for all $t \in \mathbf{R}$. Show that $\sum_{k=0}^{\infty} \frac{(-1)^k A^{2k+1}}{(2K+1)!}$ converges, for the operator norm, to a bounded (in fact, a compact) operator on ℓ^2 which we may as well call $\sin A$. Compute the spectrum of this operator.

  c. Work part (b) for the operator B.

4. Let us work with $C[0,1]$ with $\|\cdot\|_\infty$. Solve the equations:

  a. $f(s) = g(s) + \lambda \int_0^s e^{s-t} f(t) dt$, $0 \le s \le 1$
  b. $f(s) = g(s) + \lambda \int_0^s (s-t) f(t) dt$, $0 \le s \le 1$

* 5. Let T be a compact operator on an infinite dimensional Banach space B. Let $\lambda \neq 0$ be a complex number.

  a. Show that $N_0 \subseteq N_1 \subseteq \ldots$
  b. Show that $R_0 \supseteq R_q \supseteq \ldots$
  c. Suppose that λ is not an eigenvalue of T. What are the numbers p, q in that case.

4

SPECTRAL THEORY, THE BASIC TOOLS

Spectral theory is concerned with the interrelations between an operator and its spectrum, and these connections are closest for certain classes of operators on a Hilbert space. There are two main "reasons" for this. First we have available a "great many" special operators, called projections; a fact which is closely related to the geometry of a Hilbert space. Second, any bounded, linear operator on such a space has an "adjoint" (defined below) which, among other things, helps us identify the classes of operators for which we can hope to get a good spectral theory. We shall see that a Banach space which is not a Hilbert space has a rather different geometry and hence "fewer" projection operators. Also, the so-called adjoint of a bounded, linear operator on a Banach space has inherent limitations which restrict its usefulness. These, admittedly vague, remarks will become clear as our discussion progresses.

4.1 Some Infinite Dimensional Geometry, Projections

We begin with the closely related concepts of a projection operator and a pair of supplementary linear manifolds. It is useful to start by looking at these concepts in a very general setting.

Definition 1

Let $\overline{X}$ be a vector space over $\mathcal{K}$. A linear operator P on $\overline{X}$ is called a projection operator if $P^2 = P$.

Note that, on the space $\mathbf{R}^2$, the operator $P[(x, y)] = (x, 0)$ for all $(x, y) \in \mathbf{R}^2$, is a projection operator. The range of this operator is the x-axis and its null space is the y-axis. Similarly, on $\mathbf{R}^3$, the map $P[(x, y, z)] = (x, y, 0)$ for all (x, y, z) is a projection operator with range equal to the xy-plane and null space equal to the z-axis. Observe that in each of these cases the range and null space have only the zero vector in common and any vector in the space can be written as the sum of two vectors one in the range of our projection and the other in its null space; any $(x, y) \in \mathbf{R}^2$ can be written $(x, y) = (x, 0) + (0, y)$, and any (x, y, z) in $\mathbf{R}^3$ can be written as $(x, y, 0) + (0, 0, z)$. These two properties are characteristic of the range and null space of any projection operator as we shall show. First we need some more terminology.

Definition 2

Let $\overline{X}$ be a vector space over $\mathcal{K}$. Two linear manifolds Y, Z of $\overline{X}$ are said to be supplementary linear manifolds of $\overline{X}$ (we also say that each is a supplement of the other in $\overline{X}$) if their intersection contains only the zero vector, and their sum is $\overline{X}$.

More explicitly, we must have: (a) $Y \cap Z = \{0\}$; (b) $Y + Z = \{y + z | y \in Y, z \in Z\} = \overline{X}$.

Theorem 1

Let $\overline{X}$ be a vector space over $\mathcal{K}$. Then the range and null space of any projection operator on $\overline{X}$ are supplementary linear manifolds of $\overline{X}$. Conversely given a pair Y, Z of supplementary linear manifolds of $\overline{X}$ there is a unique projection operator P_y on $\overline{X}$ whose range is Y and whose null space is Z, and a unique projection operator P_z on $\overline{X}$ whose range is Z and whose null space is Y. Furthermore, $P_y = I - P_z$.

Proof. Let P be a projection operator on $\overline{X}$, let Ker P be the null space of P and let $R(P)$ be the range of this operator. If y is in both these linear manifolds then two things must be true. First, $Py = 0$, and second, there must be an x in $\overline{X}$ such that $y = Px$. But then $0 = Py = P(Px) = P^2(x) = Px = y$ and we have shown that $(\text{Ker} P) \cap R(P) = \{0\}$.

Now let w be any element of $\overline{X}$ and consider $w - Pw$. This is in Ker$P$ and since $w = (w - Pw) + Pw$ we see that w can be written as the sum of two vectors one, $w - Pw$, in Ker$P$, and the other Pw, in $R(P)$.

Let us note that this representation of the vectors in $\overline{X}$ as sums of vectors in the range and null space of P is unique. To see this suppose that $w \in \overline{X}$

and that $w = u + v = u' + v'$ where $u, u' \in \text{Ker} P$ and $v, v' \in R(P)$. Then $u - u' = v' - v$ and so $v' - v$, which is certainly is $R(P)$ is also in $\text{Ker} P$. Hence $v' - v = 0$ giving us $v' = v$. However $u - u' = v' - v$ so $u = u'$ also.

Now suppose that we are given two supplementary linear manifolds Y, Z of $\overline{X}$. Define $P_y : \overline{X} \to Y$ as follows: Write any given $x \in \overline{X}$, uniquely, as $y + z$ where $y \in Y$ and $z \in Z$, and set $P_y x = y$. It is clear that P_y is a linear operator on $\overline{X}$ whose range is Y and whose null space is Z. Furthermore, $P_y^2 = P_y$ because: For any $x \in \overline{X}$, $x = y + z$ so $P_y x = y$ and $y = y + 0$ so $P_y(P_y x) = P_y(y) = y$ showing that $P_y x = P_y^2 x$ for every $x \in \overline{X}$.

Finally, for any $x \in \overline{X}$ we write $x = y + z$ where $y \in Y$ and $z \in Z$ and we set $P_y x = y$, $P_z x = z$. Clearly then $P_y x = x - (z) = (y + z) - z = Ix - P_z x = (I - P_z)x$.

We shall call P_y the projection of $\overline{X}$ onto Y determined by the pair Y, Z. We leave it to the reader to prove that if P is any projection on $\overline{X}$, then $I - P$ is also a projection and, furthermore, $\text{Ker} P = R(I - P)$, $R(P) = \text{Ker}(I - P)$; see problem 1.

Our remarks above show that a linear manifold in $\overline{X}$ has a supplement if, and only if, it is the range of a projection operator on $\overline{X}$. It is in fact true that any linear manifold in $\overline{X}$ has a supplement but this fact is of little use. Our discussion thus far has been too general. The concepts we have introduced become much more interesting when we look at them in a Banach or Hilbert space and modify them so as to have them related to the topological as well as the algebraic structure of these spaces.

Definition 3

Let B be a Banach space and let G, H be two supplementary linear manifolds in B. If both of these manifolds are subspaces (i.e., if they are both closed) then we shall say that they are complementary subspaces of B (we shall also say that each is a complement of the other in B).

It is easy to see that the range and null space of any *continuous* projection operator on B are complementary subspaces of B; for if P is our projection then $\text{Ker} P$ is closed because P is continuous and $R(P)$ is closed because it is the null space of the continuous projection operator $I - P$. Thus the first statement in Theorem 1 has a valid analogue in the situation under discussion. The analogue of the second statement is a little more difficult to prove.

Theorem 2

Let B be a Banach space and let G be a linear subspace of B. Then G has a complement if, and only if, there is a continuous projection operator on B whose range is G.

Proof. The sufficiency of our condition is immediate. Let us suppose now that G has a complement H in B. Then any x in B can be written, in just one way, as the sum of an element $x_g \in G$ and an element $x_h \in H$. Setting $Px = x_g$ we certainly obtain a projection on B whose range is G (see the proof of Theorem 1 above), but is P continuous?

First consider $G \times H = \{(g,h) | g \in G, h \in H\}$. This is a vector space under the coordinate-wise operations; i.e., $(g_1, h_1) + (g_2, h_2) = (g_1 + g_2, h_1 + h_2)$ and $\lambda(g, h) = (\lambda g, \lambda h)$. For each (g, h) in this vector space we define $\|(g, h)\|_p = \max(\|g\|, \|h\|)$. This actually is a norm on $G \times H$ (problem 2a). Furthermore, since G and H are Banach spaces, because they are closed in B and B is a Banach space, $(G \times H, \|\cdot\|_p)$ is also a Banach space (problem 2b).

Now we define a map $\sigma : G \times H \to B$ by $\sigma[(g, h)] = g + h$ for all (g, h) in $G \times H$. It is clear that σ is linear, and since G and H are complementary subspaces, both one-to-one and onto. Furthermore, $\|\sigma(g, h)\| = \|g + h\| \leq \|g\| + \|h\| \leq 2\max(\|g\|, \|h\|) = 2\|(g, h)\|_p$ and so σ is continuous. The open-mapping theorem, actually the special case mentioned above (section 2.4, Corollary 1 to Theorem 3), tells us that σ has a continuous, linear inverse $\sigma^{-1} : B \to G \times H$. At this point we must define one more map $\pi : G \times H \to G$ by setting $\pi[(g, h)] = g$ for all $(g, h) \in G \times H$. Clearly π is linear, onto and, since $\|\pi(g, h)\| = \|g\| \leq \max(\|g\|, \|h\|) = \|(g, h)\|_p$, continuous. Finally, observe that $P = \pi \circ \sigma^{-1}$ showing, at last, that P is continuous.

Theorem 2 gives us an "answer" to the question of which linear subspaces of a Banach space have complements. In some sense we can do no better because of the following:

Theorem 3

A Banach space is a Hilbert space if, and only if, each of its linear subspaces has a complement.

The "easy" half of this theorem, the fact that every linear subspace of a Hilbert space has a complement was proved long before the other, more difficult, part. We shall prove the first part below and refer the reader to the literature for the second part [8].

Theorem 3 tells us that any Banach space that is not a Hilbert space, $(C[0, 1], \|\cdot\|_\infty)$ say, must have at least one linear subspace which has no complement. This is not easy to visualize. Such a subspace is not the range, or the null space, of any projection operator and this is what we meant when we said earlier that there are "fewer" continuous projection operators on a Banach space then there are on a Hilbert space.

An example of a linear subspace of a Banach space which has no complement can be found in [3; p. 37].

Exercises 1

* 1. Let $\overline{X}$ be a vector space over $\mathcal{K}$ and let P be a projection operator on $\overline{X}$. Show that $I - P$ is also a projection operator (just compute $(I - P)^2$) and show that $\text{Ker} P = R(I - P)$, $R(P) = \text{Ker}(I - P)$.

* 2. Let $(E_1, \|\cdot\|_1)$ and $(E_2, \|\cdot\|_2)$ be two normed spaces over the same field and let $E_1 \times E_2$ be the vector space $\{(x_1, x_2) | x_1 \in E_1, x_2 \in E_2\}$.

    a. Define, for all $(x_1, x_2) \in E_1 \times E_2$, $\|(x_1, x_2)\|_p = \max(\|x_1\|_1, \|x_2\|_2)$. Show that $\|\cdot\|_p$ is a norm on $E_1 \times E_2$.

    b. If $(E_1, \|\cdot\|_1)$ and $(E_2, \|\cdot\|_2)$ are both Banach spaces shows that $(E_1 \times E_2, \|\cdot\|_p)$ is a Banach space. Hint: Show that a sequence $\{(x_n, y_n)\} \subseteq E_1 \times E_2$ is a Cauchy sequence for $\|\cdot\|_p$ if, and only if, $\{x_n\} \subseteq E_1$ is a Cauchy sequence for $\|\cdot\|_1$, and $\{y_n\} \subseteq E_2$ is a Cauchy sequence for $\|\cdot\|_2$.

    c. Show that the maps $\pi_1 : E_1 \times E_2 \to E_1$, $\pi_1[(x, y)] = x$, and $\pi_2 : E_1 \times E_2$, $\pi_2[(x, y)] = y$, are linear, continuous, and onto.

3. In the Banach space $(C[-1, 1], \|\cdot\|_\infty)$ define two sets E and O as follows: $E = \{f | f(x) = f(-x) \text{ for all } x \text{ in } [-1, 1]\}$ and $O = \{f | f(-x) = -f(x)$ for all x in $[-1, 1]\}$.

    a. Show that E and O are linear subspaces of $(C[-1, 1], \|\cdot\|_\infty)$. First show that they are linear manifolds and then show that they are closed.

    b. Show that the subspaces E and O are complementary in $C[-1, 1]$.

    c. For each f define $(Pf)(x) = \frac{f(x)+f(-x)}{2}$, and let P be the operator on $C[-1, 1]$ which takes each f to (Pf) as just defined. Show that P is a continuous projection operator. What is its null space and what is its range?

4. a. Show that any linear subspace of $\mathbb{R}^n$ has a complement as follows: Let G be a linear subspace of $\mathbb{R}^n$ and let $\vec{v}_1, \ldots, \vec{v}_k$ be a basis for G. Then this set is, as a subset of $\mathbb{R}^n$, linearly independent. Thus we can find vectors $\vec{w}_1, \ldots, \vec{w}_{n-k}$ in $\mathbb{R}^n$ such that $\{\vec{v}_1, \ldots, \vec{v}_k, \vec{w}_1, \ldots, \vec{w}_{n-k}\}$ is a basis for $\mathbb{R}^n$.

   b. Show, by giving examples in $\mathbb{R}^2$ and $\mathbb{R}^3$, that the complement of a linear subspace is not unique.

4.2 Projections on a Hilbert Space

Our task here is to investigate projections, and all the related matters mentioned in Section 1, on a fixed Hilbert space. In this case the notion of orthogonality plays an important role.

Theorem 1

Let $(H, \langle,\rangle)$ be a Hilbert space, let M be a linear subspace of H, and let $y \in H \setminus M$. Then there is a unique vector $Py \in M$ such that $\|y - Py\| = \inf\{\|y - x\| | x \text{ in } M\}$.

Proof. Let $d = \inf\{\|y - x\| | x \text{ in } M\}$, and choose a sequence $\{x_n\} \subseteq M$ such that
$$d = \lim \|y - x_n\|. \tag{1}$$
Then we may write (Exercises 1.4, problem 1(c)),
$$\|(y-x_m)+(y-x_n)\|^2 + \|(y-x_m)-(y-x_n)\|^2 = 2\|y-x_m\|^2 + 2\|y-x_n\|^2 \tag{2}$$
A little algebra yields
$$\|x_n - x_m\|^2 = 2\|y - x_m\|^2 + 2\|y - x_n\|^2 - \|2y - (x_n + x_m)\|^2 \tag{3}$$
The very last term in this expression can be written as
$$4\|y - \frac{(x_n + x_m)}{2}\|^2 \tag{4}$$
Now $\frac{(x_n+x_m)}{2} \in M$ because x_n and x_m are in M, and M is a subspace. Consequently, our last expression (i.e., (4)) is $\geq 4d^2$. Thus
$$\|x_n - x_m\|^2 \leq 2\|y - x_m\|^2 + 2\|y - x_n\|^2 - 4d^2 \tag{5}$$
and it follows from this, via (1), that $\{x_n\}$ is a Cauchy sequence in the Hilbert space H; in particular, it is in the subspace M of H. Hence $\{x_n\}$ converges to, say, $Py \in M$. Thus
$$\|y - Py\| = \lim \|y - x_n\| = d \tag{6}$$

Finally, let us show that Py is unique. Suppose that z_1, z_2 are in M and that $\|y - z_1\| = d = \|y - z_2\|$. Then
$$\|(y - z_1) + (y - z_2)\|^2 + \|(y - z_1) - (y - z_2)\|^2 = 2\|y - z_1\|^2 + 2\|y - z_2\|^2 \tag{7}$$
or
$$\|z_2 - z_1\|^2 = 4d^2 - 4\|y - \frac{(z_1 + z_2)}{2}\|^2 \tag{8}$$

Again, reasoning as we did above, the very last term in (8) is $\geq 4d^2$ giving us $\|z_2 - z_1\|^2 \leq 0$. Clearly then $z_1 = z_2$.

Corollary 1

Let $(H, \langle,\rangle)$ be a Hilbert space, let M be a linear subspace of H and let $y \in H \setminus M$. Then $\langle y - Py, x \rangle = 0$ for every x in M.

Proof. We have just seen that $y_0 \equiv y - Py$ satisfies the equation

$$\|y_0\| = d = \inf\{\|y - x\| | x \text{ in } M\} \tag{9}$$

Now for any $z \in M$ and any scalar α we know that $Py + \alpha z$ is in M, and so

$$\begin{aligned} d^2 &= \|y_0\|^2 \leq \|y - (Py + \alpha z)\|^2 = \|y_0 - \alpha z\|^2 = \\ &\quad \langle y_0 - \alpha z, y_0 - \alpha z \rangle \\ &= \|y_0\|^2 - \alpha \langle z, y_0 \rangle - \overline{\alpha} \langle y_0, z \rangle + |\alpha|^2 \|z\|^2 \end{aligned} \tag{10}$$

Hence

$$0 \leq -\alpha \langle z, y_0 \rangle - \overline{\alpha} \langle y_0, z \rangle + |\alpha|^2 \|z\|^2 \tag{11}$$

Suppose that we could find $z \in M$ such that $\langle y_0, z \rangle \neq 0$. Then, in particular, z is not the zero vector and so we could set $\alpha = \langle y_0, z \rangle / \|z\|^2$. Putting this into (11) would give

$$0 \leq -2 \frac{|\langle y_0, z \rangle|}{\|z\|^2} + \frac{|\langle y_0, z \rangle|}{\|z\|^2} = -\frac{|\langle y_0, z \rangle|}{\|z\|^2} \tag{12}$$

which is an obvious contradiction.

At this point we need a little more terminology and it isn't entirely consistent with our earlier definitions.

Definition 1

Let $(H, \langle,\rangle)$ be a Hilbert space and let M be a linear manifold in H. The set $\{y \in H | \langle x, y \rangle = 0 \text{ for all } x \text{ in } M\}$ is denoted by $M^\perp$ and is called the orthogonal complement of M in H.

As we have said this terminology isn't consistent with our earlier use because M need not be closed. However, in the proof of the next result we shall have to use $M^\perp$ when M is not a subspace and so we thought it best to live with a little inconsistency in our definition. This should not cause any trouble.

Lemma 1

Let H be a Hilbert space and let M be a linear manifold in H. Then:

a. $\overline{M} = (M^\perp)^\perp$;

b. $M^\perp = \{0\}$ if, and only if, $\overline{M} = H$.

Proof. We recall that $\overline{M}$ denotes the closure of M (Section 3.1, Definition 2) and that, when $\overline{M} = H$, we say that M is dense in H.

Let us make some observations about $M^\perp$. These are easily proved and we leave them to the exercises (problems 1(a), (b)); hints are provided there. First note that $M^\perp$ is always a linear subspace even when M is only a linear manifold. Also, M and $\overline{M}$ have the same orthogonal complement; i.e., $\langle x, y \rangle = 0$ for all x in M if, and only if, $\langle z, y \rangle = 0$ for all $z \in \overline{M}$. Finally, it is obvious from the definitions that

$$M \subseteq (M^\perp)^\perp$$

This last observation combined with our second observation shows that

$$\overline{M} \subseteq (M^\perp)^\perp$$

Suppose that $y \in (M^\perp)^\perp$ and $y \notin \overline{M}$. Then, by Theorem 1 and its corollary, there is a unique vector $Py \in \overline{M}$ such that

$$y_0 \equiv y - Py \in (\overline{M})^\perp$$

Now both y and Py are in $(M^\perp)^\perp$, and this is a subspace, so y_0 is in here as well. Also, $y \notin \overline{M}$ while $Py \in \overline{M}$ so y_0 is not the zero vector. Since $y_0 \in (\overline{M})^\perp$, $y_0 \notin (\overline{M})^{\perp\perp}$ because if it were in here it would be self-orthogonal and the only such vector is the zero vector. But since $(\overline{M})^{\perp\perp} = (M^\perp)^\perp$ we have reached a contradiction. This proves (a).

To prove (b) we first suppose that M is dense in H. Then $(\overline{M})^\perp = H^\perp = \{0\}$ and since $\overline{M}$ and M have the same orthogonal complements we have $M^\perp = \{0\}$. Conversely, if $M^\perp = \{0\}$, then $\overline{M} = (M^\perp)^\perp$ (by (a)) $= \{0\}^\perp = H$.

We can now show that any linear subspace of a Hilbert space has a complement and we can identify the projection of the Hilbert space onto the given subspace. So let H be a fixed, Hilbert space and let M be a linear subspace of H. Consider the linear subspace $M^\perp$. Clearly $M \cap M^\perp = \{0\}$ since any vector in this intersection would be self-orthogonal. By Theorem 1 there is, for any $z \in H$, a unique vector Pz such that

$$\|z - Pz\| = \inf\{\|z - x\| \mid x \in M\}$$

Moreover, by Corollary 1, $z - Pz$ is in $M^\perp$. Now

$$z = (z - Pz) + Pz$$

showing that $M^\perp + M = H$. Thus $M^\perp$ is a complement for M in H.

Our last equation and Theorem 1 of Section 1 tell us that the map P which takes each $z \in H$ to the vector Pz, and the map $P^\perp$ which takes each $z \in H$ to $z - Pz$, are the projections of H onto M and $M^\perp$ respectively; $P : H \to M$, $P(z) = Pz$ and $P^\perp : H \to M^\perp$, $P^\perp(z) = z - Pz$. That theorem tells us nothing about the continuity of these maps. However,

$$z = P^\perp(z) + P(z) \text{ and } \langle P^\perp(z), P(z) \rangle = 0$$

hence (Exercises 1.4, problem 1(e))

$$\|z\|^2 = \|P^\perp(z)\|^2 + \|P(z)\|^2$$

showing that $\|P(z)\| \leq \|z\|$. Thus $\|P\| \leq 1$ and this map is continuous. Moreover, if $z \in M$, then $z = Pz = P(z)$ and so

$$\|z\| = \|P(z)\| \leq \|P\|\|z\|$$

giving us $\|P\| = 1$. Similarly, $\|P^\perp\| = 1$.

Let us not lose sight of the fact that the vector Pz is that point in M which is nearest the given vector z. Thus the projection P of H onto M determined by the pair M, $M^\perp$ is the map which takes each $z \in H$ to the unique point of M which is nearest to z.

Note one more thing about the projection P. For any x, y in H we have

$$\begin{aligned}\langle P(x), y \rangle &= \langle P(x), P(y) + P^\perp(y) \rangle = \langle P(x), P(y) \rangle \\ &= \langle P(x) + P^\perp(x), P(y) \rangle = \langle x, P(y) \rangle\end{aligned}$$

The significance of this observation is this: A linear subspace M of a Hilbert space H has, in general, lots of complements. Consider the xy-plane in $\mathbb{R}^3$ for instance. Any line through $(0,0,0)$ which does not lie in the xy-plane is a complement for this subspace. Thus there are lots of projections of H onto M. Of all these which is the one determined by the pair M, $M^\perp$? It is that projection P of H onto M such that

$$\langle P(x), y \rangle = \langle x, P(y) \rangle$$

for all x, y in H; we shall prove this below (Theorem 2). This leads us to an extremely important concept.

Definition 2

Let H be a Hilbert space and let A be a bounded, linear operator on this space. A bounded, linear operator B on H is called the adjoint of A if

$$\langle Ax, y \rangle = \langle x, By \rangle$$

for all x, y in H. We shall denote the adjoint of A by A^*, and we shall say that A is a self-adjoint operator in case $A = A^*$.

We shall prove in Section 3 that every bounded, linear operator on H has an adjoint. The fact that, given A, A^* is unique is easy to prove: Suppose that B, C are bounded, linear operators on H which satisfy
$$\langle x, By\rangle = \langle Ax, y\rangle = \langle x, Cy\rangle$$
for all x, y in H. Then for all x, y in H we must have
$$\langle x, (B - C)y\rangle = 0$$
In this equation we put $x = (B - C)y$, where y is arbitrary but fixed. Clearly this then gives $\|(B - C)y\|^2 = 0$ or $By = Cy$. However, y was arbitrary, so $B = C$.

Now let us look at an example. On the space ℓ^2 we recall the right-shift operator A_r (Exercises 1.5, problem 2)
$$A_r(\vec{z}) = A_r(\{z_1, z_2, \ldots\}) = \{0, z_1, z_2, \ldots\} = \{z_{j-1}\}_{j=1}^{\infty}, \quad z_0 \equiv 0.$$
For any $\vec{z} = \{z_j\}_{j=1}^{\infty}$, $\vec{w} = \{w_j\}_{j=1}^{\infty}$ in ℓ^2 we have
$$\begin{aligned}\langle A_r \vec{z}, \vec{w}\rangle &= \langle \{z_{j-1}\}_{j=1}^{\infty}, \{w_j\}_{j=1}^{\infty}\rangle = \sum_{j=1}^{\infty} z_{j-1} \overline{w}_j \\ &= \sum_{\ell=1}^{\infty} z_\ell \overline{w}_{\ell+1} = \langle \{z_\ell\}_{\ell=1}^{\infty}, \{w_{\ell+1}\}_{\ell=1}^{\infty}\rangle \\ &= \langle \vec{z}, A_\ell \vec{w}\rangle\end{aligned}$$
where we recall that $A_\ell(\{w_j\}_{j=1}^{\infty}) = \{w_{j+1}\}_{j=1}^{\infty}$. Thus
$$A_r^* = A_\ell$$

Theorem 2

Let P be a bounded, projection operator on a Hilbert space H. Then the subspaces $R(P)$, Ker$P$ are orthogonal complements in H if, and only if, P is a self-adjoint operator.

Proof. Our discussion before Definition 2 shows that the projection determined by a pair of orthogonal complements is bounded and self-adjoint. Let us now suppose that $P = P^*$. Choose any $x \in \text{Ker}P$ and any $y \in R(P)$ and observe that
$$y = P(z)$$
for some $z \in H$. Thus
$$\langle x, y\rangle = \langle x, P(z)\rangle = \langle P(x), z\rangle = \langle 0, z\rangle = 0$$
showing that $R(P)$ and Ker$P$ are orthogonal subspaces.

Exercises 2

* 1. Let H be a Hilbert space and let M be a linear manifold in H.

    a. Show that $M^\perp = \{y \in H | \langle x, y \rangle = 0 \text{ for all } x \in M\}$ is a linear subspace of H. Hint: First show that $M^\perp$ is a linear manifold. Next suppose that $\{y_n\} \subseteq M^\perp$ converges to z. Now use Exercises 1.4, problem 1(f) to show that $z \in M^\perp$.

    b. There is a theorem from classical analysis which tells us that if f is any real or complex valued function on $S \subseteq H$ which is uniformly continuous (Section 2.1, Lemma 1(d)) on S, then f can be extended in one, and only one, way to a function $\overline{f}$ which is uniformly continuous on $\overline{S}$ (the closure of S). For a fixed vector $y \in H$ let $f(x) = \langle x, y \rangle$ for all $x \in M$. Show that f is uniformly continuous on M (Section 1.4, Theorem 1). Next suppose that $f(x) = 0$ for all $x \in M$ (i.e., suppose that $y \in M^\perp$) and use the theorem stated to conclude that $f(z) = 0$ for all $z \in \overline{M}$. Conclude that $y \in (\overline{M})^\perp$.

* 2. A projection P on H such that $R(P)$ and $\text{Ker}(P)$ are orthogonal complements is called an orthogonal projection. Let P, Q be two orthogonal projections. Then:

    a. $R(P)^\perp = R(I - P)$;

    b. The three following conditions are equivalent:

        i. $R(P) \subseteq R(Q)$;
        ii. $P = PQ$;
        iii. $P = QP$;

    c. PQ is an orthogonal projection if, and only if, $PQ = QP$.

    d. If $PQ = QP$ (i.e., if these operators commute), then $R(PQ) = R(P) \cap R(Q)$. Furthermore, $P + Q - PQ$ is an orthogonal projection whose range is $R(P) + R(Q)$.

    e. Let $\{P_j\}_{j=1}^n$ be a finite set of orthogonal projections and let $P = \sum_{j=1}^n P_j$. Show that P is an orthogonal projection if, and only if, $P_k P_\ell = 0$ for $k \neq \ell$; we sometimes say that the projections are orthogonal. When this is the case show that $R(P_k) \cap R(P_\ell) = \{0\}$ for $k \neq \ell$ and that every $x \in R(P)$ can be written in one, and only one, way as a sum $\sum_{j=1}^n y_j$ where each $y_j \in R(P_j)$, $1 \leq j \leq n$.

* 3. Show that the eigenvalues of a self-adjoint operator must be real numbers.

* 4. Recall the operator P (the notation is unfortunate since this operator is not a projection) on $L^2[a,b]$ (Section 2.4, (b)) defined by $P(f) = tf(t)$. Use the fact that t is real to show that P is a self-adjoint operator.

* 5. We defined A on ℓ^2 by $A(\{z_n\}) = \{\frac{z_n}{n}\}$ (Exercises 3.2, problem 3). Find A^*.

* 6. We defined B on ℓ^2 by $B(\{z_n\}) = \{\frac{z_{n-1}}{n}\}$, $z_0 = 0$ (Exercises 3.2, problem 3). Find B^* and show that $BB^* \neq B*B$ (i.e., these operators do not commute). Show that the only eigenvalue of the operator B^* is zero.

7. Any 2×2 matrix, say $M = \begin{pmatrix} \alpha & \beta \\ \gamma & \delta \end{pmatrix}$, define a linear operator T_M on $\mathbb{C}^2$ as follows:

$$T_M \begin{pmatrix} z_1 \\ z_2 \end{pmatrix} = \begin{pmatrix} \alpha & \beta \\ \gamma & \delta \end{pmatrix} \begin{pmatrix} z_1 \\ z_2 \end{pmatrix} = \begin{pmatrix} \alpha z_1 + \beta z_2 \\ \gamma z_1 + \delta z_2 \end{pmatrix}.$$

   a. Show that the adjoint of T_M, T_M^*, is the operator deferred by the matrix $\begin{pmatrix} \overline{\alpha} & \overline{\gamma} \\ \overline{\beta} & \overline{\delta} \end{pmatrix}$. In general if $M = (\alpha_{ij})$, an $n \times n$ matrix then the adjoint of T_M is given by $(\overline{\alpha}_{ji})$.

* b. Let $K(s,t) \in L^2(R)$, where $R = [a,b] \times [a,b]$, and for each $f \in L^2[a,b]$ let

$$K(f)(s) = \int_a^b K(s,t) f(t) dt.$$

Then $K(f) \in L^2[a,b]$ and the map T_K which takes f to $K(f)$ is a bounded (in fact, a compact) linear operator on $L^2[a,b]$; we call it the Hilbert-Schmidt operator with kernel $K(s,t)$.

   i. Set $K^*(s,t) = \overline{K(t,s)}$ and compute $\|K^*(s,t)\|_2$.

   ii. Show that the adjoint of T_K, T_K^*, is the Hilbert-Schmidt operator with kernel K^*; i.e., $T_k^* = T_{K^*}$.

4.3 The Adjoint Exists!

The proof that every bounded, linear operator on a Hilbert space has an adjoint involves some very useful theorems and important concepts. However, some of these are a little bit technical so let us briefly outline the proof. Suppose that H is a Hilbert space and that A is any bounded, linear operator on H. The map σ from $H \times H$ into $\mathbb{C}$ which takes each pair (x,y) to the number $\langle x, Ay\rangle$, is very much like an inner product (Section 1.4, Definition 1). Specifically, σ is linear in the first variable, conjugate linear in the second variable, and "bounded" in the sense that $|\sigma(x,y)| \leq \|A\|\|x\|\|y\|$. We shall prove that every map from $H \times H$ into $\mathbb{C}$ which has these three properties arises just as σ did from some bounded, linear operator on H. So given B, a fixed, bounded, linear operator on H, we consider the map $(x,y) \to \langle Bx, y\rangle$ and show that is has the three properties mentioned. Thus there is an operator C such that the map $(x,y) \to \langle Bx, y\rangle$ is equal to the map $(x,y) \to \langle x, Cy\rangle$. Thus $\langle Bx, y\rangle = \langle x, Cy\rangle$ for all x, y in H showing that C is the adjoint of the given operator B. Now all we have to do is fill in the details of this argument.

Definition 1

Let $(E, \|\cdot\|)$ be a normed space over $\mathcal{K}$. A linear map from E into $\mathcal{K}$ is called a linear functional (or a linear form) on E. A bilinear functional (or bilinear form) Ω on E is a map from $E \times E$ into $\mathcal{K}$ such that:

   i. $\Omega(\alpha x + \beta y, z) = \alpha \Omega(x, z) + \beta \Omega(y, z)$ for all x, y, z in E and all α, β in $\mathcal{K}$,

   ii. $\Omega(x, \alpha y + \beta z) = \overline{\alpha}\Omega(x, y) + \overline{\beta}\Omega(x, z)$ for all x, y, z in E and all α, β in $\mathcal{K}$.

A linear functional f on E is called a bounded, linear functional if it is continuous on E; i.e., if $\sup\{\frac{|f(x)|}{\|x\|}|x \neq 0\}$ is finite (Section 2.1, Theorem 1); of course this supremum is just $\|f\|$. A bilinear form Ω on E is said to be a bounded, bilinear form on E if

$$\sup\{\frac{|\Omega(x,x)|}{\|x\|\|y\|}|x \neq 0 \neq y\}$$

is finite. When this is the case we set $\|\Omega\|$ equal to this supremum. It is instructive to note that the inner product on a Hilbert space is a bilinear form on that space which is, by the C.S.B. inequality (Section 1.4, Theorem 1), a bounded, bilinear form.

Lemma 1

Let H be a Hilbert space and let $\mathcal{L}(H)$ be the set of all bounded, linear operators on H. Then we have:

   a. For any fixed $A \in \mathcal{L}(H)$ the function $\sigma(x,y) = \langle x, Ay\rangle$ for all $x, y \in H$, is a bounded, bilinear form on H with $\|\sigma\| = \|A\|$;

   b. Given any bounded, bilinear form σ on H there is a unique $A \in \mathcal{L}(H)$ such that $\sigma(x,y) = \langle x, Ay\rangle$ for all $x, y \in H$.

Proof. (a) A straightforward calculation shows that the function σ defined in (a) is a bilinear form on H. Moreover,

$$|\sigma(x,y)| = |\langle x, Ay\rangle| \leq \|x\|\,\|Ay\| \leq \|A\|\,\|x\|\,\|y\|$$

for all $x, y \in H$ by the C.S.B. inequality (Section 1.4, Theorem 1) and the fact that A is bounded (Section 2.1, Theorem 1). Thus σ is a bounded, bilinear form and $\|\sigma\| \leq \|A\|$. In the course of proving (b) we shall show that these norms are equal. (b) We are given a bounded, bilinear form σ on H. For $y \in H$, arbitrary but, once chosen, fixed, let us set $\Omega_y(x) = \sigma(x, y)$. Then Ω_y is clearly a linear functional on H (Definition 1(i)). Moreover, Ω_y is bounded because $|\Omega_y(x)| = |\sigma(x, y)| \leq [\|\sigma\|\|y\|]\,\|x\|$, so $\|\Omega_y\| \leq \|\sigma\|\|y\|$. Now there is an important theorem which we shall prove below (Theorem 2) but we want to use now. It states that to any bounded, linear functional f on H there corresponds a unique vector $z \in H$ such that $f(x) = \langle x, z\rangle$ for all $x \in H$ and $\|f\| = \|z\|$. Applying this result to Ω_y we see that there is a unique vector $z \in H$ such that $\Omega_y(x) = \langle x, z\rangle$ for all $x \in H$ and $\|\Omega_y\| = \|z\|$. Note however that in constructing Ω_y we chose y and fixed it and so the vector z here really depends on the y chosen; i.e., if we changed y we would get a different z, so $z = A(y)$ for some function A. But now

$$\Omega_y(x) = \langle x, z\rangle = \langle x, A(y)\rangle \text{ and } \|\Omega_y\| = \|z\| = \|A(y)\|$$

so we are almost done. We have yet to show that the function A is linear, that it is bounded and that it is unique.

For $\alpha \in \mathbb{C}$ consider $\alpha y \in H$ and $\Omega_{\alpha y}$. We have

$$\begin{aligned}\langle x, A(\alpha y)\rangle &= \Omega_{\alpha y}(x) = \sigma(x, \alpha y) = \overline{\alpha}\sigma(x, y) = \overline{\alpha}\Omega_y(x)\\ &= \overline{\alpha}\langle x, A(y)\rangle = \langle x, \alpha A(y)\rangle\end{aligned}$$

for all $x \in H$. Thus $A(\alpha y) = \alpha A(y)$ for any scalar α. Similarly,

$$\begin{aligned}\langle x, A(y+z)\rangle &= \Omega_{y+z}(x) = \sigma(x, y+z) = \sigma(x, y) + \sigma(x, z)\\ &= \Omega_y(x) + \Omega_z(x) = \langle x, A(y)\rangle + \langle x, A(z)\rangle\\ &= \langle x, A(y) + A(z)\rangle\end{aligned}$$

for all $x \in H$. So we have $A(y+z) = A(y) + A(z)$ showing that A is linear; it is a linear operator on H. This operator is bounded because

$$\|A(y)\| = \|\Omega_y\| \leq \|\sigma\|\|y\|$$

giving us $\|A\| \leq \|\sigma\|$.

At this point we have shown that to any given bounded, bilinear form σ on H there corresponds an $A \in \mathcal{L}(H)$ such that $\sigma(x, y) = \langle x, Ay\rangle$ and $\|A\| \leq \|\sigma\|$. But in the proof of (a) we saw that this first equation implies $\|\sigma\| \leq \|A\|$. Hence

$$\|\sigma\| = \|A\|$$

Finally, let us show that A is unique. Suppose that $B \in \mathcal{L}(H)$ and that $\sigma(x,y) = \langle x, By \rangle$ for all $x, y \in H$. Then

$$\langle x, Ay - By \rangle = 0$$

for all $x \in H$ giving us $Ay = By$ for all $y \in H$; so $A = B$.

We have defined the norm of a bounded, bilinear form Ω to be $\|\Omega\| = \sup \left\{ \frac{|\Omega(x,y)|}{\|x\|\|y\|} \big| x \neq 0 \neq y \right\}$. As with the norm of a linear operator we can characterize $\|\Omega\|$ in several equivalent ways (Section 2.1, just after Definition 1). In fact, one can show that $\|\Omega\| = \sup \{|\Omega(x,y)| \| x\| = 1 = \|y\|\}$.

Theorem 1

Every bounded, linear operator on a Hilbert space has a unique adjoint and, moreover, the operator and its adjoint have the same norm.

Proof. Let H be a Hilbert space over $\mathbb{C}$, let $A \in \mathcal{L}(H)$ and consider the bounded, bilinear form $\sigma(x,y) = \langle Ax, y \rangle$ for all $x, y \in H$. Then

$$\begin{aligned}\|\sigma\| &= \sup\{|\langle Ax, y\rangle| \|x\| = 1 = \|y\|\} = \sup\{|\overline{\langle y, Ax\rangle}| \|x\| = 1 = \|y\|\} \\ &= \sup\{|\langle y, Ax\rangle| \|x\| = 1 = \|y\|\} = \|A\|\end{aligned}$$

by Lemma 1(a); our very last step comes from Lemma 1(a). However, part (b) of this same lemma says that there is a unique bounded, linear operator B such that

$$\sigma(x,y) = \langle x, By \rangle$$

for all $x, y \in H$ and, furthermore, $\|\sigma\| = \|B\|$. Clearly B is the adjoint of A (Section 2, Definition 2) and the norms of these operators are the same.

In the proof of Lemma 1 we made use of the so-called Riesz representation theorem for a Hilbert space. Let us prove this result now.

Theorem 2

Let H be a Hilbert space and let σ be any bounded, linear functional on H. Then there is a unique element $z \in H$ such that $\|\sigma\| = \|z\|$ and $\sigma(x) = \langle x, z \rangle$ for every $x \in H$.

Proof. Let $G = \{x \in H | \sigma(x) = 0\}$ and note that, since σ is linear and continuous, G is a linear subspace of H. We may assume that $G \neq H$ since, otherwise, we need only take z to be the zero vector. Let e be a unit (i.e., norm one) vector in $G^{\perp}$ (Section 2, Definition 1). Then $\sigma(e) \neq 0$ and the vector $\sigma(e)e$ is also in $G^\perp$. We claim that we can take $z = \overline{\sigma(e)}e$. All we need do now is show that this vector has the properties stated in the theorem.

For any $x \in H$ we may write

$$x = \left(x - \frac{\sigma(x)z}{|\sigma(e)|^2}\right) + \frac{\sigma(x)}{|\sigma(e)|^2}z$$

Let us examine each summand here in turn. First

$$\sigma\left[x - \frac{\sigma(x)z}{|\sigma(e)|^2}\right] = \sigma(x) - \frac{\sigma(x)}{|\sigma(e)|^2}\sigma(z) = \sigma(x) - \frac{\sigma(x)}{|\sigma(e)|^2}\overline{\sigma(e)}\sigma(e) = 0$$

which shows that $[x - \frac{\sigma(x)}{|\sigma(e)|^2}z]$ is in the subspace G. The second summand is, recalling the definition of z, just

$$\frac{\sigma(x)}{|\sigma(e)|^2}z = \frac{\sigma(x)\overline{\sigma(e)}}{|\sigma(e)|^2}e$$

which is a scalar multiple of e and hence belongs to $G^\perp$. So we have written x as the sum of two vectors one in G, and the other in $G^\perp$. Using this we obtain

$$\langle x, z \rangle = \langle x - \frac{\sigma(x)}{|\sigma(e)|^2}z, z \rangle + \frac{\sigma(x)}{|\sigma(e)|^2}\langle z, z \rangle$$
$$= 0 + \frac{\sigma(x)}{|\sigma(e)|^2}\overline{\sigma(e)}\sigma(e)\langle e, e \rangle = \sigma(x)$$

because $\|e\| = 1$. So we have shown that $\sigma(x) = \langle x, z \rangle$ for all $x \in H$.

Next we observe that $|\sigma(x)| = |\langle x, z \rangle| \leq \|x\|\|z\|$ (Section 1.4, Theorem 1) and so $\|\sigma\| \leq \|z\|$ (Section 2.1, Definition 2). However, $|\sigma(z)| = |\langle z, z \rangle| = \|z\|^2$ and so

$$\|z\| = \frac{|\sigma(z)|}{\|z\|} \leq \sup\{\frac{|\sigma(x)|}{\|x\|} | x \neq 0\} = \|\sigma\|$$

Thus $\|z\| = \|\sigma\|$.

Finally, we must show that the vector z is unique. If $w \in H$ satisfies the relation $\sigma(x) = \langle x, w \rangle$ for all $x \in H$, then $\langle x, z - w \rangle = 0$ for all x in H. But taking $x = z - w$ gives us $\|z - w\|^2 = 0$ and so $z = w$.

Now what exactly does Theorem 2 tell us? In order to answer that question we need some more terminology.

Definition 2

Let $(E, \|\cdot\|)$ be a normed space over $\mathcal{K}$. The vector space of all bounded, linear functionals on E (i.e., all continuous, linear maps from E into $\mathcal{K}$) is called the dual of E and will be denoted by E'. Unless the contrary is explicitly stated we shall suppose that E' has the "operator" norm (Section 2.1, Definition 2).

Observe that, since $\mathcal{K}$ is a Banach space, E' is a Banach space even when E is not (Section 2.3, Theorem 1). Theorem 2 says that there is a one-to-one, norm preserving map from H' onto H. Thus we may say that Hilbert spaces are self-dual. This is not true, in general, for a Banach space; it could not be true for a non-complete normed space E because, as we have just seen, E' is complete. Incidently, theorems which characterize, in some "concrete" way, the duals of specific spaces are all called Riesz representation theorems. We shall be able to give such a result for the space $(C[a,b], \|\cdot\|_\infty)$ later on and we shall see that this space is not self-dual. Thus the norm $\|\cdot\|_\infty$ could not come from an inner product like $\|\cdot\|_2$ does (Section 1.4, just after Definition 1).

Exercises 3

* 1. Let H be a Hilbert space and let A, B be two bounded, linear operators on H.

    a. Show that $(A^*)^* = A$, and that $(A+B)^* = A^* + B^*$.

    b. Show that $(\lambda A)^* = \overline{\lambda} A^*$ for any scalar λ, and show that $(AB)^* = B^* A^*$.

    c. If A is invertible show that A^* is invertible; hence by (a), A is invertible if, and only if, A^* is invertible. Show that $(A^{-1})^* = (A^*)^{-1}$.

    d. Show that $(A - \lambda I)^* = A^* - \overline{\lambda} I$ and conclude that if A is self-adjoint and λ is real, then $A - \lambda I$ is self-adjoint.

    e. Show that $\sigma(A^*) = \{\overline{\lambda} | \lambda \in \sigma(A)\}$.

* 2. Let σ be a bilinear form on the Hilbert space H and define $\hat{\sigma}(x) = \sigma(x, x)$ for all $x \in H$. We call $\hat{\sigma}$ the quadratic form associated with σ.

    a. Show that $\hat{\sigma}(x+y) - \hat{\sigma}(x-y) = 2\sigma(x,y) + 2\sigma(y,x)$.

    b. In the expression derived in (a) set $y = iy$ to obtain
    $$\hat{\sigma}(x+iy) - \hat{\sigma}(x-iy) = -2i\sigma(x,y) + 2i\sigma(y,x).$$

    c. Multiply the result in (b) by i and add this to (a) to get
    $$\sigma(x,y) = \frac{1}{4}[\hat{\sigma}(x+y) - \hat{\sigma}(x-y) + i\hat{\sigma}(x+iy) - i\hat{\sigma}(x-iy)]$$

    We call this the polar identity (compare with Exercises 1.4, problem 1(d)).

d. Use the polar identity to prove that σ is symmetric (i.e., $\sigma(x,y) = \overline{\sigma(y,x)}$) if, and only if, $\hat{\sigma}(x)$ is real for all x.

e. Let A be a bounded, linear operator on H. Show that the following are equivalent: (i) A is self-adjoint; (ii) $\sigma(x,y) = \langle Ax, y \rangle$ for all x, y in H, is symmetric; (iii) $\hat{\sigma}(x) = \langle Ax, x \rangle$ is real for all $x \in H$.

* 3. Let φ, ψ be two bilinear forms on H show that $\varphi(x,y) = \psi(x,y)$ for all x, y in H if, and only if, $\hat{\varphi}(x) = \hat{\psi}(x)$ for all x in H.

* 4. Let σ be a bilinear form on H.

a. Show that σ is bounded if, and only if, $\hat{\sigma}$ is bounded. Assume that σ is bounded and prove that $\|\hat{\sigma}\| \le \|\sigma\| \le 2\|\hat{\sigma}\|$. Note: We say that $\hat{\sigma}$ is bounded if there is a number M such that $|\hat{\sigma}(x)| \le M\|x\|^2$ for all $x \in H$, and if $\hat{\sigma}$ is bounded then $\|\hat{\sigma}\|$ is taken to be $\sup_{x \ne 0} \frac{\hat{\sigma}(x)}{\|x\|}$. Hint: Use the polar identity and the Parallelogram identity (Exercises 1.4, problem 1(c)).

b. If σ is bounded and symmetric show that $\|\sigma\| = \|\hat{\sigma}\|$. Hint: We need only show, because of (a), that $|\sigma(x,y)| \le \|\hat{\sigma}\|$ for any two unit vectors x, y in H. Now write $\sigma(x,y) = \rho e^{i\alpha}$, ρ and α real and $\rho \ge 0$, and let x' be the unit vector $e^{-i\alpha}x$. Now use the polar identity.

c. If A is a bounded, self-adjoint operator on H show that $\|A\| = \sup_{\|x\|=1} |\langle Ax, x \rangle|$. Hint: Use 2(e), use (b), and use Lemma 1.

* 5. Let A be any bounded, linear operator on H.

a. Show that there exist two self-adjoint operators B, C such that $A = B + iC$. Show that B and C are uniquely determined by these requirements.

b. Referring to (a) show that B and C commute if, and only A is normal; i.e., $BC = CB$ if, and only if, $AA^* = A^*A$.

6. The Riesz representation theorem sets up a one-to-one, norm preserving correspondence between H' and H; here H is a fixed Hilbert space. Denote this one-to-one map by γ.

a. Show that $\gamma(\sigma_1 + \sigma_2) = \gamma(\sigma_1) + \gamma(\sigma_2)$ for any $\sigma_1, \sigma_2 \in H'$.

b. Show that $\gamma(\alpha\sigma) = \overline{\alpha}\gamma(\sigma)$ for any $\sigma \in H'$ and any $\alpha \in \mathcal{K}$.

7. Let E, F be two normed spaces over the same field and let $T : E \to F$ be a bounded, linear map. We can define a map T^* from F' into E' (see Definition 2 above) as follows: For any $f \in F'$, $T^*(f)$ is to be a linear

functional on E. We set $T^*(f)(x) = f(Tx)$ for each x in E. The map T^* is sometimes called the adjoint of T. It really is the adjoint when $E = F = H$ a Hilbert space. Prove this. Also, show that T^* is always a bounded and linear map from F' into E'.

4.4 The Adjoint of a Compact Operator

The main result of this section is that the adjoint of a compact operator is also a compact operator. Our proof uses the weak topology of a Hilbert space and this is a tricky thing to work with. The Banach-Steinhaus theorem arises naturally here and we shall give a proof of this important result. Some readers may find it helpful to go through the entire section reading only the definitions and the statements of the theorems, and leave the proofs for another time.

Definition 1

A sequence $\{x_k\}$ of points of H is said to be a weak Cauchy sequence, or we sometimes say that it is weakly Cauchy, if for each $y \in H$ the sequence $\{\langle x_k, y \rangle\}$ of complex numbers is a Cauchy sequence in $\mathbb{C}$. We shall say that $\{x_k\}$ is weakly convergent to $x_0 \in H$ if for every $y \in H$, $\lim \langle x_k, y \rangle = \langle x_0, y \rangle$ in $\mathbb{C}$.

When discussing weak convergence some authors will call a sequence which converges for the norm of H a strongly convergent sequence. We shall not do this. For us the terms convergent, Cauchy, etc. always refer to the norm of H. If something else is intended we make this intention explicit by modifying these terms with the words "weak" or "weakly."

Observe that if $\{x_k\} \subseteq H$ is convergent to x_0 then it is weakly convergent to x_0 also. This is easy to see since

$$|\langle x_k, y \rangle - \langle x_0, y \rangle| = |\langle x_k - x_0, y \rangle| \leq \|x_k - x_0\|\|y\|$$

by the C.S.B. inequality (Section 1.4, Theorem 1). Let us now construct a sequence which is weakly convergent but not convergent. Suppose that H is infinite dimensional and choose any infinite, linearly independent sequence in H. Then by applying the Gram-Schmidt process to this sequence we can obtain an infinite orthonormal sequence $\{e_n\}$ in H (Section 1.9, just after Lemma 1). Now for any $y \in H$ we have, by Bessel's inequality (Section 1.6, Theorem 2)

$$\sum_{n=1}^{\infty} |\langle y, e_n \rangle|^2 \leq \|y\|^2$$

But since the nth term of any convergent series of complex numbers must tend to zero we see that

$$\lim \langle y, e_n \rangle = \lim \langle e_n, y \rangle = 0$$

for every $y \in H$. This says that the sequence $\{e_n\}$ is weakly convergent to the zero vector. However, $\|e_n\| = 1$ for every n so the sequence could not converge to the zero vector (Exercises 1.3, problem 1).

So now we know that weak convergence is well named (i.e., convergence implies weak convergence) and that it is distinct from ordinary (norm) convergence. One of the theorems we shall prove is this: A bounded, linear operator T on H is a compact operator if, and only if, it maps any weakly convergent sequence onto a convergent sequence; i.e., if $\{x_k\}$ is any weakly convergent sequence, then the sequence $\{Tx_k\}$ is convergent. This, and the fact that the adjoint of a compact operator is compact, are nice results which are often useful, unfortunately it takes some work to prove them. Let us start by discussing the Banach-Steinhaus theorem. This innocent looking theorem has some remarkable applications [3; Section 6.2]. Suppose that E is a normed space and that $\mathcal{O}$ is a non-empty subset of E' (Section 3, Definition 2); so $\mathcal{O}$ is a set of bounded, linear functions on E. Observe that $\mathcal{O}$ can be "bounded" in two ways at least. First it could happen that $\|f\| \leq M$ for all $f \in \mathcal{O}$, here M is a fixed number, and when this is the case we shall say that $\mathcal{O}$ is norm bounded. On the other hand we could have a number M for each fixed $x \in E$ such that $|f(x)| \leq M$ for all $f \in \mathcal{O}$; i.e., $M = M(x)$, $|f(x)| \leq M(x)$ all $f \in \mathcal{O}$ and as we change $x \in E$ the number $M(x)$ changes. In this case we shall say that $\mathcal{O}$ is pointwise bounded on E.

Theorem 1

Let B be a Banach space. A subset of B' is norm bounded if, and only if, it is pointwise bounded in B.

Proof. It is easy to see that a norm bounded set is pointwise bounded and this does not depend on the fact that B is a Banach space (see problem 2). So let $\mathcal{O}$ be pointwise bounded on B and let us show that it is norm bounded. First observe that we may suppose that $\mathcal{O}$ is countable; i.e., $\mathcal{O} = \{f_n | n = 1, 2, \ldots\} \subseteq B'$. We prove this as follows: If $\mathcal{O}$ is not norm bounded then, for each n, we can find $f_n \in \mathcal{O}$ with $\|f_n\| > n$. Clearly $\{f_n | n = 1, 2, \ldots\}$ as just chosen is not bounded. Thus $\mathcal{O}$ unbounded implies some countable subset of $\mathcal{O}$ is unbounded or, to turn it around, if every countable subset of $\mathcal{O}$ is bounded, $\mathcal{O}$ must be bounded.

Now assume, as we have just shown we may, that $\mathcal{O}$ is $\{f_n | n = 1, 2, \ldots\}$ and that this set is pointwise bounded on B. We shall now show that to prove that $\{f_n\}$ is norm bounded it suffices to prove that there is an x_0 in B, a $\delta > 0$, and a $k > 0$ such that $|f_n(x)| < k$ for all n and all x in $\{x \in B | \|x - x_0\| \leq \delta\}$;

i.e., it suffices to show that our set is uniformly bounded on some ball. It is convenient to denote our last set by $\mathcal{B}(\delta, x_0)$. Suppose that our condition is satisfied for an x_0 in B and numbers δ, k. Let x be a point of B with $\|x\| \leq \delta$. Then

$$|f_n(x)| \leq |f_n(x + x_0) - f_n(x_0)| \leq |f_n(x + x_0)| + |f_n(x_0)| \leq 2k$$

for every n because $\|(x + x_0) - x_0\| = \|x\| \leq \delta$. Thus $|f_n(x)| \leq 2k$ for every n and every x in the set $\mathcal{B}(\delta, 0)$. Now if y is any non-zero point of B, then $\delta y \|y\|^{-1}$ is in $\mathcal{B}(\delta, 0)$. So $|f_n(y)| \leq 2k \|y\| \delta^{-1}$ for all n. Then clearly $\|f_n\| \leq 2k\delta^{-1}$ for every n.

Now we argue by contradiction. Suppose that $\{f_n\}$ is not norm bounded. Then, as we just proved, this set is unbounded on every closed ball. More explicitly, for any closed ball $\mathcal{S}$ (i.e., any set of the form $\mathcal{B}(\delta, x_0)$) and any $k > 0$ there is an integer n and a point x in $\mathcal{S}$ such that $|f_n(x)| \geq k$. Choose a ball $\mathcal{S}$, a point $x_1 \in \mathcal{S}$, and an integer $n(1)$ such that $|f_{n(1)}(x_1)| > 1$. Since $f_{n(1)}$ is continuous this inequality is satisfied at each point of a ball $\mathcal{S}_1$ contained in $\mathcal{S}$ and containing x_1 whose diameter is less than 2^{-1}. By our assumption there is a point $x_2 \in \mathcal{S}_1$ and an integer $n(2)$ such that $|f_{n(2)}(x_2)| > 2$. As before, this must hold at each point of some ball $\mathcal{S}_2$ contained in $\mathcal{S}_1$ and containing x_2 whose diameter is less than 2^{-2}. Now choose $x_3 \in \mathcal{S}_2$ and an integer $n(3)$ such that $|f_{n(3)}(x)| > 3$. Using the continuity of $f_{n(3)}$ we find a ball $\mathcal{S}_3$ contained in $\mathcal{S}_2$ and containing x_3 such that $|f_{n(3)}(x)| > 3$ for all $x \in \mathcal{S}_3$ and the diameter of $\mathcal{S}_3$ is less than 2^{-3}. Continue in this way. After $\mathcal{S}_1, \mathcal{S}_2, \ldots, \mathcal{S}_{k-1}$ have been chosen, choose $x_k \in \mathcal{S}_{k-1}$ and an integer $n(k)$ such that $|f_{n(k)}(x_k)| > k$. By the continuity of $f_{n(k)}$ this inequality must hold at each point of a ball $\mathcal{S}_k$ contained in $\mathcal{S}_{k-1}$ and containing x_k whose diameter is less than 2^{-k}.

Now use the fact that $(B, \|\cdot\|)$ is a Banach space. Because of this, the set $\bigcap_{k=1}^{\infty} \mathcal{S}_k$ is not empty (see problem 3). But if y is a point in this intersection, then $|f_{n(k)}(y)| > k$ for $k = 1, 2, \ldots$, contradicting the fact that $\mathcal{O} = \{f_n\}$ is pointwise bounded on B.

We mention without proof that the theorem need not be true for normed spaces which are not complete [3; p. 50].

Corollary 1

Any weak Cauchy sequence in a Hilbert space H is a bounded set.

Proof. Let $\{x_k\} \subseteq H$ be a weak Cauchy sequence. Then for each k let $\sigma_k(x) = \langle x, x_k \rangle$. We recall (Section 3, Theorem 2) that each σ_k is a continuous, linear functional on H and that $\|\sigma_k\| = \|x_k\|$ for each k. Now for each fixed $y \in H$ the sequence $\{\langle x_k, y \rangle\}_{k=1}^{\infty}$ is a Cauchy sequence in $\mathbb{C}$ (Definition 1) and any such sequence is a bounded set in $\mathbb{C}$. Thus $\{|\langle x_k, y \rangle| = |\langle y, x_k \rangle| = |\sigma_k(y)|\}_{k=1}^{\infty}$ is a bounded subset of $\mathbb{C}$ for each fixed y; i.e., $\{\sigma_k\}$ is pointwise bounded on H. Since every Hilbert space is a Banach space we see that for some M, $\|\sigma_k\| \leq M$ for all k. However, $\|\sigma_k\| = \|x_k\|$ for each k and so $\|x_k\| \leq M$ for all k.

Theorem 2

Any weak Cauchy sequence in a Hilbert space H is weakly convergent to a point of H.

Proof. Let $\{x_k\} \subseteq H$ be a weak Cauchy sequence and, for each k, define $\sigma_k(x) = \langle x, x_k \rangle$ for all $x \in H$. As we saw in Corollary 1 the sequence $\{\sigma_k\}$ is bounded; i.e., there is an M such that $\|x_k\| = \|\sigma_k\| \leq M$ for all k. Define one more linear functional on H as follows:

$$\sigma(x) = \lim \sigma_k(x) = \lim \langle x, x_k \rangle = \lim \overline{\langle x_k, x \rangle}$$

By hypothesis $\{\langle x_k, x \rangle\}$ converges in $\mathbb{C}$ for each fixed $x \in H$ so σ is well-defined and, it is easy to check, linear. But

$$|\sigma(x)| = |\lim \sigma_k(x)| = \lim |\sigma_k(x)| \leq M\|x\|$$

showing that σ is bounded. Thus (Section 3, Theorem 2) there is a unique $y \in H$ such that $\sigma(x) = \langle x, y \rangle$ for all $x \in H$. However, this says that

$$\langle x, y \rangle = \sigma(x) = \lim \sigma_k(x) = \lim \langle x, x_k \rangle$$

or

$$\langle y, x \rangle = \overline{\sigma(x)} = \lim \langle x_k, x \rangle$$

showing that $\{x_k\}$ converges weakly to y (Definition 1).

We need one more fact about weak convergence in a Hilbert space and then we shall be able to prove our main results.

Theorem 3

Any bounded sequence in a Hilbert space has a subsequence which is weakly convergent.

Proof. Let H be a Hilbert space and let $\{x_n\} \subseteq H$ be a bounded sequence. We may assume that $\|x_n\| \leq 1$ for all n. Let G be the closed, linear span of $\{x_n\}$; i.e., the closure of $lin\{x_n\}$ (Section 1.1, Definition 2 and Section 3.1, Definition 2). If G is finite dimensional, recall (Exercises 3.1, problem 4(d)) that G is a linear subspace of H, then $\{x_n\}$ will have a convergent subsequence; convergent for the norm and hence weakly convergent

(Section 3.1, Theorem 1 and Theorem 2). Hence we may assume that G is infinite dimensional. Now G is clearly a separable space (Section 1.9, Definition 1 and Exercises 1.9, problem 2(b)) and so it must have a countable orthonormal basis which we shall denote by $\{e_n | n = 1, 2, \ldots\}$ (Section 1.9, Theorem 1).

Consider now $\{\langle x_n, e_1 \rangle | n = 1, 2, \ldots\}$. This is a bounded sequence of complex numbers and so there is a subsequence $\{x_{1n}\}$ of $\{x_n\}$ such that $\{\langle x_{1n}, e_1 \rangle\}$ is convergent. Next consider $\{\langle x_{1n}, e_2 \rangle | n = 1, 2, \ldots\}$. Again we may choose a subsequence $\{x_{2n}\}$, a subsequence this time of $\{x_{1n}\}$, such that $\{\langle x_{2n}, e_2 \rangle\}$ is convergent. Observe that $\{\langle x_{2n}, e_1 \rangle\}$ is also convergent. Continue in this way. After sequences $\{x_{1n}\}, \{x_{2n}\}, \ldots, \{x_{kn}\}$, each a subsequence of the one before it, have been chosen we note that $\{\langle x_{kn}, e_{k+1} \rangle\}$ is bounded and hence that there is a subsequence $\{x_{k+1\,n}\}$ of $\{x_{kn}\}$ such that $\lim_n \langle x_{k+1\,n}, e_{k+1} \rangle$ exists. Clearly, $\lim_n \langle x_{k+1\,n}, e_j \rangle$ exists for each $j \leq k+1$. Finally, let us set $y_1 = x_n$, $y_2 = x_{22}, \ldots, y_k = x_{kk}, \ldots$. Then $\{y_n\}$ is a subsequence of $\{x_n\}$, it is also a subsequence of $\{x_{1n}\}$ showing that $\{\langle y_n, e_1 \rangle\}$ is convergent. Also, $\{y_n\}_{n \geq k}$ is a subsequence of $\{x_{kn} | n = 1, 2, \ldots\}$ showing that $\{\langle y_n, e_k \rangle | n = 1, 2, \ldots\}$ is convergent for each k. We shall finish the proof by showing that $\{y_n\}$ is weakly convergent.

First observe that if $z = \sum_{j=1}^p a_j e_j$, $p < \infty$, then we clearly that $\{\langle y_n, z \rangle\}$ convergent. Suppose that $z \in G$ is arbitrary. Then given $\varepsilon > 0$ we may choose p so that for $z_1 = \sum_{j=1}^n \langle z, e_j \rangle e_j$, $\|z - z_1\| < \varepsilon$ (Section 1.6, Theorem 3). Then

$$|\langle y_m, z \rangle - \langle y_n, z \rangle| = |\langle y_m - y_n, z \rangle| = |\langle y_m - y_n, z - z_1 \rangle + \langle y_m - y_n, z_1 \rangle|$$
$$\leq |\langle y_m - y_n, z - z_1 \rangle| + |\langle y_m - y_n, z_1 \rangle| \leq 2\varepsilon + |\langle y_m - y_n, z_1 \rangle|$$

Now this last term tends to zero as $m, n \to \infty$ so for m, n large enough we have

$$|\langle y_m, z \rangle - \langle y_n, z \rangle| \leq 3\varepsilon$$

and this proves that $\{y_n\}$ is a weak Cauchy sequence in G. By Theorem 2, $\{y_n\}$ has a limit $y \in G$; weak limit. The only thing left to prove is that $\{y_n\}$ is weakly convergent to y in H; i.e., given any $x \in H$ we must show that

$$\lim \langle y_n, x \rangle = \langle y, x \rangle$$

To do this we make use of the orthogonal complement of G, $G^\perp$ (Section 2, Definition 1 and the discussion after Lemma 1). We have unique elements $y_0 \in G^\perp$, $z \in G$ such that $x = y_0 + z$. Then

$$\lim \langle y_n, x \rangle = \lim \langle y_n, y_0 + z \rangle = \lim \langle y_n, y_0 \rangle + \lim \langle y_n, z \rangle$$
$$= \lim \langle y_n, z \rangle = \langle y, z \rangle$$

However, we also have $\langle y, x \rangle = \langle y, y_0 + z \rangle = \langle y, y_0 \rangle + \langle y, z \rangle = \langle y, z \rangle$.

And now, at last, our two main results.

Theorem 4

A bounded, linear operator on a Hilbert space is a compact operator if, and only if, it maps any weakly convergent sequence onto a convergent sequence.

Proof. Let H be a Hilbert space and let T be a bounded, linear operator on H. Suppose first that $\{Tx_n\}$ is convergent whenever $\{x_n\}$ is weakly convergent. Let $\mathcal{B}_1 = \{x \in H \,|\, \|x\| \leq 1\}$. We shall show that T is a compact operator by showing that $T(\mathcal{B}_1)$ is a relatively compact set (Section 3.1, Corollary 1 to Theorem 2). Take any sequence $\{y_n\} \subseteq T(\mathcal{B}_1)$ and choose, for each n, an element $x_n \in \mathcal{B}_1$ such that $Tx_n = y_n$. Now $\{x_n\}$ is a bounded sequence in H and so, by Theorem 3, it has a subsequence $\{x'_n\}$ which is weakly convergent. But then $\{y'_n\}$ is a subsequence of $\{y_n\}$ ($y'_n = Tx'_n$ for each n) which is convergent, showing that $T(\mathcal{B}_1)$ is relatively compact (Section 3.1, Theorem 2).

Now suppose that T is a compact operator and let $\{x_n\}$ be any weakly convergent sequence in H. If $\{Tx_n\}$ does not converge then it must have two subsequences $\{Tx'_n\}$ of $\{Tx''_n\}$ which converge to y', y'' respectively with $y' \neq y''$ (Section 3.1, Corollary 1(c)) to Theorem 2 and Theorem 2). But then for any $z \in H$ we would have

$$\begin{aligned} \langle y', z \rangle &= \lim \langle Tx'_n, z \rangle = \lim \langle x'_n, T^*z \rangle = \lim \langle x_n, T^*z \rangle \\ &= \lim \langle x''_n, T^*z \rangle = \lim \langle Tx''_n, z \rangle = \langle y'', z \rangle \end{aligned}$$

because $\{x_n\}$ is weakly convergent and so all of its subsequences are also weakly convergent and to the same limit. Since z was arbitrary we must conclude that $y' = y''$ contradicting our assumption. Thus $\{Tx_n\}$ must be convergent.

Note how we needed to know that T had an adjoint to carry out this proof (Section 3).

Corollary 2

If T is a compact operator on a Hilbert space H and if $\{x_n\} \subseteq H$ is weakly convergent to x_0, then the sequence $\{Tx_n\}$ is convergent (norm convergent) to Tx_0.

Proof. By the theorem $\{Tx_n\}$ converges to, say, $y \in H$. But then $\{Tx_n\}$ is also weakly convergent to y. Thus all we have to do is show that $\{Tx_n\}$ is weakly convergent to Tx_0. But

$$\langle Tx_0 - Tx_k, x \rangle = \langle T(x_0 - x_k), x \rangle = \langle x_0 - x_k, T^*x \rangle$$

for any $x \in H$ and our result follows from this.

Theorem 5

The adjoint of a compact operator on a Hilbert space is a compact operator on this space.

Proof. We shall use Theorem 4. Suppose that T is a compact operator on the Hilbert space H and that $\{x_n\} \subseteq H$ is weakly convergent. We may assume that $\|x_n\| \leq 1$ for every n (Corollary 1 to the Banach-Steinhaus theorem). Then

$$\begin{aligned} \|T^*x_m - T^*x_n\|^2 &= \langle T^*(x_m - x_n), T^*(x_m - x_n)\rangle = \langle TT^*(x_{m*} - x_n), \\ &\quad (x_m - x_n)\rangle \\ &\leq \|TT^*(x_m - x_n)\|\|x_m - x_n\| \leq 2\|TT^*(x_m - x_n)\| \end{aligned}$$

Now TT^* is a compact operator (Exercises 3.1, problem 2(a)) and so, since $\{(x_m - x_n)\}$ converges weakly to zero,

$$\lim \|TT^*(x_m - x_n)\| = 0$$

by Corollary 1 to Theorem 4. Combining this with our first inequality we see that

$$\lim \|T^*x_m - T^*x_n\| = 0$$

Thus $\{T^*x_n\}$ is convergent, it is a Cauchy sequence in a Hilbert space, and this says that T^* is compact because $\{x_n\}$ was an arbitrary weakly convergent sequence.

Exercises 4

* 1. Let H be a Hilbert space and let $\{x_n\}$ be a weakly convergent sequence with limit x_0 in H.

    a. Show that if $\{x_n\}$ is also weakly convergent to $y_0 \in H$ then $x_0 = y_0$.

    b. Let A be any bounded, linear operator on H. Show that the sequence $\{Ax_n\}$ is weakly convergent to Ax_0.

* 2. Let E be a normed space and let $\mathcal{O}$ be a subset of E' which is norm bounded, i.e., there is a fixed member M such that $\|f\| \leq M$ for every $f \in \mathcal{O}$. Show that $\mathcal{O}$ is pointwise bounded on E; i.e., given $x \in E$ find a constant K such that $|f(x)| \leq K$ for every $f \in \mathcal{O}$. Hint: K is M times something.

152 CHAPTER 4 SPECTRAL THEORY, THE BASIC TOOLS

* 3. Let B be a Banach space and let $\{S_n\}_{n=1}^{\infty}$ be a decreasing (i.e., $S_n \supseteq S_{n+1}$ for each n) sequence of closed subsets of B such that $\lim_n (\text{dia } S_n) = 0$; here dia $S_n = \max\{\|x - y\| | x, y \in S_n\}$. Show that $\cap_{n=1}^{\infty} S_n \neq \emptyset$. Hint: Choose $x_n \in S_n$ for $n = 1, 2, \ldots$ and use the fact that the diameters tend to zero to show that the sequence $\{x_n\}$ is Cauchy. Show that the limit of this sequence must be in every S_n. Where does the fact that S_n is closed come in?

4. Give a definition of weak convergence for a sequence in an arbitrary normed space E which, in case E happens to be a Hilbert space, reduces to Definition 1. Hint: Section 3, Theorem 2. (b) Show that any weakly convergent sequence in E is a bounded set. Hint: Use the Banach-Steinhaus theorem and the fact that E' is always a Banach space.

5
THE SPECTRAL THEOREM, PART I

We shall prove the spectral theorem for certain compact operators on a Hilbert space, and then apply these results to the study of Fredholm integral equations. It is wise to begin by reviewing some linear algebra.

5.1 Matrices

When considering systems of ordinary differential equations one often has to work with $\exp(A)$ where A is a square matrix. This is one instance in which one must consider the question of defining a function of a linear operator; there is, of course, a close connection between square matrices and operators on a finite dimensional space. Let us try now to identify a 'reasonably large' class of functions $\mathcal{F}$ and 'reasonably large' class of linear operators $\mathcal{D}$ (on a finite dimensional space) such that $f(T)$ has meaning for all $f \in \mathcal{F}$ and all $T \in \mathcal{D}$.

One approach to this problem is suggested by our mention of the exponential function. Suppose that $f(z)$ can be expanded in a power series about

$z = 0$ which is convergent for, say, $|z| < r$. So

$$f(z) = \sum_{k=0}^{\infty} b_k z^k, \quad |z| < r \tag{1}$$

Then, since A^k has meaning for every non-negative integer k, we might set

$$f(A) = \sum_{k=0}^{\infty} b_k A^k \tag{2}$$

provided, of course, that the series in (2) is convergent. Now if A is large it may not be easy to estimate the size of the various numbers in A^k and so it becomes difficult to say whether or not (2) has any meaning. However, if we place some restrictions on A, then the problem becomes tractable.

An $n \times n$ matrix A will be denoted by (a_{ij}), $1 \le i, j \le n$. The numbers a_{ii}, $1 \le i \le n$, constitute the main diagonal of A. When we say that A is a diagonal matrix we shall mean that all of the entries of A which are not on the main diagonal are zero; i.e., $a_{ij} = 0$ whenever $i \ne j$. Now if $D = (d_{ij})$ is a diagonal matrix then, for any non-negative integer k, $\mathcal{D}^k$ is also a diagonal matrix and its entries along its main diagonal are the corresponding entries of D raised to the kth power. Thus:

$$D = \begin{pmatrix} d_{11} & 0 & \cdots & 0 \\ 0 & d_{22} & \cdots & 0 \\ 0 & 0 & \cdots & 0 \\ 0 & 0 & \cdots & d_{nn} \end{pmatrix}, \quad \mathcal{D}^k = \begin{pmatrix} d_{11}^k & 0 & \cdots & 0 \\ 0 & d_{22}^k & \cdots & 0 \\ \cdots & \cdots & \cdots & \cdots \\ 0 & 0 & \cdots & d_{nn}^k \end{pmatrix} \tag{3}$$

Hence the finite sum

$$\sum_{k=0}^{\ell} b_k \mathcal{D}^k \tag{4}$$

is just the diagonal matrix which has the numbers

$$\sum_{k=1}^{\ell} b_k d_{ii}^k \tag{5}$$

along its main diagonal. Now if, for every i between 1 and n, we have $|d_{ii}| < r$ then, letting $\ell \to \infty$ in (4) we get the diagonal matrix that has $f(d_{ii})$ along its main diagonal. We take this $n \times n$ matrix to be $f(D)$. So

$$f(D) = \begin{pmatrix} f(d_{11}) & 0 & 0 & \cdots & 0 \\ 0 & f(d_{22}) & 0 & \cdots & 0 \\ \cdots & \cdots & \cdots & \cdots & \cdots \\ 0 & 0 & 0 & \cdots & f(d_{nn}) \end{pmatrix} \tag{6}$$

The diagonal matrices are, of course, rather special. We can, however, extend the discussion above to a much wider class of matrices; the so-called diagonalizable matrices. First recall that the $n \times n$ diagonal matrix that has all of its diagonal entries equal to 1 is called the $n \times n$ identity matrix. We shall denote this by I. Given an $n \times n$ matrix A there may exist another (necessarily unique) $n \times n$ matrix A^{-1} such that $AA^{-1} = A^{-1}A = I$. When this is the case we shall say that A is invertible and that A^{-1} is its inverse.

Definition 1

A square matrix A is said to be a diagonalizable matrix if there is a diagonal matrix D and an invertible matrix B such that $A = BDB^{-1}$.

When $A = BDB^{-1}$ we have, for every non-negative integer k, $A^k = (BDB^{-1})(BDB^{-1})\cdots(BDB^{-1}) = BD^kB^{-1}$. Thus, the discussion given above for D yields in this case the matrix $Bf(D)B^{-1}$. We take this to be $f(A)$.

Let us look at a simple example to help us recall how one goes about diagonalizing a matrix. Consider

$$A = \begin{pmatrix} 1 & 2 \\ 4 & 3 \end{pmatrix} \tag{7}$$

and suppose that we want to define $\exp(A)$. Of course

$$\exp(z) = \sum_{n=0}^{\infty} z^n/n!, \quad |z| < \infty \tag{8}$$

so we have no problem about convergence. The first thing we do is find the eigenvalues of A. These are the numbers λ for which the equations (Section 2.3, Definition 3)

$$(A - \lambda I)\vec{v} = 0 \tag{9}$$

have a non-trivial solution. Equivalently (see the exercises) they are the solutions to the polynomial equation

$$\det(A - \lambda I) = 0 \tag{10}$$

where det denotes the determinant. Equation (10) is the characteristic equation of the matrix A. Now in this case, the case of A given by (7), the eigenvalues turn out to be -1 and $+5$. The equations (9) when $\lambda = -1$ have the solution $(-1, 1)$ and when $\lambda = 5$ they have the solution $(1, 2)$. We know then that we can write $A = BDB^{-1}$ where

$$D = \begin{pmatrix} -1 & 0 \\ 0 & 5 \end{pmatrix} \tag{11}$$

and B is the matrix whose first column is $(-1,1)$ and second column is $(1,2)$. So

$$B = \begin{pmatrix} -1 & 1 \\ 1 & 2 \end{pmatrix} \qquad (12)$$

Finally, $\exp(A) = B\exp(D)B^{-1}$ and this is

$$\begin{pmatrix} -1 & 1 \\ 1 & 2 \end{pmatrix} \begin{pmatrix} \exp(-1) & 0 \\ 0 & \exp(5) \end{pmatrix} \begin{pmatrix} -1 & 1 \\ 1 & 2 \end{pmatrix}^{-1} \qquad (13)$$

One should note that the eigenvalues of A give us the entries of our diagonal matrix and that the corresponding eigenvector (i.e., the solutions to (9)) give us the columns of our matrix B. Thus we may write down a solution to the problem we started with as follows:

Let f be any complex-valued function which is analytic at $z = 0$ and let A be any diagonalizable matrix. If every eigenvalue of A is within the radius of convergence of the Maclaurin series for f, then $f(A)$ is defined. Furthermore, if $A = BDB^{-1}$, D a diagonal matrix whose diagonal entries are the eigenvalues of A, then $f(A) = Bf(D)B^{-1}$.

Exercises 1

1. Compute the eigenvalues of the matrix (7) above. Solve the equations (9) when $\lambda = -1$ and again when $\lambda = 5$. Find the inverse of the matrix (12) above and show that $A = BDB^{-1}$.

2. We assume that the reader has some familiarity with determinants. Recall that for any two $n \times n$ matrices A, B, $\det(AB) = (\det A)(\det B)$. Using this show that:

    a. If A is invertible, then $\det(A^{-1}) = \frac{1}{\det(A)}$

    b. A is invertible if, and only if, $\det(A) \neq 0$

    c. The eigenvalues of A, see equation (9), are the complex numbers λ for which $\det(A - \lambda I) = 0$.

* 3. Two $n \times n$ matrices A and C are said to be similar if there is an invertible matrix B such that $A = BCB^{-1}$.

    a. If A and C are similar and A is invertible, then C is invertible.

    b. Show that similar matrices have the same eigenvalues (2(c)).

THE SPECTRAL THEOREM, FINITE-DIMENSIONAL CASE   157

4. (See Exercises 4.3, problem 1) Let A be an $n \times n$ matrix, $A = (a_{ij})$. We define the adjoint of A, A^*, to be $(\overline{a}_{ji})$. So we interchange the rows and the columns of A and we take the complex conjugate of each entry.

   a. If B is another $n \times n$ matrix show that $(A+B)^* = A^* + B^*$ and that $(AB)^* = B^* A^*$.

   b. For any scalar λ show that $(\lambda A)^* = \overline{\lambda} A^*$. If A is invertible show that $(A^{-1})^* = (A^*)^{-1}$.

   c. Show that the adjoints of two matrices which are similar (problem 3) are also similar; use 4b.

   d. We say that A is self-adjoint if $A = A^*$. Show that the diagonal entries, and also the eigenvalues, of a self-adjoint matrix are real numbers.

   e. We say that an $n \times n$ matrix U is a unitary matrix if $U^* = U^{-1}$. Show that the eigenvalues of any such matrix have absolute value 1.

5. (see Exercises 4.3, problem 5) Given an arbitrary $n \times n$ matrix A write: $A = B + iC$ where $i = \sqrt{-1}$. Take the adjoint of A, use 4(a) and 4(b), and use the equation for A and the equation for A^* to solve for B, C under the assumption that B and C are self-adjoint (4(c)). Thus, for any A, we can find self-adjoint matrices B, C such that $A = B + iC$. Show that B, C are unique. Finally, show that $BC = CB$, we say B and C commute, if, and only if, A commutes with A^*. A matrix which commutes with its adjoint is called a normal matrix. Clearly, all self-adjoint (4c) and all unitary (4c) matrices are normal.

5.2   The Spectral Theorem, Finite-dimensional Case

In the last paragraph we discussed diagonizable matrices but we didn't answer the question of which matrices are of this kind. We shall discuss this question here and in the next section. First, however, we must recall the connection between a matrix and a linear operator.

Suppose that T is a linear operator on a finite dimensional space V, and let $v_1, \ldots, v_n$ be a Hamel basis for V (this is sometimes called an ordered basis because we have designated one vector as the first basis element, another as the second, and so on). Then for each j, $1 \leq j \leq n$, there are scalars a_{ij} such that

$$T(v_j) = a_{1j} v_1 + a_{2j} v_2 + \cdots + a_{nj} v_n \qquad (1)$$

The matrix of T for the given basis, call it $M(T)$, is taken to be (a_{ij}), $1 \leq i \leq n$ and $1 \leq j \leq n$. We recall that, if $M'(T)$ dentoes the matrix of T in any other basis for V, the matrices $M(T)$ and $M'(T)$ are similar; i.e., there is an invertible matrix C such that

$$M(T) = C[M'(T)]C^{-1} \qquad (2)$$

It follows from this that the matrix of T for some basis of V is diagonalizable if, and only if, the matrix of T for any basis of V is diagonalizable; for if $M'(T) = BDB^{-1}$ for some diagonal matrix D, then $M(T) = (CB)D(CB)^{-1}$ by (2). Thus we may define a linear operator on V to be a diagonalizable operator if its matrix representation is a diagonalizable matrix.

We have already observed (Exercises 1, problem 3b) that similar matrices have the same eigenvalues. Thus the eigenvalues of a linear operator T are the same as the eigenvalues of any matrix representing T. In particular, if f is analytic at $z = 0$ and if all of the eigenvalues of T lie within the radius of convergence of the Maclaurin series for f and if $M(T)$ is the matrix representation of T in some given basis for V, then we may define $f(T)$ to be the operator an V whose matrix for the given basis of V is the matrix $f[M(T)]$.

There is one case when it is clear that a given operator is diagonalizable. Suppose that T is a linear operator on V and that there is a basis $v_1, \ldots, v_n$ for V consisting of eigenvectors of T. Then

$$\begin{aligned} Tv_1 &= \lambda_1 v_1 + 0v_2 + \cdots + 0v_n \\ Tv_2 &= 0v_1 + \lambda_2 v_2 + \cdots + 0v_n \\ &\cdots \\ Tv_n &= 0v_1 + 0v_2 + \cdots + \lambda_n v_n \end{aligned} \qquad (3)$$

and so the matrix of T for this basis is

$$\begin{pmatrix} \lambda_1 & 0 & 0 & \cdots & 0 \\ 0 & \lambda_2 & 0 & \cdots & 0 \\ \cdots & \cdots & \cdots & \cdots & \cdots \\ 0 & 0 & 0 & \cdots & \lambda_n \end{pmatrix} \qquad (4)$$

In general, the λ's will not all be distinct. In fact if T happens to be λI, then they are all the same. The general situation can be illustrated as follows: Suppose that V has dimension seven and that T is a linear operator on V with three distinct eigenvalues λ, μ, ν where $M_\lambda = \{v \in V | (T - \lambda I)v = 0\}$ has dimension three and M_μ, M_ν each has dimension two. What happens is that we have a basis u_1, u_2, u_3 for M_λ, a basis v_4, v_5 for M_μ and a basis w_6, w_7 for M_ν and then the set $u_1, u_2, u_3, v_4, v_5, w_6, w_7$ is a basis for V. So any vector $v \in V$ can be written

$$v = (\alpha_1 u_1 + \alpha_2 u_2 + \alpha_3 u_3) + (\beta_4 v_4 + \beta_5 v_5) + (\gamma_6 w_6 + \gamma_7 w_7) \qquad (5)$$

Note that $\alpha_1 u_1 + \alpha_2 u_2 + \alpha_3 u_3$ is in M_λ, call this vector u_λ. Similarly, set $v_\mu = \beta_4 v_4 + \beta_5 v_5$, $w_\nu = \gamma_6 w_6 + \gamma_7 w_7$. Then each $v \in V$ can be written

$$v = u_\lambda + v_\mu + w_\nu \tag{6}$$

In other words, $V = M_\lambda + M_\mu + M_\nu$. Moreover, this representation of v is unique; so, in some sense, M_λ, M_μ, M_ν are independent.

Now what have we done? We have observed that for a linear operator having a basis consisting of its eigenvectors, the space V splits into linear subspaces in such a way that when T is restricted to any of these subspaces it acts as a scalar times the identity; for $T|M_\lambda$ is just λI and $T|M_\nu$ is just νI, and T restricted to M_μ is μI. So we have obtained a nice decomposition of our operator T into very simple pieces. In order to do this we assumed that there was a basis for V consisting of eigenvectors of T, in other words we assumed that some matrix representation of T is diagonal. So the operators we have been discussing are diagonalizable. The reader can easily check that if we are given a diagonalizable operator T then it must be true that there is a basis for V consisting of eigenvectors of T; it is the basis for V in which the matrix representation of T is a diagonal matrix. So we have proved:

Theorem 1

A linear operator T on a finite dimensional vector space V is diagonalizable if, and only if, there is a Hamel basis for V each element of which is an eigenvector for T.

We have said that the spaces M_λ, M_μ, M_ν are in some sense independent. Before continuing with our discussion of diagonalizable operators let us make this idea more precise. Let us say that a finite set $W_1, \ldots, W_k$ of non-trivial linear subspaces of V is linearly independent if the following is true: Whenever w_j is a non-zero vector in W_j, $1 \leq j \leq k$, the set $\{w_j\}_1^k$ is linear independent (Section 1.1, Definition 3).

Lemma 1

Let T be any linear operator on a finite dimensional space V and let $\lambda_1, \lambda_2, \ldots, \lambda_k$ be the distinct eigenvalues of T. Then the eigenspaces $M_j = \{v \in V | (T - \lambda I)v = 0\}$, $1 \leq j \leq k$, are linearly independent.

Proof. Let $w_j \in M_j$ be a non-zero vector, $1 \leq j \leq k$, and suppose that

$$\sum_{j=1}^k w_j = 0 \tag{7}$$

Choose and fix i, $1 \leq i \leq k$ and let $U_i = \Pi_{j \neq i}(T - \lambda_j I)$. This is just a polynomial in the operator T and so the factors in the product defining U certainly commute. Hence

$$U_i w = 0 \text{ all } w \in M_j \text{ all } j \neq i \tag{8}$$

Furthermore

$$U_i w_i = \Pi_{j \neq i}(\lambda_i - \lambda_j) w_i \tag{9}$$

and this is not the zero vector since the λ's are distinct. Thus applying U_i to both sides of (7) gives $w_i = 0$, and since i was arbitrary we see that (7) implies $w_j = 0$ for all j.

It is easy to show that, when $W_1, W_2, \ldots, W_k$ are independent linear subspaces of V their sum $W_1 + \cdots + W_k = \{\sum_{j=1}^{k} w_j | w_j \in W_j \text{ for } 1 \leq j \leq k\}$ is a linear manifold in V (see problem 5). We shall denote this sum by the symbol $W_1 \oplus \cdots \oplus W_k$ and we shall call it the direct sum of the spaces $W_1, \ldots, W_k$. Suppose now that $W_1, \ldots, W_k$ are independent and that their direct sum is V. Then for each i, $1 \leq i \leq k$, the direct sum $(*) W_1 \oplus \cdots \oplus \hat{W}_i \oplus \cdots \oplus W_k$, where $\hat{W}_i$ means we leave out this subspace, is a complement for W_i in V (Section 4.1, Definition 2). Thus there is a projection P_i on V whose range is W_i and whose null space is $(*)$ (Section 4.1, Theorem 1).

Theorem 2

Let T be a diagonalizable operator on V and let $\lambda_1, \ldots, \lambda_k$ be the distinct eigenvalues of T. Then there are proections $\{P_j\}_{j=1}^{k}$ on V such that:

a. $T = \sum_{j=1}^{k} \lambda_j P_j$

b. $I = \sum_{j=1}^{k} P_j$

c. $P_i P_j = 0$ for $i \neq j$

Proof. For each j, $1 \leq j \leq k$, let W_j be the eigenspace of the eigenvalue λ_j. Then these spaces are independent and nontrivial and, since T is diagonalizable

$$V = W_1 \oplus W_2 \oplus \cdots \oplus W_k \tag{10}$$

by Theorem 1. For each j let P_j be the projection of V onto W_j whose null space is $W_1 \oplus \cdots \oplus \hat{W}_j \oplus \cdots \oplus W_k$. For any $v \in V$ we have, by (10),

$$v = P_1 v + \cdots + P_k v \tag{11}$$

In other words

$$I = \sum_{j=1}^{k} P_j \tag{12}$$

which proves (b). The way we defined P_j shows that $P_i P_j = 0$ for $i \neq j$ and so we have (c) verified. Finally, applying T to both sides of (11) and recalling that $P_j v \in W_j$ which is the eigenspace of T corresponding to the eigenvalue λ_j, we can write:

$$Tv = \sum_{j=1}^{k} T(P_j v) = \sum_{j=1}^{k} \lambda_j (P_j v) = (\sum_{j=1}^{k} \lambda_j P_j) v \qquad (13)$$

Since $v \in V$ is arbitrary this says

$$T = \sum_{j=1}^{k} \lambda_j P_j. \qquad (14)$$

Observe that the discussion given in the last section concerning functions of a matrix can be given here using Theorem 2 for a diagonalizable operator. By (a) and (c) of that Theorem we see that $T^p = \sum_{j=1}^{k} \lambda_j^p P_j$ and so $f(T)$ will be the operator $\sum_{j=1}^{k} f(\lambda_j) P_j$ whenever f is an analytic function whose circle of convergence contains all of the eigenvalues of T. We will discuss this later.

Theorem 3

Let T be a linear operator on V and suppose that there are distinct scalars $\lambda_1, \ldots, \lambda_k$ and non-zero projections $P_1, \ldots, P_k$ on V which satisfy conditions (a), (b) and (c) of Theorem 2. Then T is diagonalizable, the λ's are the distinct eigenvalues of T and each P_j is the projection of V onto the eigenspace W_j of T corresponding λ_j whose null space is $W_1 \oplus \cdots \oplus \hat{W}_j \oplus \cdots \oplus W_k$.

Proof. Since $T = \sum \lambda_j P_j$, $TP_i = \lambda_i P_i$, because $P_j P_i = 0$ when $i \neq j$. This says that the range of P_i is contained in the null space of $(T - \lambda_i I)$. Since P_i is not the zero operator we see that λ_i is an eigenvalue of T. Furthermore, if

$$(T - \lambda I)v = 0 \qquad (15)$$

for some scalar λ and some non-zero vector v, then

$$(\sum_{j=1}^{k} \lambda_j P_j - \lambda I)v = (\sum_{j=1}^{k} \lambda_j P_j - \lambda \sum_{j=1}^{k} P_j)v = \sum_{j=1}^{k} (\lambda_j - \lambda) P_j v = 0 \qquad (16)$$

Now v is not the zero vector so some $P_j v \neq 0$ giving us $\lambda_j - \lambda = 0$. Hence the only distinct eigenvalues of T are the given λ's.

We have noted that the range of P_j is contained in the null space of $(T - \lambda_j I)$. Since $I = \sum P_j$ the sum $M_1 + \cdots + M_k$, where $M_j = \text{Ker}(T - \lambda_j I)$, spans V. But, by Lemma 1, these spaces are independent. Hence $V = M_1 \oplus \cdots \oplus M_k$. So once we show that $M_j = \text{Range } P_j$ all j, then we shall have shown that there is a basis for V consisting of eigenvectors of T; i.e., by Theorem 1, we shall have shown that T is diagonalizable.

Suppose $Tv = \lambda_i v$. Then $\sum_{j=1}^{k}(\lambda_j - \lambda_i)P_j v = 0$ and hence $(\lambda_j - \lambda_i)P_j v = 0$ for each j. Thus $P_j v = 0$ for $j \neq i$. Since $v = Iv = \sum P_j v$ and $P_j v = 0$ for $j \neq i$, we must have $v = P_i v$ showing that every eigenvector of T is in the range of some P_j.

We have shown that diagonalizable operators have some "nice" properties (Theorem 2) and that these "nice" properties characterize diagonalizable operators. What we have not yet done is find a reasonably large class of operators all elements of which are diagonalizable. We really will do that in the next section.

Exercises 2

1. Compute the matrix of the orthogonal projection of $\mathbf{R}^3$ onto the xy-plane. What are the eigenvalues of this operator?

* 2. Show that every linear operator on a finite dimensional space has an eigenvalue.

3. Let P_{xy} be the orthogonal projection of $\mathbf{R}^3$ onto the xy-plane and let P_z be the orthogonal projection of $\mathbf{R}^3$ onto the z-axis.

   a. Show that $P_{xy} P_y = 0$

   b. Show that $I = P_{xy} + P_z$

   c. Let $T = 3P_{xy} + 2P_z$. Find the matrix of T for the standard basis in $\mathbf{R}^3$.

   d. Compute $\exp(T)$.

* 4. Let $(H, \langle,\rangle)$ be a Hilbert space, let $\{M_j\}_{j=1}^{n}$ be a finite family of linear subspaces of H. We shall say that this family is pairwise orthogonal if, for any i, j with $i \neq j$, and any $x \in M_i$, $y \in M_j$, $\langle x, y \rangle = 0$.

   a. Show that a family of pairwise orthogonal linear subspaces is linearly independent (Section 1.1, Defintion 3).

   b. Show that the direct sum of a family of pairwise orthogonal subspaces is a linear subspace (i.e., show that it is closed).

c. For each j let P_j be the orthogonal projection of H onto M_j. Here $\{M_j\}_{j=1}^n$ is a family of pairwise orthogonal subspaces. By (b) we know that $M_1 \oplus \cdots \oplus M_n$ is a linear subspace of H and hence is the range of an orthogonal projection P on H. Show that $P = P_1 + P_2 + \cdots + P_n$.

5. Let V be a finite dimensional space and let $M_1, \ldots, M_k$ be a set of linearly independent linear manifolds in V. Show that $M_1 \oplus \cdots \oplus M_k$ is linear manifold in V.

6. Show that the matrix
$$A = \begin{pmatrix} 0 & 0 & 1 \\ 0 & 0 & 3 \\ 0 & 0 & 0 \end{pmatrix}$$
is not diagonalizable. Hint: If A is a diagonalizable then $A = BDB^{-1}$ where D is a diagonal matrix. Show that the diagonal entries of D must be the eigenvalues of A. Next complete these values.

5.3 Invariant Subspaces

Our investigation of diagonalizable matrices and operators has led us very far already. However, we must go further still in order to find a nice large class of operators each element of which is diagonalizable. We shall do that here. First, however, we need some terminology. As usual, $(H, \langle, \rangle)$ denotes a fixed Hilbert space.

Definition 1

Let T be a bounded, linear operator on H. A linear subspace M of H is said to be invariant under T if $T(M) \subseteq M$ (i.e., $T(x) \in M$ whenever $x \in M$). We shall say that the linear subspace M reduces the operator T if both M and $M^\perp$ (Section 4.2, Definition 1) are invariant under T.

Of course H and the zero subspace reduce any bounded linear operator. What one wants is a non-trivial subspace which is invariant under or, better yet, reduces T. We might mention that the question of whether any bounded, linear operator on a Hilbert space has a non-trivial invariant subspace remains, at this time, open and has been so for a long time. The reader may also want to reread Remark 1(b) of Section 3.3.

Observe that when M reduces T then the study of T becomes the study of $T|M$ and $T|M^\perp$, since $M \oplus M^\perp = H$ (Section 4.2, just after the proof of lemma 1).

Theorem 1

Let T be a bounded, linear operator on H. A linear subspace M of H is invariant under T if, and only if, the linear subspace $M^\perp$ if invariant under T^*.

Proof. Suppose that M is invariant under T. Let $x \in M$, $y \in M^\perp$ and note that $\langle Tx, y \rangle = 0$ because $Tx \in M$. However, this says that $\langle x, T^*y \rangle = 0$ and since $x \in M$, $y \in M^\perp$ were arbitrary we conclude that $T^*(M^\perp) \subseteq M^\perp$.

Now suppose that $M^\perp$ is invariant under T^*. Then, by what we have just proved, $M^{\perp\perp} = M$ is invariant under $T^{**} = T$ (Section 4.2, Lemma 1 and Section 4.3, Exercises 4, problem 1(a)).

Corollary 1

A linear subspace of H reduces the operator T if, and only if, it is invariant under both T and T^*.

Theorem 2

Let M be a linear subspace of H and let P be the orthogonal projection of H onto M. Then M is invariant under the bounded, linear operator T if, and only if, $TP = PTP$.

Proof. First suppose that M is invariant under T and let $x \in H$. Then $Px \in M$ of course and so $T(Px) \in M$. But then $P[T(Px)] = T(Px)$ which shows, since x was arbitrary, that $PTP = TP$.

Now suppose that $TP = PTP$. For any $y \in M$ we have $Py = y$ so $TPy = Ty$. But then $Ty = (PTP)y$ which is clearly in M. Thus $T(M) \subseteq M$ and we have shown that M is invariant under T.

Corollary 2

Let M be a linear subspace of H and let P be the orthogonal projection of H onto M. Then M reduces the bounded, linear operator T if, and only if, P and T commute (i.e., $PT = TP$).

Proof. Suppose that M reduces T. Then by Corollary 1 above M is invariant under both T and T^*. It follows from this and Theorem 2 that $TP = PTP$ and $T^*P = PT^*P$. Now P is an orthogonal projection so it is self-adjoint (Section 4.2, Theorem 2, and Exercises 4.2, problem 2). Thus $(T^*P)^* = (PT^*P)^*$ gives us $PT = PTP$, and since we already have $PTP = TP$ we see that $PT = TP$.

Conversely, let us suppose that $PT = TP$. Then we have $P(PT) = P(TP)$ or, since $P^2 = P$, $PT = PTP$. Taking the adjoint of both sides of this and again using the fact that P is self-adjoint, we get $PT^*P = T^*P$. By Theorem 2 we conclude that M is invariant under T^*. On the other hand $PT = TP$ gives also $(PT)P = (TP)P$ or $PTP = TP^2 = TP$. Again by Theorem 2 we see that M is invariant under T. Finally, by Corollary 1, M reduces T.

The eigenspaces of a linear operator are obviously invariant under that operator; for if $M_\lambda = \{v \in H | (T - \lambda I)v = 0\}$, then $T|M_\lambda$ is just λI on M_λ. When T is a normal operator, recall that T is said to be normal if T commutes with T^*, then we shall see that each eigenspace reduces the operator. Before we can prove this we need some preliminary results.

Lemma 1

Suppose that T is a linear operator on H such that $\langle Tx, x \rangle = 0$ for all $x \in H$. Then T is the zero operator.

Proof. Let us first show that if $\langle Tx, y \rangle = 0$ for all x and all y, then T is the zero operator. If not then there is some $x \in H$ such that Tx is not the zero vector. But $\langle Tx, y \rangle = 0$ for all y and, in particular, we can take $y = Tx$. Then we have $\|Tx\|^2 = 0$ which is impossible. Thus it suffices to prove that $\langle Tx, x \rangle = 0$ for all x implies $\langle Tx, y \rangle = 0$ for all x and all y.

The following identity is easily checked by direct calculation.

$$\langle T(\alpha x + \beta y), \alpha x + \beta y \rangle - |\alpha|^2 \langle Tx, x \rangle - |\beta|^2 \langle Ty, y \rangle = \alpha \overline{\beta} \langle Tx, y \rangle + \overline{\alpha} \beta \langle Ty, x \rangle \quad (1)$$

By assumption the left-hand side of (1) is zero, hence so also is the right-hand side; and this is for all scalars α, β and any vectors x, y. Thus

$$\alpha \overline{\beta} \langle Tx, y \rangle + \overline{\alpha} \beta \langle Ty, x \rangle = 0 \quad (2)$$

Set $\alpha = 1 = \beta$ in (2) to get

$$\langle Tx, y \rangle + \langle Ty, x \rangle = 0 \quad (3)$$

and then set $\alpha = i$ and $\beta = 1$ in (2) to get

$$i \langle Tx, y \rangle - i \langle Ty, x \rangle = 0 \quad (4)$$

We may divide (4) by i and add the result to (3) to get $\langle Tx, y \rangle = 0$ for all x, y in H.

Corollary 3

An operator T on H is a normal operator if, and only if, $\|T^*x\| = \|Tx\|$ for every $x \in H$.

Proof. Suppose that $\|T^*x\| = \|Tx\|$ for all $x \in H$. Then $\|T^*x\|^2 = \|Tx\|^2$ and so $\langle T^*x, T^*x \rangle = \langle Tx, Tx \rangle$. These yield $\langle TT^*x, x \rangle = \langle T^*Tx, x \rangle$ which is equivalent to $\langle (TT^* - T^*T)x, x \rangle = 0$ for all x in H. By our Lemma we see that $TT^* = T^*T$; i.e., T is normal.

Conversely, if T is known to be normal then we can write $\langle (TT^* - T^*T)x, x \rangle = 0$ for all x and now follow the reasoning above in reverse.

Corollary 4

If T is a normal operator on H with eigenvalue λ then $x \in H$ is an eigenvector for T corresponding to λ if, and only if, x is an eigenvector of T^* corresponding to $\overline{\lambda}$.

Proof. This is really a corollary of corollary 1. Since T is normal so also is $T - \lambda I$. Thus, by Corollary 1, $\|(T-\lambda I)^*x\| = \|(T-\lambda I)x\|$ for every x. However, $(T - \lambda I)^* = T^* - \overline{\lambda}I$ (Exercises 4.3, problem 1(d)). Thus $(T - \lambda I)x = 0$ if, and only if, $(T^* - \overline{\lambda}I)x = 0$.

Theorem 3

Any eigenspace of a normal operator reduces that operator. Furthermore, the eigenspaces of a normal operator are pairwise orthogonal (Exercises 2, problem 4).

Proof. Let T be a normal operator on H and let W_λ be an eigenspace of T corresponding to the eigenvalue λ. It is clear that W_λ is invariant under T and so, in order to show that this space reduces T, it suffices to show that W_λ is invariant under T^* (Corollary 1 to Theorem 1). But if $x \in W_\lambda$ then $Tx = \lambda x$ and $T^*x = \overline{\lambda}x$, by Corollary 2 of Lemma 1, hence $T^*x \in W_\lambda$ because $x \in W_\lambda$ and $\overline{\lambda}x$ certainly also belongs to W_λ.

Next suppose that λ, μ are two distinct eigenvalues of T with corresponding eigenspaces W_λ, W_μ respectively. Let $x \in W_\lambda, y \in W_\mu$ so that $Tx = \lambda x$, $Ty = \mu y$. Then

$$\lambda \langle x, y \rangle = \langle \lambda x, y \rangle = \langle Tx, y \rangle = \langle x, T^*y \rangle = \langle x, \overline{\mu}y \rangle = \mu \langle x, y \rangle$$

again by Corollary 2 of Lemma 1. It follows that $(\lambda - \mu)\langle x, y \rangle = 0$ and since $\lambda \neq \mu$, $\langle x, y \rangle = 0$.

Let us apply our results to partially answer a question raised several times above.

We have seen (Theorem 2.2) that if an operator T on V is diagonalizable, if $\lambda_1, \ldots, \lambda_k$ are the distinct eigenvalues of T and if, for each j, W_j is the eigenspace corresponding to λ_j, then there are projections $P_1, \ldots, P_k$ such that

a. $T = \sum_{j=1}^k \lambda_j P_j$,
b. $I = \sum_{j=1}^k P_j$,
c. $P_i P_j = 0$ for $i \neq j$

here each P_j is the projection of V onto W_j whose null space is $W_1 \oplus W_2 \oplus \cdots \oplus \hat{W}_j \oplus \cdots \oplus W_k$. One sometimes says, because of (c), that the projections P_j are mutually orthogonal. This is a little bit misleading since the P_j's need not be orthogonal projections. There is one case, however, when we can be sure that they are orthogonal projections:

Theorem 4

Let V be a finite dimensional space and let T be a normal operator on V. Then T is a diagonalizable operator and, furthermore, the projections appearing in the representation of T, (a), (b), and (c) above, are orthogonal projections.

Proof. Let $\lambda_1, \lambda_2, \ldots, \lambda_m$ be the distinct eigenvalues of T and let $W_1, \ldots, W_m$ be the eigenspaces corresponding to these eigenvalues. By Theorem 3, the spaces W_j, $1 \leq j \leq m$, are pairwise orthogonal and so $W = W_1 \oplus \cdots \oplus W_m$ is a linear manifold (= subspace since V is finite dimensional) in V. Now if P_j is the orthogonal projection of V onto W_j for $j = 1, 2, \ldots, m$, then $P = P_1 + \cdots + P_m$ is the orthogonal projection of V onto W (Exercises 2, problem 4 parts (b) and (c)). By Theorem 3 the operator T commutes with every P_j and so it must commute with P. Thus, By Corollary 1 of Theorem 2, the space W reduces T; i.e., $T|W$ is a linear operator on W and $T|W^\perp$ is a linear operator on $W^\perp$. However, all of the eigenvectors of T are in W, hence $T|W^\perp$ has no eigenvalues because it has no eigenvectors. But V, and hence $W^\perp$, is a finite dimensional space and so we cannot have a nontrivial operator on this space that has no eigenvalues (Exercises 2, problem 2); note $T|W^\perp$ could not be the zero operator since that would mean $\lambda = 0$ is an eigenvalue of T and we would then have counted it among the λ_j's. It follows that $W^\perp$ must be the zero subspace and so

$$V = W_1 \oplus \cdots \oplus W_m$$

clearly then (b) $I = \sum_{j=1}^m P_j$ and we certainly have (c) $P_i P_j = 0$, $i \neq j$. Now $TI = T(\sum_{j=1}^m P_j) = \sum_{j=1}^m TP_j$ and for any $v \in V$

$$Tv = (TI)v = \sum_{j=1}^m T(P_j v) = \sum_{j=1}^m \lambda_j (P_j v) = (\sum_{j=1}^m \lambda_j P_j) v$$

thus we have (a) $T = \sum_{j=1}^{m} \lambda_j P_j$. By Theorem 3 of Section 2 we conclude that T is a diagonalizable operator.

Theorem 5

Let T be a linear opeartor on a finite dimensional space V and suppose that there are distinct scalars $\lambda_1, \lambda_2, \ldots, \lambda_k$ and non-zero *orthogonal* projections $P_1, \ldots, P_k$ on V which satisfy conditions (a), (b) and (c) stated above. Then T is a diagonalizable operator, the λ_j's are the distinct eigenvalues of T and, furthermore, T is a normal operator.

Proof. Most of our statements were proved in Theorem 2.3. The only thing left to do here is to show that, under the conditions stated, T is a normal operator. We have (a) $T = \sum_{j=1}^{k} \lambda_j P_j$, hence

$$T^* = \sum_{j=1}^{k} \overline{\lambda}_j P_j^* = \sum_{j=1}^{k} \overline{\lambda}_j P_j$$

(Section 4.2, Theorem 2 and Exercises 4.2, problem 2) because the P_j's are orthogonal projections. But then

$$TT^* = (\sum_{j=1}^{k} \lambda_j P_j)(\sum_{j=1}^{k} \overline{\lambda}_j P_j) = \sum_{j=1}^{k} |\lambda_j|^2 P_j^2 = \sum_{j=1}^{k} |\lambda_j|^2 P_j$$

because $P_i P_j = 0$ when $i \neq j$ and $P_j^2 = P_j$. Also

$$T^*T = (\sum_{j=1}^{k} \overline{\lambda}_j P_j)(\sum_{j=1}^{k} \lambda_j P_j) = \sum_{j=1}^{k} |\lambda_j|^2 P_j$$

showing that $T^*T = TT^*$; i.e., showing that T is a normal operator.

Let us sum up some of what we have done. We worked with a fixed, finite dimensional Hilbert space $(V, \langle,\rangle)$. Given a Hamel basis for V any linear operator on V has a matrix relative to that basis. The matrices of the same operator relative to two different bases for V are similar. A square matrix is called diagonalizable if it is similar to a diagonal matrix and we defined a linear operator on V to be diagonalizable if its matrix is diagonalizable. We found that a linear operator T on V is of this type if, and only if, we could find distinct scalars $\lambda_1, \ldots, \lambda_k$ and projections $P_1, \ldots, P_k$ such that

a. $T = \sum_{j=1}^{k} \lambda_j P_j$,

b. $I = \sum_{j=1}^{k} P_j$,

c. $P_i P_j = 0$ for $i \neq j$.

The P_j's are orthogonal projections if, and only if, T is a normal operator.

Exercises 3

1. Let A denote the right-shift operator on ℓ^2; so $A(\{x_n\}_{n=1}^\infty) = \{x_{n+1}\}_{n=0}^\infty$, $x_1 \equiv 0$. Let $W = \{\{x_n\} \in \ell^2 | x_1 = 0\}$. Show that W is a linear subspace of ℓ^2 which is invariant under A. Show that W does not reduce A.

* 2. Let A be a bounded, linear operator on H which is self-adjoint. Show that any subspace of H which is invariant under A reduces this operator.

3. Let T be a normal operator on a finite dimensional Hilbert space. Suppose that B is another operator on this space which commutes with T. Show that B commutes with $f(T)$ for any f for which $f(T)$ has meaning.

4. Let T be a normal operator on $(F, \langle , \rangle)$ and write $T = \sum \lambda_j P_j$. If $f(T)$ has meaning show that $f(T) = \sum f(\lambda_j) P_j$. Is $f(T)$ a normal operator?

* 5. Let A be a bounded, linear operator on H and let M be a linear subspace of H which is invariant under A. To avoid triviality let us suppose that $\{0\} \subsetneq M \subsetneq H$.

   a. Show that $A|M$, the restriction of A to M, is a bounded, linear operator on M.

   b. If A is a self-adjoint operator (respectively a normal operator), show that $A|M$ is a self-adjoint (respectively normal) operator.

   c. If A is a compact operator, show that $A|M$ is a compact operator on M. One way to do this is as follows:

      i. Show that $\{x_n\} \subset M$ is weakly convergent to $y \in M$ if, and only if, $\lim \langle x_n, y \rangle = \langle y, z \rangle$ for all z in M (not H but M).

      ii. Now recall that A is compact if, and only if, it maps any weakly convergent sequence to one which is convergent for the norm.

5.4 The Spectral Theorem: Compact Operators

We shall now begin our discussion of the spectral theorem for an operator on an infinite dimensional space. The natural place to start is with compact operators because they have such simple spectra. Recall that the spectrum of a compact operator is either a finite set or a sequence converging to zero. The non-zero points of the spectrum are eigenvalues of the operator and each of the corresponding eigenspaces is finite dimensional (Section 3.3, Corollary 2 to Theorem 3 and Remark 1(a)). Before going any further let us clarify where the number zero fits in this discussion (see also the second paragraph of Section 3.3).

Theorem 1

Every compact opeartor on an infinite dimensional Hilbert space has the number zero in its spectrum and, furthermore, this number is a generalized eigenvalue of the operator.

Proof. Let $(H, \langle, \rangle)$ be an infinite dimensional Hilbert space and let A be a compact, linear operator on H. There is an infinite orthonormal sequence $\{e_n\} \subseteq H$; this can be constructed by first choosing an infinite, linearly independent subset of H, there must be one since H is infinite dimensional, and then applying the Gram-Schmidt process (Section 1.9 just after Lemma 1) to it. Since A is a compact operator the sequence $\{Ae_n\}$ has a convergent subsequence with limit $y \in H$. We shall denote this subsequence, redundantly, by $\{Ae_n\}$ also. Now by Bessel's inequality (Section 1.6, Theorem 2)

$$\sum_{n=1}^{\infty} |\langle x, e_n \rangle|^2 \leq \|x\|^2 \text{ for each } x \in H \qquad (1)$$

It follows from this that, for each $x \in H$,

$$\lim_{n \to \infty} \langle e_n, x \rangle = 0 \qquad (2)$$

because the limit of the nth term of any convergent series is zero. But then

$$\langle y, x \rangle = \lim \langle Ae_n, x \rangle = \lim \langle e_n, A^*x \rangle = 0 \qquad (3)$$

for any $x \in H$ showing that y is the zero vector. Hence

$$\lim \|(A - 0I)e_n\| = 0 \qquad (4)$$

which says that the number zero is a generalized eigenvalue (Section 2.4, Definition 1) of the operator A.

We have seen that, even in the finite dimensional case, not every operator can be diagonalized (Exercises 2, problem 6). Let us give an example of a compact operator on ℓ^2 which cannot be diagonalized.

Example 1. Recall the infinite dimensional Hilbert space $\ell^2 = \{\{z_n\}_{n=1}^{\infty} \mid \sum_{n=1}^{\infty} |z_n|^2 < \infty\}$ and the linear operator B defined as follows:

$$(*) \qquad B(\{z_n\}_{n=1}^{\infty}) = \{\frac{z_{n-1}}{n}\}_{n=1}^{\infty}, \quad z_0 \equiv 0$$

(Exercises 3.2, problem 3) hence

$$B[(z_1, z_2, \ldots, z_n, \ldots)] = (0, \frac{z_1}{2}, \frac{z_2}{3}, \ldots, \frac{z_n}{n+1}, \ldots)$$

Clearly

$$\|B(\{z_n\})\|^2 = \sum_{n=1}^{\infty} \frac{|z_{n-1}|^2}{n^2} \leq \sum_{n=1}^{\infty} |z_{n-1}|^2 = \|\{z_n\}\|^2$$

and so B is a bounded, linear operator. Let us show that it is a compact operator. For each p let B_p be the linear operator on ℓ^2 defined by:

(**) $\quad B_p(\{z_n\}_{n=1}^{\infty}) = \{\frac{z_{n-1}}{n}\}_{n=1}^{p}, \quad z_0 \equiv 0$

hence

$$B_p[(z_1, z_2, \ldots, z_p, z_{p+1}, \ldots)] = (0, \frac{z_1}{2}, \frac{z_2}{3}, \ldots, \frac{z_{p-1}}{p}, 0, 0, 0, \ldots)$$

It is clear that B_p has finite rank and that it is a bounded operator, thus B_p is a compact operator (Exercises 3.1, problem 5). Now if $\vec{z} = \{z_n\} \in \ell^2$ has norm one or less then

$$\|(B - B_p)\vec{z}\|^2 = \|(B - B_p)\{z_n\}\|^2 = \sum_{n=p}^{\infty} |\frac{z_n}{n+1}|^2 \leq \sum_{n=p}^{\infty} (\frac{1}{n+1})^2$$

and the latter sum tends to zero as $p \to \infty$. Thus

$$\lim \|B - B_p\| = 0$$

showing that B is a compact operator (Section 3.2, Theorem 2).

Now Theorem 1 tells us that the number zero is in the spectrum of B and that it is a generalized eigenvalue of this operator. We claim that $\sigma(B) = \{0\}$. To prove this we suppose that $\lambda \neq 0$ is in $\sigma(B)$. Then λ would have to be an eigenvalue of B. However

$$(B - \lambda I)\{z_n\} = \{\frac{z_{n-1}}{n}\} - \{\lambda z_n\} = (-\lambda z_1, \frac{z_1}{2} - \lambda z_2, \frac{z_2}{3} - \lambda z_3, \ldots)$$

Suppose that $\{z_n\}$ is an eigenvector corresponding to λ. Then the vector just obtained must be the zero vector giving us the equations

$$\lambda z_1 = 0, \quad \frac{z_1}{2} - \lambda z_2 = 0, \quad \frac{z_2}{3} - \lambda z_3 = 0, \quad \text{etc.}$$

Since $\lambda \neq 0$ our first equation gives $z_1 = 0$ and this, combined with the next equation, gives $z_2 = 0$ which in turn gives $z_3 = 0$, etc. Hence $\{z_n\}$ is the zero vector which contradicts the fact that it is an eigenvector corresponding to λ. We conclude that B has no eigenvalues.

It is clear that there is no hope of being able to write $B = \sum_{j=1}^{\infty} \lambda_j P_j$ where the P_j's are projections such that $P_i P_j = 0$ for $i \neq j$ and $I = \sum_{j=1}^{\infty} P_j$ because, if we could do this, the λ_j would have to be the eigenvalues of B and B doesn't have any. One might note (Exercises 4.2, problem 6) that B is not a normal operator.

Example 1 shows that we must make some assumptions about our compact operator if we are to be able to prove a spectral theorem; for we must be able to guarantee that the operator has "sufficiently" many eigenvalues. The next result is in this direction.

Theorem 2

Let A be a compact self-adjoint operator on $(H, \langle,\rangle)$. Then there is an eigenvalue λ of A such that $|\lambda| = \|A\|$.

Proof. For any $\lambda \in \sigma(A)$ we have $|\lambda| \leq \|A\|$ (Section 2.3, just after Definition 2). Let us assume, as we may since otherwise the result is trivial, that $\|A\| > 0$. Now since A is self-adjoint $\langle Ax, x \rangle$ is real for all x in H and

$$\|A\| = \sup\{|\langle Ax, x \rangle| \,|\, \|x\| = 1\} \tag{5}$$

(see Exercises 4.3, problem 2(e)(iii) and 4(c)). Thus, using (1), we can select a sequence $\{x_n\}$ of unit vectors in H such that

$$\|A\| = \lim |\langle Ax_n, x_n \rangle| \tag{6}$$

Now $\{x_n\}$ is a bounded sequence and so it has a subsequence which is weakly convergent (Section 4.4, Theorem 3) and we'll call this subsequence $\{x_n\}$ also. Since the operator A is compact the sequence $\{Ax_n\}$ is convergent for the norm of H (Section 4.4, Theorem 4). Thus

$$\lim \langle Ax_n, x_n \rangle = \lambda, \text{ say, where } |\lambda| = \|A\| \tag{7}$$

$$\lim Ax_n = y, \text{say}, y \in H \tag{8}$$

and convergence is for the norm. From (4) and (Exercises 1.3, problem 1) we have

$$\|y\| = \lim \|Ax_n\| \leq \|A\| \tag{9}$$

because each x_n is a unit vector. Also

$$\|Ax_n - \lambda x_n\|^2 = \langle Ax_n - \lambda x_n, Ax_n - \lambda x_n \rangle = \|Ax_n\|^2 - 2\lambda \langle Ax_n, x_n \rangle + |\lambda|^2$$

because λ is real. From this we get

$$\begin{aligned} 0 \leq \lim \|Ax_n - \lambda x_n\|^2 &= \lim \|Ax_n\|^2 - 2\lambda \lim \langle Ax_n, x_n \rangle + |\lambda|^2 \\ &= \|y\|^2 - 2\lambda^2 + \lambda^2 = \|y\|^2 - \|A\|^2 \leq 0 \end{aligned}$$

THE SPECTRAL THEOREM: COMPACT OPERATORS

Thus
$$\|y\| = \|A\| > 0$$

so $y \neq 0$ and $\lim \lambda x_n$ must exist and equal $\lim A x_n = y \neq 0$. Since $|\lambda| = \|A\| > 0$, $\lambda \neq 0$, and so $\lim x_n = y/\lambda$ and

$$(A - \lambda I)(y/\lambda) = (A - \lambda I)(\lim x_n) = \lim(A - \lambda I)x_n = 0$$

This says that λ is an eigenvalue of A with eigenvector y/λ and since we already have $|\lambda| = \|A\|$ we are done.

Let us now prove the spectral theorem for compact, self-adjoint operators. The proof is fairly long and involved. We begin by setting up our notation and making some simple observations. Next we prove the theorem in a simple case. After that we digress into a discussion of infinite dimensional geometry to obtain a result we need in order to finish the proof. Finally, we complete the proof and state the theorem. The matter does not end there, for after obtaining a useful corollary we use our theorem to prove an analagous result for compact, normal operators.

Suppose now that A is a fixed, compact, self-adjoint operator on H. Then the non-zero eigenvalues of A are real numbers, there are countably many of them, and we may label them so that

$$|\lambda_1| \geq |\lambda_2| \geq |\lambda_3| \geq \ldots \geq |\lambda_n| \geq \ldots \qquad (10)$$

For each n let $M_n = \{x \in H | (A - \lambda_n I)x = 0\}$ be the eigenspace corresponding to λ_n and let P_n be the orthogonal projection of H onto M_n; recall also that each of these spaces is finite dimensional. Notice that, since A is a normal operator (self-adjoint implies normal), the spaces M_n are pairwise orthogonal and so $P_m P_n = 0$ for $m \neq n$ (Section 3, Theorem 3). At this point we distinguish two cases:

Case 1. Suppose that A has only finitely many eigenvalues

$$|\lambda_1| \geq |\lambda_2| \geq \ldots \geq |\lambda_k| > 0$$

say. We set $M = M_1 \oplus M_2 \oplus \cdots \oplus M_k$, let us write this $\sum_{j=1}^{k} \oplus M_j$, and let M_0 be the orthogonal complement of M; here we have used Exercises 2, problem 4(b). It is clear that M is invariant under A and so M_0 is also invariant under this operator (Exercises 3, problem 2). Thus $A|M_0$ is a self-adjoint, compact operator on M_0 (Section 4.4, Theorem 4). Now use Theorem 2 to conclude that $A|M_0$ has an eigenvalue λ such that

$$|\lambda| = \|A|M_0\|.$$

Note that the eigenvectors corresponding to λ lie in M_0. But any eigenvalue of $A|M_0$ is also an eigenvalue of A. Thus λ is an eigenvalue of A and so its corresponding eigenvectors lie in some M_p. This is impossible because these vectors would then lie in both $M_p \subseteq \sum_{j=1}^{k} \oplus M_j$ and M_0, and these subspace are orthogonal complements. It follows that $A|M_0$ has no non-zero eigenvalues and so, by Theorem 2, this must be the zero operator.

Now let P_0 be the orthogonal projection of H onto M_0. Then for any $x \in H$ we must have

$$x = \sum_{j=0}^{k} P_j x \qquad (11)$$

(note that the sum starts with zero) and so

$$I = \sum_{j=0}^{k} P_j \qquad (12)$$

Since this sum is finite we can write

$$Ax = A(\sum_{j=0}^{k} P_j x) = \sum_{j=0}^{k} A(P_j x) = \sum_{j=1}^{k} \lambda_j (P_j x)$$

because $P_j x \in M_j$ for $1 \leq j \leq k$ showing that these are eigenvectors of A, and $P_0 x \in M_0$ where A reduces to the zero operator. We conclude that

$$A = \sum_{j=1}^{k} \lambda_j P_j \qquad (13)$$

This proves the spectral theorem for this case. Observe that the range of A is $M = \sum_{j=1}^{k} \oplus M_j$ and since k is finite and each M_j is finite dimensional ($1 \leq j \leq k$, not M_0), we see that A has finite rank (Exercises 3.1, problem 5).

The second case is that in which A has infinitely many eigenvalues hence infinitely many eigenspaces $M_1, M_2, \ldots$. These are pairwise orthogonal, finite dimensional spaces and before we can complete the proof of the spectral theorem we must first discuss the meaning of the infinite sum $\sum_{j=1}^{\infty} \oplus M_j$.

Let us set $\mathcal{M} = \cup_{j=1}^{\infty} M_j$ and let $lin\mathcal{M}$ be the linear manifold generated by this set (Section 1.1, Definition 2). So $lin\mathcal{M}$ is the set of all *finite* sums $\sum_{k=1}^{n} \alpha_k x_k$ where each x_k is in $\mathcal{M}$ and each α_k is a scalar. We recall that the closure of $lin\mathcal{M}$, we shall denote it by $\overline{lin\mathcal{M}}$, is a linear subspace of H (Exercises 3.1, problem 4(d)). Now we would like to be able to say that, since the M_j's are pairwise orthogonal (the fact that they are finite dimensional is irrelevant here), each $x \in \overline{lin}\mathcal{M}$ can be written as an infinite series

$$x = \sum_{j=1}^{\infty} x_j, \quad x_j \in M_j \text{ for each } j.$$

This is true and we shall need this fact in our proof of the spectral theorem, but proving it takes some work.

The first thing we do is set

$$\sum_{j=1}^{\infty} \oplus M_j = \{x \in H | x = \sum_{j=1}^{\infty} x_j, x_j \in Mj \text{ all } j \text{ and convergence is for the norm}\}$$

Observe that any such x is the limit of the sequence

$$\{\sum_{j=1}^{n} x_j\}_{n=1}^{\infty} \subseteq lin\mathcal{M}$$

and so any such x is in $\overline{lin}\mathcal{M}$ (Section 2.4, Theorem 1). Combining this observation with the fact that, trivially, $lin\mathcal{M} \subseteq \sum_{j=1}^{\infty} \oplus M_j$ we may write

$$lin\mathcal{M} \subseteq \sum_{j=1}^{\infty} \oplus M_j \subseteq \overline{lin}\mathcal{M}$$

Hence to prove that the latter two sets are identical it suffices to show that our infinite sum is closed (Exercises 3.1, problem 4(c)). In doing this the following result is useful.

Lemma 1

Let $\{y_k\}_{k=1}^{\infty}$ be a sequence of pairwise orthogonal vectors in H. Then:

  i. $\sum_{k=1}^{\infty} y_k$ is convergent if, and only if, $\sum_{k=1}^{\infty} \|y_k\|^2$ is convergent;

  ii. If $\sum_{k=1}^{\infty} y_k$ converges to the vector $y \in H$ then $\|y\|^2 = \sum_{k=1}^{\infty} \|y_k\|^2$.

Proof. Suppose that $\sum_{k=1}^{\infty} \|y_k\|^2 < \infty$. Then for any m, n, with $m \leq n$, we can write

$$\|\sum_{k=m}^{n} y_k\|^2 = \langle \sum_{m}^{n} y_k, \sum_{m}^{n} y_k \rangle = \sum_{k=m}^{n} \|y_k\|^2 \tag{14}$$

and the latter sum tends to zero as $m, n \to \infty$; we have used the fact that our vectors are pairwise orthogonal. It follows that $\{\sum_{k=1}^{\ell} y_k\}_{\ell=1}^{\infty}$ is a Cauchy sequence in the Hilbert space H and so our series of vectors must converge. This proves half of (i).

Now suppose that $\sum_{k=1}^{\infty} y_k = y$. Then we have

$$\|y\|^2 = \langle y, y \rangle = \langle \lim_n \sum_{k=1}^{n} y_k, \lim_n \sum_{k=1}^{n} y_k \rangle \qquad (15)$$

$$= \lim_n \langle \sum_{k=1}^{n} y_k, \sum_{k=1}^{n} y_k \rangle = \lim_n \sum_{k=1}^{n} \|y_k\|^2$$

$$= \sum_{k=1}^{\infty} \|y_k\|^2$$

This proves (ii) and also the other half of (i).

We might think of (ii) as a kind of infinite dimensional Pythagorean theorem. Let us use it now to show that

$$\sum_{j=1}^{\infty} \oplus M_j$$

is closed. Suppose that $\{x_n\}$ is a sequence of points in this sum which converges to, say, x. Then for each n we have, from the way our sum was defined,

$$x_n = \sum_{k=1}^{\infty} y_{n,k}, \quad y_{n,k} \in M_k \text{ for each } k \qquad (16)$$

Now using (ii) of Lemma 1, we may write

$$\|x_m - x_n\|^2 = \sum_{k=1}^{\infty} \|y_{m,k} - y_{n,k}\|^2 \qquad (17)$$

Since $\{x_n\}$ was given to be a convergent sequence we can, for any given $\varepsilon > 0$, choose N so that

$$\|x_m - x_n\| < \varepsilon \text{ whenever } m, n \geq N. \qquad (18)$$

Thus, using (12),

$$\sum_{k=1}^{\infty} \|y_{m,k} - y_{n,k}\|^2 < \varepsilon^2 \text{ whenever } m, n \geq N. \qquad (19)$$

It follows from this that, for each fixed k, $\{y_{n,k}\}_{n=1}^{\infty}$ is a Cauchy sequence in M_k; for

$$\|y_{m,k} - y_{n,k}\|^2 \leq \sum_{k=1}^{\infty} \|y_{m,k} - y_{n,k}\|^2 < \varepsilon^2 \text{ for } m, n \geq N. \qquad (20)$$

Since M_k is closed, it is a subspace, in H and H is complete $\{y_{n,k}\}_{n=1}^{\infty}$ converges to, say, $y_k \in M_k$ for each k.

Now we are trying to show that our infinite sum is closed. We chose a sequence in this sum, $\{x_n\}$, and we assumed that it converged to x. What we must do is show that $x \in \sum_{k=1}^{\infty} \oplus M_j$; i.e., we must show that we can write x as the sum, infinite sum, of vectors one from each M_j. In the last paragraph we constructed vectors y_k such that $y_k \in M_k$ for each k. The fact is that the sum of these vectors is convergent and its limit is the vector x. Let us prove this now. Using (14) we may write, for any fixed ℓ,

$$\sum_{k=1}^{\ell} \|y_{m,k} - y_{n,k}\|^2 < \varepsilon^2 \text{ for } m, n \geq N \tag{21}$$

Letting $m \to \infty$ in this finite sum we get

$$\sum_{k=1}^{\ell} \|y_k - y_{n,k}\|^2 \leq \varepsilon^2 \text{ for } n \geq N \text{ and each } \ell. \tag{22}$$

Thus, letting $\ell \to \infty$ in (17), we obtain

$$\sum_{k=1}^{\infty} \|y_k - y_{n,k}\|^2 \leq \varepsilon^2 \text{ for } n \geq N \tag{23}$$

This last inequality together with (i) of Lemma 1 tells us that

$$\sum_{k=1}^{\infty} (y_k - y_{n,k})$$

is convergent. Since $\sum_{k=1}^{\infty} y_{n,k}$ is also convergent ((11)) we see that $\sum_{k=1}^{\infty} y_k$ converges; for

$$\sum_{k=1}^{\infty} y_k = \sum_{k=1}^{\infty} (y_k - y_{n,k}) + \sum_{k=1}^{\infty} y_{n,k} = \sum_{k=1}^{\infty} (y_k - y_{n,k}) + x_n$$

again by (11). Since the ε in (18) is arbitrary, changing $\varepsilon > 0$ only increases N, we see that

$$\sum_{k=1}^{\infty} y_k = \lim x_n = x \tag{24}$$

This proves that our infinite sum of spaces is closed and that

$$\overline{\text{lin}} \, \mathcal{M} = \sum_{j=1}^{\infty} \oplus M_j.$$

We have just seen that any $x \in \sum_{j=1}^{\infty} \oplus M_j$ can be written as the sum of an infinite series one term from each M_j. Let us show here that this can be done in only one way. Suppose that

$$x = \sum_{j=1}^{\infty} x_j \text{ and } x = \sum_{j=1}^{\infty} y_j$$

where $x_j, y_j \in M_j$ for each j. Then clearly $\sum_{j=1}^{\infty}(x_j - y_j) = 0$ and so, by (ii) of Lemma 1,

$$\sum_{j=1}^{\infty} \|x_j - y_j\|^2 = 0$$

But this is a sum of non-negative terms the only way it can be zero is if $\|x_j - y_j\| = 0$ for every j; i.e., $x_j = y_j$ for each j.

Let us now return to the proof of the spectral theorem. Recall that A was a fixed, compact, self-adjoint operator on H whose non-zero eigenvalues $\{\lambda_j\}$ were labeled so that

(*) $|\lambda_1| \geq |\lambda_2| \geq \ldots \geq |\lambda_k| \geq \ldots$

Case 2. Suppose that A has infinitely many eigenvalues so that the sequence in (*) is infinite and $\lim \lambda_k = 0$. Again we let $M_j = \{x \in H | (A - \lambda_j I)x = 0\}$ be the eigenspace corresponding to the eigenvalue λ_j. These are finite dimensional, pairwise orthogonal spaces and, as we have just laboriously proved, $\sum_{j=1}^{\infty} \oplus M_j$ is a linear subspace of H. Again let M_0 be the orthogonal complement of our direct sum and, exactly as in Case 1, we see that $A|M_0$ is the zero operator. So if P_0 is the orthogonal projection of H onto M_0 then any $x \in H$ can be written

$$x = P_0 x + y, \qquad y \in \sum_{j=1}^{\infty} \oplus M_j$$

Thus, since $y = \sum_{j=1}^{\infty} P_j x$ where P_j is the orthogonal projection of H onto M_j, we have

$$\text{(a) } Ix = \sum_{j=0}^{\infty} P_j x \quad \text{(b) } Ax = \sum_{j=0}^{\infty} A(P_j x) = \sum_{j=1}^{\infty} \lambda_j (P_j x) \tag{25}$$

for every $x \in H$; note that $P_i P_j = 0$ for $i \neq j$.

Now, for any $x \in H$ we have

$$(A - \sum_{j=1}^{n} \lambda_j P_j)x = \sum_{j=1}^{\infty} \lambda_j (P_j x) - \sum_{j=1}^{n} \lambda_j (P_j x) = \sum_{j=n+1}^{\infty} \lambda_j (P_j x)$$

and so by taking norms we get

$$\|(A - \sum_{j=1}^{n} \lambda_j P_j)x\|^2 = \sum_{n+1}^{\infty} \lambda_j^2 \|P_j x\|^2 \leq \lambda_{n+1}^2 \sum_{j=0}^{\infty} \|P_j x\|^2 = \lambda_{n+1}^2 \|x\|^2$$

where we have used (*); i.e., $|\lambda_{n+1}| \geq |\lambda_j|$ for all $j \geq n+1$. Hence

$$\|A - \sum_{j=1}^{n} \lambda_j P_j\| \leq |\lambda_{n+1}| \tag{26}$$

and since $\lim \lambda_{n+1} = 0$ we see that, for the operator norm,

$$A = \sum_{j=1}^{\infty} \lambda_j P_j \tag{27}$$

Finally, let us observe that (21) can be sharpened. For if we take $x \in M_{n+1}$ then $\sum_{j=n+1}^{\infty} P_j x = \lambda_{n+1} x$ and (21) becomes

$$\|A - \sum_{j=1}^{n} \lambda_j P_j\| = |\lambda_{n+1}|$$

Let us now sum up the results of all this argument.

Theorem 3

(Spectral Theorem for Compact, Self-adjoint Operators) Let A be a compact, self-adjoint operator on a Hilbert space H. Let $\{\lambda_j\}$ be the distinct, non-zero eigenvalues of A arranged in such a way that

$$|\lambda_1| \geq |\lambda_2| \geq \ldots, \quad \lambda_i \neq \lambda_j \text{ for } i \neq j$$

For each j let M_j be the eigenspace corresponding to λ_j and let P_j be the orthogonal projection of H onto M_j. Then:

(I) If $\{\lambda_j\}$ is a finite set, then A is an operator of finite rank and

$$(9) \quad A = \sum_{j=1}^{k} \lambda_j P_j$$

k being the number of elements in $\{\lambda_j\}$.

(II) If $\{\lambda_j\}$ is an infinite sequence, then $\lim \lambda_j = 0$ and

$$A = \sum_{j=1}^{\infty} \lambda_j P_j \tag{28}$$

where the series converges for the operator norm. Furthermore, $\|A - \sum_{j=1}^{n} \lambda_j P_j\| = |\lambda_{n+1}|$ for every n. We also note that $P_i P_j = 0$ for $i \neq j$ in both (I) and (II) and these projections have finite rank.

We call equation (13) in (I) and (22) in (II) the spectral decomposition of A. These equations show that a compact, self-adjoint operator is completely determined by its non-zero eigenvalues and their corresponding eigenspaces.

Theorem 4

Let A be a compact, self-adjoint operator on a Hilbert space H and let

$$A = \sum_{j=1}^{\infty} \lambda_j P_j$$

be its spectral decomposition. Then a bounded, linear operator on H commutes with A if, and only if, it commutes with every P_j, $j = 1, 2, \ldots$.

Proof. Let B be a bounded, linear operator on H and suppose that $BP_j = P_j B$ for $j = 1, 2, \ldots$. Then

$$\begin{aligned}
BA &= B(\sum_{j=1}^{\infty} \lambda_j P_j) = B[\lim_n \sum_{j=1}^{n} \lambda_j P_j] = \lim_n B(\sum_{j=1}^{n} \lambda_j P_j) \\
&= \lim_n \sum_{j=1}^{n} \lambda_j (BP_j) = \lim_n \sum_{j=1}^{n} \lambda_j (P_j B) = (\lim_n \sum_{j=1}^{n} \lambda_j P_j)B \\
&= AB
\end{aligned}$$

Next suppose that we know that $AB = BA$. Then if M_j is the eigenspace corresponding to λ_j we have, for $x \in M_j$,

$$(A - \lambda_j I)Bx = B(A - \lambda_j I)x = 0$$

and, since $B^*A = AB^*$ (because A is self-adjoint)

$$(A - \lambda_j I)B^*x = B^*(A - \lambda_j I)x = 0$$

These equations show that both B and B^* map M_j into itself; i.e., M_j is invariant under both B and B^*. Thus M_j reduces B (Section 3, Corollary 1 to Theorem 1 and Definition 1). We conclude from this that (Section 3, Corollary 1 to Theorem 2)

$$P_j B = B P_j$$

and this is true for each j.

In the next section we shall use Theorem 2 to obtain a spectral theorem for compact, normal operators.

Exercises 4

* 1. Recall the compact operator A on ℓ^2 defined by

$$A(\{z_k\}_{k=1}^\infty) = \{\frac{z_k}{k}\}_{k=1}^\infty$$

(Exercises 3.2, problem 3)

   a. Show that A is a self-adjoint operator (Exercises 4.2, problem 5)

   b. Find the spectral decomposition of A; i.e., find the eigenvalues of this operator, find the corresponding eigenspaces and the projections of ℓ^2 onto these spaces, write A as in (27).

   c. For any fixed, positive integer n show that A^n is a compact, self-adjoint operator (Exercises 4.3, problem 1(b) and Exercises 3.1, problem 2(a)). Find the spectral decomposition of A^n.

   d. Let $p(x)$ be a fixed polynomial with real coefficients and no constant term. Show that $p(A)$ is a compact, self-adjoint operator and find its spectral decomposition.

* 2. Let A be a compact, self-adjoint operator on H, let $\{\lambda_j\}$ be its non-zero, distinct, eigenvalues and let, for each j, M_j be the eigenspace corresponding to λ_j.

   a. Show that $\overline{A(H)}$, the closure of $\{Ax | x \in H\}$, is equal to $\sum_{j=1}^\infty \oplus M_j$. Show also that this is a separable subspace of H which reduces A.

   b. Show that the orthogonal complement of $\overline{A(H)}$ is $\{x \in H | Ax = 0\}$.

   c. Show that $\overline{A(H)}$ has an orthonormal basis $\{e_j\}_{i=1}^\infty$ consisting of eigenvectors of A and that

$$Ax = \sum \lambda_i \langle x, e_i \rangle e_i$$

   for all $x \in H$; here the λ_i's are no longer distinct (each eigenvalue appears as many times its multiplicity).

   d. Let M_0 be the subspace defined in (b). Show that if B is a bounded, linear operator on H and if $AB = BA$ then $BP_0 = P_0B$ where P_0 is the orthogonal projection of H onto M_0.

5.5 The Spectral Theorem: Compact Operators (continued)

Here we shall extend Theorem 4.2 to compact, normal operators. Let A be such an operator on a Hilbert space H and write

$$A = B + iC, \quad B = \frac{1}{2}(A + A^*), \quad C = \frac{1}{2i}(A - A^*)$$

It is clear that $B + iC$ is A and that B and C are self-adjoint operators. Since A is a compact operator so also is A^* (Section 4.4., Theorem 5) and hence B and C are compact, self-adjoint operators. Finally, recall that since A is a normal operator $BC = CB$ (Exercises 4.3, problem 5(b)). Now use the spectral theorem to write

$$B = \sum_{j=1}^{\infty} \alpha_j P_j, \quad C = \sum_{k=1}^{\infty} \beta_k P_k$$

where these are the spectral decompositions of the compact, self-adjoint operators B and C respectively; the cases in which one or both of these sums is finite will be left to the reader. Let M_j be the eigenspace corresponding to α_j and let N_k be the eigenspace corresponding to β_k, so P_j is the orthogonal projection of H onto M_j and Q_k is the orthogonal projection of H onto N_k; $j = 1, 2, \ldots$ and $k = 1, 2, \ldots$. We know that

$$\sum_{j=1}^{\infty} \oplus M_j \text{ and } \sum_{k=1}^{\infty} \oplus N_k$$

are subspaces of H. Let M_0, N_0 denote, respectively, their orthogonal complements and let P_0, Q_0 be, respectively, the orthogonal projections onto these subspaces.

Claim: $P_j Q_k = Q_k P_j$ for $j = 0, 1, 2, \ldots$ and $k = 0, 1, 2, \ldots$.

Proof. Since $BC = CB$ we must have, by Theorem 3 of Section 4, $CP_j = P_j C$ for $j = 0, 1, 2, \ldots$; because the P_j are the projections appearing in the spectral decomposition of B. However, this last set of equations says that each P_j commutes with every Q_k again using Theorem 3 and now the fact that the Q_k are the projections appearing in the spectral decomposition of C.

It follows from the claim we have just proved that $P_j Q_k$ is an orthogonal projection and that its range is the space $M_j \cap N_k$ (Exercises 4.2, prob-

THE SPECTRAL THEOREM: COMPACT OPERATORS (CONTINUED)

lem 2(c)). It also follows that $(P_j Q_k) \circ (P_\ell Q_m) = P_j \circ Q_k \circ Q_m \circ P_\ell = 0$ for $k \neq m$ or $j \neq \ell$. Now since

$$H = \sum_{k=0}^{\infty} \oplus N_k$$

we have, for any $x \in H$,

$$Bx = B(\sum_{k=0}^{\infty} Q_k x) = (\sum_{j=0}^{\infty} \alpha_j P_j)(\sum_{k=0}^{\infty} Q_k x) = \sum_{j=0}^{\infty}(\sum_{k=0}^{\infty} \alpha_j (P_j Q_k) x)$$

Similarly,

$$Cx = \sum_{k=0}^{\infty} \sum_{j=0}^{\infty} \beta_k (Q_k P_j) x = \sum_{k=0}^{\infty} (\sum_{j=0}^{\infty} \beta_k (P_j Q_k) x)$$

because the projections commute. Now we can interchange the order of summation in this last double sum because:

$$Cx = \sum_{j=0}^{\infty} P_j (Cx) = \sum_{j=0}^{\infty} P_j (\sum_{k=0}^{\infty} \beta_k Q_k x) = \sum_{j=0}^{\infty} \sum_{k=0}^{\infty} \beta_k (P_j Q_k) x$$

here we have used the fact that

$$H = \sum_{j=0}^{\infty} \oplus M_j$$

Combining our results:

$$Ax = (B + iC)x = \sum_{j=0}^{\infty} \sum_{k=0}^{\infty} \alpha_j (P_j Q_k) x + i \sum_{j=0}^{\infty} \sum_{k=0}^{\infty} \beta_k (P_j Q_k) x \quad (1)$$

$$= \sum_{j=0}^{\infty} (\sum_{k=0}^{\infty} (\alpha_i + i\beta_k) P_j Q_k x)$$

Of course some of the products $P_j Q_k$ may be zero. Those numbers $\alpha_j + i\beta_k$ for which the subspace $M_j \cap N_k$ is not the zero subspace are the non-zero eigenvalues of A (omitting $\alpha_0 + i\beta_0 = 0$) and we can arrange them in a sequence $\{\lambda_n\}_{n=1}^{\infty}$ such that

$$|\lambda_1| \geq |\lambda_2| \geq \ldots \quad \text{and} \quad \lim \lambda_n = 0$$

Let $\{R_n\}_{n=1}^{\infty}$ be the orthogonal projections onto the eigenspaces corresponding to the λ_n's, each $R_n = P_j Q_k$ for suitable j and k and each R_n has range $L_n = M_j \cap N_k$. Also, let R_0 be the orthogonal projection on $L_0 = (\sum_{n=1}^{\infty} \oplus L_n)^\perp$.

Then from (1) we have, for any $x \in H$, $Ax \in \sum_{n=1}^{\infty} L_n$, $R_n(Ax) = \lambda_n(R_n x)$ for $1 \leq n < \infty$

$$Ax = \sum_{n=1}^{\infty} R_n(Ax) = \sum_{n=1}^{\infty} \lambda_n(R_n x)$$

$$\|(A - \sum_{n=1}^{m} \lambda_n R_n)x\|^2 = \|\sum_{m+1}^{\infty} \lambda_n R_n x\|^2 = \sum_{m+1}^{\infty} |\lambda_n|^2 \|R_n x\|^2$$

$$\leq |\lambda_{m+1}|^2 \sum_{n=0}^{\infty} \|R_n x\|^2 = |\lambda_{m+1}|^2 \|x\|^2$$

If we choose x to be an element of R_{m+1} then

$$\|(A - \sum_{n=1}^{m} \lambda_n R_n)x\|^2 = \|A\|^2 = |\lambda_{m+1}|^2 \|x\|^2$$

Thus

$$\|A - \sum_{n=1}^{m} \lambda_n R_n\| = |\lambda_{m+1}|, \quad A = \sum_{n=0}^{\infty} \lambda_n R_n$$

There is one point we have to prove. Suppose that x is an eigenvector of A corresponding to the eigenvalue λ. Then

$$x = \sum_{n=0}^{\infty} R_n x \neq 0$$

and

$$(A - \lambda I)x = \sum_{n=0}^{\infty} (\lambda_n - \lambda) R_n x = 0$$

Now we cannot have $R_n x = 0$ for all n since our first sum is $\neq 0$. It follows that for this n we must have $\lambda_n - \lambda = 0$; for $R_n x \in L_n$ and these spaces are pairwise orthogonal. Thus $\lambda = \lambda_n$ and $R_m x = 0$ for $m \neq n$, $R_n x = \lambda_n x \in L_n$. This shows that $\{\lambda_n\}_{n=1}^{\infty}$ contains all non-zero eigenvalues of A and that $\{L_n\}_{n=1}^{\infty}$ is the sequence of eigenspaces.

We may now state:

Theorem 1

(Spectral Theorem for Compact, Normal Operators). Let A be a compact, normal operator on a Hilbert space H. Let $\{\lambda_n\}_{n=1}^{\infty}$ be the sequence of non-zero eigenvalues of A labeled in such a way that

$$|\lambda_1| \geq |\lambda_2| \geq |\lambda_3| \geq \ldots, \quad \lambda_j \neq \lambda_k \text{ for } j \neq k, \quad \lim_n \lambda_n = 0$$

THE SPECTRAL THEOREM: COMPACT OPERATORS (CONTINUED)

and let P_n be the orthogonal projection of H onto M_n (the eigenspace corresponding to λ_n). Then each P_n has finite rank, $P_n P_m = 0$ for $n \neq m$, and

$$A = \sum_{n=1}^{\infty} \lambda_n P_n$$

where the series converges for the operator norm. Furthermore, $\|A - \sum_{n=1}^{m} \lambda_n P_n\| = |\lambda_{m+1}|$ for all m.

For any operator A we can set $B = \frac{1}{2}(A + A^*)$ and $C = \frac{1}{2i}(A - A^*)$ and have $A = B + iC$ where B and C are self-adjoint operators. When A is also a compact operator, then B and C are compact and so it may seem that we can get a spectral theorem for any compact operator. This is not true (see problem 1). In proving Theorem 1 we made use of the fact that all of the projections in the spectral decomposition of B commute with all of the projections in the spectral decomposition of C and this was implied by the fact that $BC = CB$. This last fact is true when, and only when, A is a normal operator. The next theorem shows that we can still obtain useful information from the representation $A = B + iC$ even when A is not normal.

Theorem 2

A bounded, linear operator on a Hilbert space is a compact operator if, and only if, it is the limit (for the operator norm) of a sequence of operators of finite rank.

Proof. Let A be a compact operator on a Hilbert space H and write $A = B + iC$ where, as we have noted several times above, B and C are compact, self-adjoint operators. By Theorem 4.2 both B and C have spectral decompositions:

$$B = \sum_{j=1}^{\infty} \alpha_j P_j \qquad C = \sum_{k=1}^{\infty} \beta_k Q_k$$

where each P_j and each Q_k is an orthogonal projection having finite rank and our series both converge for the operator norm. For any fixed integer n let set

$$A_n = \sum_{j=1}^{n} \alpha_j P_j + i \sum_{k=1}^{n} \beta_k Q_k$$

and note that each A_n has finite rank, because our sums are finite. Furthermore, $\{A_n\}$ converges to A for the operator norm. This proves that any compact operator on H is the limit of a sequence of operators having finite rank. The converse follows from Theorem 2 of Section 3.2.

Remark. An operator on a Banach space which is the limit, for the operator norm, of a sequence of operators having finite rank is certainly a compact operator. For many years, it was not known if the converse was true. Finally, in 1973, Enflo showed that a compact operator on a Banach space need *not* be the limit of a sequence of operators having finite rank [4].

Exercises 5

1. Let H be a Hilbert space and let $\{P_j\}_{j=1}^{\infty}$ be a sequence of non-zero, orthogonal projections on H each having finite rank. Suppose that $P_j P_k = 0$ for $i \neq k$, and let $\{\lambda_j\}_{j=1}^{\infty}$ be a sequence of distinct (i.e., $\lambda_j \neq \lambda_k$ if $j \neq k$) non-zero complex numbers such that $|\lambda_1| \geq |\lambda_2| \geq \ldots$ and $\lim \lambda_j = 0$.

   a. Show that $\sum_{j=1}^{\infty} \lambda_j P_j$ converges for the operator norm to a compact, normal operator (which we shall call A) on H. Thus the spectral decomposition described in Theorem 4.2 and Theorem 1 are possible only for compact normal operators.

   b. Show that $\sum_{j=1}^{\infty} \lambda_j P_j$ is the spectral decomposition of the operator A.

   c. Show that A is self-adjoint if, and only if, each λ_j is a real number.

* 2. Let A be a compact, normal operator on H and let $A = \sum_{n=1}^{\infty} \lambda_n P_n$ be its spectral decomposition.

   a. Find the spectral decomposition of the compact, normal operator A^*.

   b. Let T be a bounded, linear operator on H which commutes with every P_n. Show that T commutes with both A and A^*.

5.6 Fredholm Integral Equations

Let us recall that an integral equation in which the unknown function occurs under the integral sign and nowhere else is called an equation of the first kind. We shall prove an interesting theorem about this type of equation very soon, but for now let us look at those equations which are not of the first kind; i.e., equations of the second kind. We want to use our considerable knowledge of compact operators to study these equations. Again recall that a, b are real numbers with $a < b$ and $R = [a,b] \times [a,b]$. For any fixed $K(s,t) \in L^2[R]$ we

define
$$K(f)(s) = \int_a^b K(s,t)f(t)dt \qquad (1)$$

for any $f \in L^2[a,b]$. Then $K(f) \in L^2[a,b]$ and the operator T which takes f to $K(f)$ is a compact, linear operator on this space which we call the Hilbert-Schmidt operator with kernel $K(s,t)$. Also

$$f(s) - \lambda T(f)(s) = g(s) \qquad (2)$$

where $g \in L^2[a,b]$ is known and $\lambda \neq 0$ is a fixed scalar, is a typical Fredholm equation of the second kind (Section 3.3, Remark 2). Observe once again that the position of the λ, natural enough in this context, causes some minor problems. In the notation of operator theory (2) becomes:

$$(\lambda^{-1}I - T)f = \lambda^{-1}I(g) \qquad (3)$$

and we say that λ is a characteristic value of the equation (2) if, and only if, λ^{-1} is an eigenvalue of the operator T. Since T is compact the set of all characteristic values of (2) is either a finite set or it is a sequence of complex numbers whose absolute values tend to infinity. We might also recall that when $|\lambda|\|T\| < 1$ equation (2) has a unique solution for any given g and in fact (Section 2.3, Theorem 2)

$$f(s) = \sum_{n=0}^{\infty} \lambda^n T^n(g) \qquad (4)$$

Theorem 1

(Fredholm Alternative). Let T be a Hilbert-Schmidt operator and let $\lambda \neq 0$ be any complex number. We refer to equation (2). We have: (i) The given complex number λ is either a characteristic value of (2) or it is a regular value of this equation (i.e., λ^{-1} is either an eigenvalue of the operator T or it is a regular value of this operator); (ii) If λ is a characteristic value of (2) then the null space of $(I - \overline{\lambda}T^*)$ is a finite dimensional subspace of $L^2[a,b]$ and (2) has a solution if, and only if, the given function g is in the orthogonal complement of this null space.

Proof. Our reformulation of (i) in terms of the operator T renders this statement obvious. Suppose that λ is a characteristic value of (2) so that $\mu = \lambda^{-1}$ is an eigenvalue of the compact operator T. Thus $\overline{\mu}$ must be an eigenvalue of T^* (Exercises 4.3, problem 1(e)). But T^* is a compact operator (Section 4.4, Theorem 5) and so the null space of $(\overline{\mu}I - T^*)$, call it $N(\overline{\mu}I - T^*)$, is finite dimensional (Section 3.3, Theorem 1). However,

$$N(\overline{\mu}I - T^*) = N[\overline{\lambda}(\overline{\mu}I - T^*)] = N[\overline{\lambda}(\overline{\lambda}^{-1}I - T^*)] = N(I - \overline{\lambda}T^*)$$

and so this last space is also finite dimensional. Finally, (2) has a solution if, and only if, g is in $R(I - \lambda T)$; the range of $I - \lambda T$. Let us show that this range is equal to $N(I - \overline{\lambda}T^*)^\perp$. We shall prove, immediately below, that for any bounded, linear operator S on a Hilbert space H,

$$\overline{R(S)}^\perp = N(S^*)$$

(Section 3.1, Definition 2). Hence, since T is compact, we have

$$R(I - \lambda T) = N(I - \overline{\lambda}T^*)^\perp$$

because this range is closed (Section 3.3, Corollary 1 to Lemma 1).

Lemma 1

Let H be a Hilbert space and let S be a bounded, linear operator on H. Then $\overline{R(S)}^\perp = N(S^*)$ and $\overline{R(S)} = N(S^*)^\perp$.

Proof. Let $y \in \overline{R(S)}^\perp$. Equivalently, $0 = \langle Sx, y \rangle$ for all x and this is equivalent to $\langle x, S^*y \rangle = 0$ for all x. But the latter is true for all x if, and only if, $S^*y = 0$; i.e., if, and only if, $y \in N(S)$. The second statement follows from the first and the fact that

$$\overline{R(S)}^{\perp\perp} = \overline{R(S)}$$

(Section 4.2, Lemma 1).

Once again let us return to the integral equation

$$f(s) - \lambda T(f)(s) = g(s)$$

where T is a Hilbert-Schmidt operator with kernel $K(s,t) \in L^2[R]$. Suppose now that T is self-adjoint. Then

$$K(s,t) = \overline{K(t,s)} \quad \text{a.e.}$$

by (Exercises 4.2, problem 7(b)). Let $\{\lambda_n^{-1}\}$ be an enumeration of the eigenvalues of T, which we know must be real since $T = T^*$ (Exercises 4.2, problem 3), counted according to their multiplicity; recall that the multiplicity of an eigenvalue is the dimension of the corresponding eigenspace, and since T is compact this must be finite, so we count an eigenvalue with multiplicity p, p-times — this was suppressed in our discussion of the spectral theorem since there we considered only the distinct eigenvalues (see also the example in Section 2, especially equations (5) and (6)). Next let $\{u_n\}$ be the eigenfunctions corresponding to $\{\lambda_n^{-1}\}$ and write, as we may by the spectral theorem,

$$(*) \qquad T(f)(s) = \sum_{n=1}^\infty \lambda_n^{-1} \langle f, u_n \rangle u_n(s)$$

(see Exercises 4, problem 2(c)) where convergence is for the L^2-norm. We would like to see what we can say when we know that the function $K(s,t)$ is continuous and not just in L^2. The first thing we note is that $T(f)$ is a continuous function even when $f \in L^2[a,b]$. To see this we observe that

$$|T(f)(s_1) - T(f)(s_2)| \leq \int_a^b |K(s_1,t) - K(s_2,t)||f(t)|dt$$
$$\leq \|f\|_2 \max |K(s_1,t) - K(s_2,t)|$$

and use the fact that K is continuous. It follows from this that the eigenfunctions corresponding to any of our eigenvalues are continuous and so we may regard T as an operator on the Banach space $(C[a,b], \|\cdot\|_\infty)$.

Theorem 2

(Hilbert-Schmidt) The series $(*)$ converges to $T(f)(s)$ uniformly over $[a,b]$.

Proof. For each fixed $s \in [a,b]$ and any integers m, p we have:

$$\sum_{n=m}^{m+p} |\lambda_n^{-1} \langle f, u_n \rangle u_n(s)| \leq (\sum_{n=m}^{m+p} |\langle f, u_n \rangle|^2)^{1/2} (\sum_{n=m}^{m+p} |\lambda_n^{-1} u_n(s)|)^{1/2}$$

by (Section 1.4, Theorem 1). Now since

$$\lambda_n^{-1} u_n(s) = \int_a^b K(s,t) u_n(t) dt$$

we see that, for each fixed s, $\lambda_n^{-1} u_n(s)$ is the nth coefficient in the expansion of the L^2-function $K(s,\cdot)$, L^2-as a function of the missing variable, in terms of the orthogonal set $\{u_n\}$ (see Section 1.8, Theorem 7 and Section 3, Theorem 3). Now K is continuous on $R = [a,b] \times [a,b]$ so it is bounded on this set. Let M be a bound for it. Then, by Bessel's inequality (Section 1.6, Theorem 2) we may write

$$\sum_{n=m}^{m+p} |\lambda_n^{-1} u_n(s)|^2 \leq \int_a^b |K(s,t)|^2 dt \leq M^2(b-a)$$

Combining our two inequalities we have

$$(\sum_{n=m}^{m+p} |\lambda_n^{-1} \langle f, u_n \rangle u_n(s)|)^2 \leq M^2(b-a) \sum_{n=m}^{m+p} |\langle f, u_n \rangle|^2$$

However, again using Bessel's inequality this time on the last sum, we see that

$$lim_{m \to \infty} \sum_{n=m}^{m+p} |\langle f, u_n \rangle|^2 = 0$$

and so
$$\lim_{m\to\infty} \sum_{n=m}^{m+p} |\lambda_n^{-1}\langle f, u_n\rangle u_n(s)| = 0$$

and note that this holds uniformly in s because the right-hand side of our last inequality does not depend on s. Thus the partial sums of the series in (*) form a Cauchy sequence in $(C[a,b], \|\cdot\|_\infty)$ and hence they converge to a function in this space. But the series in (*) converges to $T(f)(s)$ for the L^2-norm and, since $C[a,b] \subseteq L^2[a,b]$, $T(f)(s)$ must be equal to this continuous function a.e., hence they must be identical; because, as we observed just above, $T(f)(s)$ is continuous.

In those cases where the eigenvalues and eigenfunctions of our self-adjoint Hilbert-Schmidt operator are known we can use (*) to solve equation (2) as follows:
$$f(s) - \lambda T(f)(s) = g(s)$$
$$(*) \quad T(f)(s) = \sum_{n=1}^{\infty} \lambda_n^{-1}\langle f, u_n\rangle u_n(s)$$

Put (*) into (2) to get
$$(**) \quad f(s) = \lambda \sum_{n=1}^{\infty} C_n u_n(s) + g(s)$$

where $C_n = \lambda_n^{-1}\langle f, u_n\rangle$ for $n = 1, 2, \ldots$. Now we are trying to find f and to do this it suffices to find each C_n. Take the inner product of both sides of (**) with the function u_m
$$\langle f, u_m\rangle = \lambda C_m + \langle g, u_m\rangle$$
and note that $\langle f, u_m\rangle = \lambda_m C_m$ so
$$(***) \quad C_m = \frac{\langle g, u_m\rangle}{\lambda_m - \lambda}, \quad m = 1, 2, \ldots$$

Here, of course, we must suppose that λ is not a characteristic value; i.e., $\lambda \neq \lambda_m$, $m = 1, 2, \ldots$. Using these value of C_m we see that (**) is the solution to (2). Moreover, if K is continuous, then (**) converges uniformly to f.

Now what happens if the λ in (2) is equal to, say, λ_i? In this case the Fredholm alternative (Theorem 1) tells us that (2) has a solution if, and only if, $g \in N(I - \lambda_i T)^\perp$ (remember T is a self-adjoint so λ_i is real). When g is in this space then in (**) we may take for the coefficients of those eigenfunctions in $N(I - \lambda_i T)$ any numbers and for the remaining eigenfunctions we take the coefficients given by (***). Thus in this case the solutions of (2) form an n-parameter family of functions, here $n = \dim(I - \lambda_i T)$, and they are given by (**).

Let us end this section with a brief discussion of Fredholm equations of the first kind. These are of the form

$$T(f)(s) = g(s) \tag{5}$$

where $g \in L^2[a,b]$ is known and T is a Hilbert-Schmidt operator with kernel $K \in L^2[R]$. Now T is a compact operator hence so also is T^*T and this last operator is self-adjoint. Also

$$\langle T^*T(f), f \rangle = \langle T(f), T(f) \rangle = \|T(f)\|^2 \geq 0$$

for all $f \in L^2[a,b]$. It follows that the eigenvalues of this product operator, which we know must be real numbers, are positive numbers. Let us arrange then, counted according to their multiplicities as a decreasing sequence:

$$\mu_1 \geq \mu_2 \geq \cdots$$

Now define, for each n, λ_n by the equation $\lambda_n^2 = \mu_n$. Then

$$\lambda_1^2 \geq \lambda_2^2 \geq \lambda_3^2 \geq \ldots, \quad (T^*T)v_n = \lambda_n^2 v_n$$

where, of course, v_n represents an eigenfunction corresponding to the eigenvalue λ_n^2 and we may, and do, assume that $\{v_n\}$ is an orthonormal set (Section 3, Theorem 3). Define

$$u_n(s) = \lambda_n^{-1} T(v_n)(s), \quad n = 1, 2, 3, \ldots \tag{6}$$

We first observe that $\{u_n\}$ is an orthonormal set because

$$\begin{aligned}
\langle u_n, u_m \rangle &= \langle \lambda_n^{-1} T(v_n), \lambda_m^{-1} T(V_m) \rangle = \\
&= (\lambda_n \lambda_m)^{-1} \langle v_n, T^*T(v_m) \rangle = (\lambda_n \lambda_m)^{-1} \langle v_n, \lambda_m^2 v_m \rangle \\
&= \lambda_n^{-1} \lambda_m \langle v_n, v_m \rangle = 0 \text{ when } n \neq m
\end{aligned}$$

and

$$\begin{aligned}
\langle u_n, u_n \rangle &= \langle \lambda_n^{-1} T(v_n), \lambda_n^{-1} T(v_n) \rangle = \lambda_n^{-2} \langle T(v_n), T(v_n) \rangle \\
&= \lambda_n^{-2} \langle v_n, T^*T(v_n) \rangle = \lambda_n^{-2} \langle v_n, \lambda_n^2 v_n \rangle = \langle v_n, v_n \rangle = 1
\end{aligned}$$

Theorem 3

(Picard) In order that equation (5) have a solution in $L^2[a,b]$ it is necessary and sufficient that $g \in N(T^*)^\perp$ and that $\sum_{n=1}^\infty \lambda_n^{-1} |\langle g, u_n \rangle|^2$ be convergent.

Proof. Suppose first that there is a function $f \in L^2[a,b]$ such that $T(f) = g$. Then clearly $g \in R(T)$ and so $g \in N(T^*)^\perp$ (see Lemma 1). We may write

$$f(s) = h(s) + \sum_{n=1}^{\infty} c_n v_n(s) \qquad (7)$$

where $h \in N(T^*T)$ and $c_n = \langle f, v_n \rangle$. Using (7), the fact that $T^*T(f) = T^*(g)$, and Bessel's inequality (Section 1.6, Theorem 2) we have:

$$\begin{aligned}
\sum_{n=1}^{\infty} \lambda_n^{-2} |\langle g, v_n \rangle|^2 &= \sum_{n=1}^{\infty} \lambda_n^{-2} |\langle g, \lambda_n^{-1} T(v_n) \rangle|^2 = \sum_{n=1}^{\infty} \lambda_n^{-4} |\langle g, T(v_n) \rangle|^2 \\
&= \sum_{n=1}^{\infty} \lambda_n^{-4} |\langle T^*(g), v_n \rangle|^2 = \sum_{n=1}^{\infty} \lambda_n^{-4} |\langle (T^*T)(f), v_n \rangle|^2 \\
&= \sum_{n=1}^{\infty} \lambda_n^{-4} |\langle T(f), T(v_n) \rangle|^2 \\
&= \sum_{n=1}^{\infty} \lambda_n^{-4} |\langle f, T^*T(v_n) \rangle|^2 = \sum_{n=1}^{\infty} \lambda_n^{-4} |\langle f, \lambda_n^2 v_n \rangle|^2 \\
&= \sum_{n=1}^{\infty} |\langle f, v_n \rangle|^2 = \sum_{n=1}^{\infty} |c_n|^2
\end{aligned}$$

$\leq \|f\|_2^2$ showing that our series is convergent.

Now suppose that our two conditions are satisfied. Equation (7) shows that the range of T is spanned by

$$T(f) = T(h) + \sum_{n=1}^{\infty} c_n T(v_n) = T(h) + \sum_{n=1}^{\infty} c_n \lambda_n u_n$$

So if $\langle f, u_n \rangle = 0$ for all n then $T^*(f) = T^*(T(h)) = 0$ so $f \in N(T^*)$; we chose $h \in N(T^*T)$. Now $\{u_n\}$ is an orthonormal set and so

$$\sum_{j=1}^{\infty} \langle g, u_j \rangle u_j$$

converges in $L^2[a,b]$ and

$$\langle g - \sum_{j=1}^{\infty} \langle g, u_j \rangle u_j, u_n \rangle = 0 \text{ for all } n$$

hence

$$g - \sum_{j=1}^{\infty} \langle g, u_j \rangle u_j$$

is an element of $N(T^*)$. But, by hypothesis, $g \in N(T^*)^\perp$ and clearly $u_j \in R(T) \subseteq N(T^*)^\perp$ for any j. Thus our difference, g minus its series expansion in $\{u_n\}$, is in both $N(T^*)$ and $N(T^*)^\perp$. It follows that

$$g = \sum_{j=1}^{\infty} \langle g, u_j \rangle u_j$$

Now we are assuming that

$$\sum_{n=1}^{\infty} \lambda_n^{-2} |\langle g, u_n \rangle|^2 < \infty$$

so we may define an element $f \in L^2[a,b]$ by setting

$$f(s) = \sum_{n=1}^{\infty} \lambda_n^{-1} \langle g, u_n \rangle v_n(s)$$

(see Exercises 1.6, problem 4). Then

$$T(f) = \sum_{n=1}^{\infty} \lambda_n^{-1} \langle g, u_n \rangle T(v_n) = \sum_{n=1}^{\infty} \lambda_n^{-1} \langle g, u_n \rangle (\lambda_n u_n) = \sum_{n=1}^{\infty} \langle g, u_n \rangle u_n = g$$

Thus f is a solution to (5) for the given g.

Exercises 6

1. Let S be a bounded, linear operator on a Hilbert space H.

   a. Show that $\overline{R(S^*)}^\perp = N(S)$ and that $\overline{R(S^*)} = N(S)^\perp$.

   b. If S is a normal operator (i.e., $SS^* = S^*S$) show that $\overline{R(S)} = \overline{R(S^*)}$.

2. Let T be a compact operator on H. Show that T^*T is a self-adjoint operator whose eigenvalues are positive numbers.

5.7 Some Further Remarks on Integral Operators

Any integral operator whose kernel is defined and continuous on $[a,b]$, where $b-a < \infty$, is a Hilbert-Schmidt operator. Thus its non-zero spectrum consists entirely of eigenvalues, there are only countably many such values and each of them has finite multiplicity. Here we shall examine two integral operators with continuous kernels which are defined on infinite intervals.

(a) Our first example is due to Picard. For any real α we may write:

$$\int_{-\infty}^{\infty} e^{-|s-t|} e^{i\alpha t} dt = \int_{-\infty}^{s} e^{-(s-t)+i\alpha t} dt + \int_{s}^{\infty} e^{-(t-s)+i\alpha t} dt$$

$$= \frac{e^{-(s-t)+i\alpha t}}{1+i\alpha} \Big|_{-\infty}^{s} + \frac{e^{-(t-s)+i\alpha t}}{-1+i\alpha} \Big|_{s}^{\infty}$$

$$= \frac{e^{(i\alpha s)}}{1+i\alpha} - \frac{e^{i\alpha s}}{-1+i\alpha} = \left(\frac{2}{1+\alpha^2}\right) e^{i\alpha s}$$

Thus if we set $K(s,t) = e^{-|s-t|}$, $-\infty < s, t < \infty$, then

$$\int_{-\infty}^{\infty} K(s,t) e^{i\alpha t} = \left(\frac{2}{1+\alpha^2}\right) e^{i\alpha t} = \lambda e^{i\alpha t} \tag{1}$$

We see here that each number

$$\lambda = \frac{2}{1+\alpha^2}, \quad \alpha \text{ real}$$

is an eigenvalue for the operator with kernel K on $(-\infty, \infty)$. This is the set $(0, 2]$ which is not a countable set.

(b) Our second example is due to Weyl. This is a little more complicated. We must refer to some well-known integration formulas. First

$$\int e^{px} \sin qx \, dx = e^{px} \frac{p \sin qx - q \cos qx}{p^2 + q^2}$$

and from this we get

$$\int_{0}^{\infty} e^{-at} \sin st \, dt = \frac{s}{a^2 + s^2} \quad a > 0 \tag{2}$$

Next, for $p > 0$,

$$\int_{0}^{\infty} \frac{\cos px}{1+x^2} dx = \frac{\pi}{2} e^{-p}$$

and differentiating this with respect to p gives us

$$\int_{0}^{\infty} \frac{t}{a^2 + t^2} \sin st \, dt = \frac{\pi}{2} e^{-as} \quad a, s > 0 \tag{3}$$

Combining (2) and (3) we obtain

$$\int_{0}^{\infty} \sqrt{\frac{2}{\pi}} \sin st \left(\sqrt{\frac{2}{\pi}} e^{-at} + \frac{t}{a^2+t^2}\right) dt = \sqrt{\frac{2}{\pi}} e^{-as} + \frac{s}{a^2+s^2} \tag{4}$$

or, letting $K(s,t) = \sqrt{\frac{2}{\pi}} \sin st$,

$$\int_0^\infty K(s,t) \left(\sqrt{\frac{2}{\pi}} e^{-at} + \frac{t}{a^2+t^2} \right) dt = \frac{\pi}{2} e^{-as} + \frac{s}{a^2+s^2} \quad (5)$$

This tells us that $\lambda = 1$ is an eigenvalue of the operator with kernel K on $[0, \infty)$, but corresponding to this eigenvalue there are uncountably many eigenvectors; one for each $a > 0$.

Let us look now at some nonlinear functional analysis and apply it to differential equations. To begin with consider $(C[0,1], \|\cdot\|_\infty)$, the Banach space of continuous functions on $[0,1]$ with the sup. norm (Section 1.2, equation (4)). Fix two numbers y_0 and ℓ and consider $C^* = \{f \in C[0,1] | \|f(x) - y_0\|_\infty \leq \ell\}$. This is not a linear manifold in $C[0,1]$; for given $f \in C^*$ the functions αf is not in C^* when α is large enough. It is a closed set however: If $\{f_n\} \subseteq C^*$ and $f_n \to f_0$ for the sup. norm, then f_0 certainly is in $C[0,1]$ and, furthermore,

$$\|f_0(x) - y_0\|_\infty = \lim \|f_n(x) - y_0\|_\infty \leq \ell$$

by problem (1) exercises 1.3 (closed sets were defined in Section 2.2, Definition 2 and we have used here Theorem 1 of Section 2.4).

Now since C^* is not a linear manifold we can't use $\|\cdot\|_\infty$ to put a norm on this set. We can use it to put a "metric" on C^* and this turns out to be quite useful.

Definition 1

Let S be a non-empty set and let ρ be a non-negative real-valued function defined on $S \times S = \{(s,t) | s,t \in S\}$. We shall say that ρ is a metric on S if: (a) $\rho(s,t) = 0$ if, and only if, $s = t$; (b) $\rho(s,t) = \rho(t,s)$ for all s,t in S; (c) $\rho(s,u) \leq \rho(s,t) + \rho(t,u)$ for all s,t,u in S.

The reader is probably familiar with many metrics. A simple example is this: Let $(E, \|\cdot\|)$ be a normed space, then $\rho(x,y) = \|x - y\|$ is a metric on E. Note that ρ gives us a metric on any subset of E. In particular, the metric defined on $C[0,1]$ by $\|\cdot\|_\infty$ gives us a metric on C^*.

Convergent sequences (Section 1.3, Definition 2) can be defined for metric spaces. A metric space in which every Cauchy sequence of points of the space converges to a point of the space is called a complete metric space.

It is easy to check that when $(E, \|\cdot\|)$ is a Banach space then E with the metric $\rho(x,y) = \|x - y\|$ is a complete metric space. Thus $(C[0,1], \|\cdot\|_\infty)$ gives us the complete metric space $(C[0,1], \rho)$ where $\rho(f,g) = \|f - g\|_\infty$. Also, since C^* is a closed subset of $C[0,1]$, (C^*, ρ) is a complete metric space (Exercises 2.4, problem 2(d)).

Definition 2

Let (S, ρ) be a metric space and let T be a mapping from S onto S. We shall say that T is a contraction of S if there is a number α, $0 < \alpha < 1$, such that

$$\rho(Ts, Tt) \leq \alpha \rho(s, t) \text{ for all } s, t \in S$$

Theorem 1

Suppose that (S, ρ) is a complete metric space and that T is a contraction of S. Then there is one, and only one, fixed point of T in S; i.e., one, and only one, point $s \in S$ such that $Ts = s$.

Proof. Choose any point $s_0 \in S$ and construct a sequence

$$s_0, s_1 = Ts_0, \quad s_2 = Ts_1 = T^2 s_0, \quad s_3 = Ts_2 = T^3 s_0, \ldots$$

Let us show that $\{s_n\}_{n=0}^{\infty}$ is a Cauchy sequence in S. For any integers m, n with $n < m$ we have

$$\begin{aligned}
\rho(s_n, s_m) &\leq \rho(s_n, s_{n+1}) + \rho(s_{n+1}, s_{n+2}) + \cdots + \rho(s_{m-1}, s_m) \\
&\leq \rho(T^n s_0, T^n s_1) + \rho(T^{n+1} s_0, T^{n+1} s_1) + \cdots + \rho(T^{m-1} s_0, T^{m-1} s_1) \\
&\leq (\alpha^n + \alpha^{n+1} + \cdots + \alpha^{m-1}) \rho(s_0, s_1) \\
&\leq \left(\sum_{k=0}^{\infty} \alpha^{n+k} \right) \rho(s_0, s_1) = \frac{\alpha^n}{1 - \alpha} \rho(s_0, s_1)
\end{aligned}$$

since $0 < \alpha < 1$ our geometric series is convergent with known sum. However, $\lim \alpha^n = 0$ as $n \to \infty$ because $0 < \alpha < 1$ and this shows that $\{s_n\}$ is a Cauchy sequence.

Now (S, ρ) is a complete metric space hence $\{s_n\}$ must converge to a point $s \in S$. We see from Definition 2 that

$$\lim Ts_n = Ts$$

or

$$\lim s_{n+1} = Ts$$

But $\lim s_{n+1} = s$ and so we must have $s = Ts$. Thus T has a fixed point.

Finally, suppose that $t \in S$ is any fixed point of T. Then

$$\rho(s, t) \leq \rho(Ts, Tt) \leq \alpha \rho(s, t)$$

which, again because $0 < \alpha < 1$, tells us that $\rho(s, t) = 0$. We conclude that $s = t$.

We shall apply Theorem 1 to show that, under reasonable conditions, the differential equation with initial condition

$$y' = f(x,y), \quad y(x_0) = y_0$$

has a unique solution. Note that $\sigma(x)$ satisfies our problem if

$$\sigma'(x) = f[x, \sigma(x)], \quad \sigma(x_0) = y_0$$

or, equivalently, if σ satisfies

$$(*) \quad \sigma(x) = y_0 + \int_{x_0}^{x} f[t, \sigma(t)] dt$$

The trick we shall apply is to arrange things to that the integral operator in (*) is a contraction of a complete metric space.

Theorem 2

Suppose that $f(x,y)$ is a function which is continuous on some rectangle G and satisfies the Lipschitz condition

$$|f(x, y_1) - f(x, y_2)| \leq M|y_1 - y_2|$$

in G; here M is a constant. Let (x_0, y_0) be a point in the interior of G. Then there is a $\delta > 0$ and a unique solution σ to (*) defined on $(x_0 - \delta_1, x_0 + \delta)$.

Proof. The function f is bounded on G (a continuous function on a compact set is bounded) and so there is a k such that $|f(x,y)| \leq k$ for all x, y in G. Choose $\delta > 0$ so that:

i. $|x - x_0| < \delta$ and $|y - y_0| < k\delta$ implies (x,y) is in the interior of G;

ii. $M\delta < 1$.

Next let C^* denote the space of all functions σ which are continuous on $|x - x_0| \leq \delta$ and satisfy $|\sigma(x) - y_0| \leq k\delta$. It is easy to see (see our discussion above) that the sup norm gives us a metric ρ on C^* and that (C^*, ρ) is a complete metric space. Define a mapping T from C^* to C^* as follows: Given $\sigma \in C^*$, $T\sigma = \chi$ where

$$\chi(x) = y_0 + \int_{x_0}^{x} f[t, \sigma(t)] dt$$

It is clear that χ, as just defined, is continuous on $|x - x_0| \leq \delta$ and, furthermore,

$$|\chi(x) - y_0| \leq \int_{x_0}^{x} |f[t, \sigma(t)]| dt \leq k|x - x_0| \leq k\delta$$

hence χ does belong to C^*; i.e., T is a mapping from C^* to C^*. Next we show that T is a contraction.

Suppose that $T\sigma_1 = \chi_1$, $T\sigma_2 = \chi_2$. Then

$$\chi_2(x) - \chi_1(x) = \int_{x_0}^{x} (f[t, \sigma_1(t)] - f[t, \sigma_2(t)])dt$$

$$|\chi_2(x) - \chi_1(x)| \leq \int_{x_0}^{x} M|\sigma_1(t) - \sigma_2(t)|dt$$

$$\leq M \max |\sigma_1(t) - \sigma_2(t)||x - x_0|$$

$$\leq (M\delta)\rho(\sigma_1, \sigma_2)$$

by our Lipschitz condition and the definition of ρ. Now $M\delta < 1$ hence this last inequality says

$$\rho(T\sigma_1, T\sigma_2) \leq \alpha\rho(\sigma_1, \sigma_2)$$

where $\alpha = M\delta < 1$; i.e., T is a contraction. By Theorem 1 there is a unique point $\sigma(x) \in C^*$ such that $T\sigma = \sigma$ which translates into

$$\sigma(x) = y_0 + \int_{x_0}^{x} f[t, \sigma(t)]dt$$

Thus σ is the unique solution to our differential equation with initial condition.

Remark. Peano has shown that our differential equation with initial condition has a solution even if we only assume that f is continuous. However, the solution is not unique in that case.

6

THE SPECTRAL THEOREM, PART II

Roughly speaking, the spectral theorem proved in the last chapter tells us that we can write any compact, normal operator as an infinite series of orthogonal projections each multiplied by an eigenvalue of the operator. We first proved the theorem for compact, self-adjoint operators because we could show that this type operator had a non-trivial, real spectrum — remember the operator whose spectrum was $\{0\}$, there is no hope of getting a spectral theorem for such an operator. Next we extended our theorem to compact, normal operators by using the fact that any such operator could be written as a linear combination of two commuting compact self-adjoint operators. We turn now to the class of bounded, linear operators — non-compact operators. Again we look first at the self-adjoint operators. It turns out that the spectrum of such an operator is non-empty and that it is contained in some closed, bounded interval on the real line. Now there are immediate complications here. First the spectrum is no longer a countable set and, furthermore, the non-zero points of the spectrum need not be eigenvalues of the operator. Still, we can write our operator as a "sum" of projections provided we understand that the infinite series obtained in the case of a compact operator (whose spectrum is a countable set) is here replaced by a process of integration (the "analogue" of summation in an uncountable situation). Once we do that, and it takes a lot of doing, we can extend the result to normal operators.

Needless to say, there are a great many complications in this, the general, case that did not arise before. So there are many technical matters that must be dealt with before we can prove our main result. However, many of these are quite interesting and useful in their own right.

6.1 The Spectrum of a Self-adjoint Operator

Let us recall (Exercises 4.2, problem 3) that the eigenvalues of a self-adjoint operator are real numbers. However, a self-adjoint operator need not have any eigenvalues. An example is the multiplication operator on $L^2[a,b]$ (Section 2.4(b)) which was shown to be self-adjoint in (Exercises 4.2, problem 4). See also (Section 2.4 Definition 1 and the discussion just after it) where generalized eigenvalues are defined and it is shown that the spectrum of the multiplication operator consists entirely of such points.

Theorem 1

If T is a bounded, self-adjoint operator on a Hilbert space H, then

$$\|T\| = \sup\{|\langle Tz, z\rangle| | z \in H \text{ and } \|z\| = 1\}$$

Proof. Recall that $\|T\|$ was defined to be (Section 2.1, Definition 2)

$$\sup\{\|Tz\| | z \in H \text{ and } \|z\| = 1\}$$

Note that, since T is self-adjoint,

$$\langle Tz, z\rangle = \langle z, T^*z\rangle = \langle z, Tz\rangle = \overline{\langle Tz, z\rangle}$$

hence $\langle Tz, z\rangle$ is a real number for every vector $z \in H$. Also, by the C.S.B. inequality (Section 1.4, Theorem 1)

$$|\langle Tz, z\rangle| \leq \|T\|$$

when $\|z\| = 1$. Thus if we denote the supremum (in the hypothesis of the theorem) by M, then

$$M \leq \|T\|.$$

In order to show the reverse inequality we first choose and fix $z \in H$ such that $\|z\| = 1$ and $Tz \neq 0$; since our result is trivial when T is the zero operator we may, and do, assume that this is not the case here. Next we set $\beta = \|Tz\|^{1/2}$, we set $x = \beta z + \frac{Tz}{\beta}$, and we set $y = \beta z - \frac{Tz}{\beta}$. Then

$$\|x\|^2 + \|y\|^2 = \frac{1}{2}\{\|x+y\|^2 + \|x-y\|^2\} = 2\beta^2 + \frac{2\|Tz\|^2}{\beta^2} = 4\|Tz\| \quad (1)$$

by (Exercises 1.4, problem 1(c)) and direct calculation.

Now for any non-zero vector u we have

$$|\langle Tu, u\rangle| = \|u\|^2 |\langle T\left(\frac{u}{\|u\|}\right), \frac{u}{\|u\|}\rangle| \leq M\|u\|^2 \qquad (2)$$

So by combining (1) and (2) we get

$$\langle Tx, x\rangle - \langle Ty, y\rangle \leq |\langle Tx, x\rangle| + |\langle Ty, y\rangle|$$
$$\leq M(\|x\|^2 + \|y\|^2) = 4M\|Tz\|$$

remember that our inner products in this case are real numbers. Now a direct calculation using the definitions of x, y and β gives us $\langle Tx, x\rangle - \langle Ty, y\rangle = 4\|Tz\|^2$. This, together with (2) shows that $\|Tz\| \leq M$ whenever $\|z\| = 1$. We conclude that $\|T\| \leq M$ as was to be shown.

For any bounded, linear operator S on H and any complex number λ the adjoint of $S - \lambda I$ is $S^* - \overline{\lambda} I$ (Exercises 4.3, problem 1(d)). Also

$$\begin{aligned} M_\lambda &= \{x \in H | Sx = \lambda x\} = \{x \in H | (S - \lambda I)x = 0\} \\ &= \{x \in H | \langle (S - \lambda I)x, y\rangle = 0 \text{ for all } y \in H\} \\ &= \{x \in H | \langle x, (S^* - \overline{\lambda} I)y\rangle = 0 \text{ for all } y \in H\} = [(S^* - \overline{\lambda} I)(H)]^\perp \end{aligned}$$

(Section 4.2, Definition 1). We shall need this fact below.

Lemma 1

The complex number λ is an eigenvalue of the self-adjoint operator T if, and only if,

$$\overline{[(T - \lambda I)(H)]} \neq H;$$

i.e., the range of $T - \lambda I$ is not dense in H (see Section 3.1 Definition 2 for our notation, dense sets were defined in Section 1.9 Definition 1 (see also Exercises 3.1, problem 4(e))).

Proof. Suppose that λ is an eigenvalue of T. Then M_λ is a non-trivial subspace of H; i.e., $M_\lambda \neq \{0\}$. However, we have observed, just before stating this lemma, that $M_\lambda = [(T^* - \overline{\lambda} I)(H)]^\perp$ and since T is self-adjoint we know λ must be real, so

$$M_\lambda = [(T - \lambda I)(H)]^\perp.$$

Now by Lemma 1 of Section 4.2 the set $[(T - \lambda I)(H)]^{\perp\perp}$ is equal to $\overline{[(T - \lambda I)(H)]}$. By that same result this last set is equal to H if, and only if, $[(T - \lambda I)(H)]^\perp = \{0\}$. But this perpendicular is M_λ and M_λ is not the zero subspace.

Next suppose that λ is not an eigenvalue of the operator T. Then the number $\bar{\lambda}$ is also not an eigenvalue of T; this is so easy that it may be difficult. But to say that $\bar{\lambda}$ is not an eigenvalue of T is to say that

$$\{0\} = \{x \in H | T^*x = \bar{\lambda}x\} = [(T - \lambda I)(H)]^\perp$$

Recalling again Lemma 1 of Section 4.2 we see that

$$H = \{0\}^\perp = [(T - \lambda I)(H)]^{\perp\perp} = \overline{[(T - \lambda I)(H)]}$$

hence the range of $T - \lambda I$ is dense in H.

Lemma 2

Let H_1, H_2 be two Hilbert spaces over the same field and let $T : H_1 \to H_2$ be a bounded, linear map. Suppose that there is a fixed positive constant k such that

$$k\|x\|_1 \leq \|Tx\|_2 \text{ for all } x \in H_1$$

Then the range of T is a closed subset of H_2.

Proof. Suppose that $\{y_n\}_{n=1}^\infty$ is a sequence in the range of T which converges to a point $w \in H_2$. We must show that w is in the range of T (Section 2.4, Theorem 1).

For each $n = 1, 2, \ldots$ we can choose $x_n \in H_1$ such that $Tx_n = y_n$. Then for any m, n we have, by our hypothesis,

$$k\|x_m - x_n\|_1 \leq \|T(x_m - x_n)\|_2 = \|Tx_m - Tx_n\|_2 = \|y_m - y_n\|_2$$

Now since $\{y_n\}$ is convergent this last difference tends to zero as $m, n \to \infty$. Thus $\{x_n\}$ is a Cauchy sequence in the Hilbert space H_1. Such a sequence must converge to, say, $v \in H_1$. But then

$$Tv = \lim Tx_n = \lim y_n = w$$

showing that w is in the range of T; we have used Section 1.3, Definitions 1 and 2 and Section 2.1 Theorem 1.

Theorem 2

Let T be a bounded, self-adjoint operator on a Hilbert space H. A complex number λ is *not* in $\sigma(T)$ if, and only if, there is a positive constant k such that $k\|x\| \leq \|(T - \lambda I)x\|$ for all x in H.

Proof. First suppose that $\lambda \notin \sigma(T)$. Then $(T - \lambda I)^{-1}$ is a bounded, linear operator on H (Section 2.3 Definition 2). Hence, for any $x \in H$ we have

$$x = (T - \lambda I)^{-1}[(T - \lambda I)x]$$

and so
$$\|x\| \leq \|(T - \lambda I)^{-1}\|\|(T - \lambda I)x\|$$
This gives us our inequality with the constant k equal to $\|(T - \lambda I)^{-1}\|^{-1}$.

Now suppose that for a given complex number λ we have found a positive constant k such that our inequality holds. Then by Lemma 2 the range of $T - \lambda I$, $R(T - \lambda I)$, is a closed subset of H; it is actually a subspace of H. Now the given number λ could not be an eigenvalue of T for, if it were, then $(T - \lambda I)x = 0$ for some non-zero vector x and our inequality would then give $k\|x\| \leq 0$ contradicting the fact that k is positive. Hence, by Lemma 1, $R(T - \lambda I)$ is dense in H. So we must then have $R(T - \lambda I) = H$ since the former set is both closed and dense in H.

Now in order to show, as we are trying to do, that $\lambda \notin \sigma(T)$ we must show that $T - \lambda I$ has a bounded linear inverse. We have just observed that this map is onto and, since λ is not an eigenvalue of T (we proved this in the last paragraph) $T - \lambda I$ is a one-to-one map. Since it is certainly bounded it has a bounded, linear inverse by the Open-Mapping Theorem (Section 2.4, Theorem 3).

Definition 1

Let T be a bounded, self-adjoint operator on a Hilbert space H. Then the two numbers
$$m_T = \inf\{\langle Tz, z\rangle | \|z\| = 1\}, \quad M_T = \sup\{\langle Tz, z\rangle | \|z\| = 1\}$$
are called the spectral bounds of T.

The next theorem will justify this terminology. Observe that, by Theorem 1, $\|T\| = \max\{|m_T|, |M_T|\}$.

Theorem 3

Let T be a self-adjoint operator on a Hilbert space H. Then $\sigma(T)$, the spectrum of T, is a non-empty subset of $\mathbf{R}$ which is contained in the interval $[m_T, M_T]$. Furthermore, the end points of this interval are in $\sigma(T)$, and $\sigma(T)$ consists entirely of generalized eigenvalues of T.

Proof. If $\lambda = \alpha + \beta i$, $\beta \neq 0$, is a given complex number then, for any $x \in H$, we have
$$\|(T - \lambda I)x\|^2 = \langle (T - \lambda I)x, (T - \lambda I)x\rangle = \|Tx - (\alpha I)x\|^2 + \beta^2\|x\|^2 \geq \beta^2\|x\|^2$$
Thus, by Theorem 2, $\lambda \notin \sigma(T)$. We conclude that $\sigma(T) \subseteq \mathbf{R}$.

Next suppose that $\lambda < m_T$ and let $x \in H$. We may write
$$\|Tx - (\lambda I)x\|\|x\| \geq \langle Tx - \lambda I x, x\rangle = \langle Tx, x\rangle - \lambda\langle x, x\rangle \geq (m_T - \lambda)\|x\|^2.$$

Since $m_T - \lambda > 0$ we conclude, again from Theorem 2, that $\lambda \notin \sigma(T)$. A similar argument shows that any $\lambda > M_T$ is not in $\sigma(T)$. Thus $\sigma(T) \subseteq [m_T, M_T]$.

In order to show that the spectrum of T is a non-empty set it suffices to show that it contains M_T. Replacing, if necessary, T by $T - m_T I$ we may suppose that $0 \leq m_T \leq M_T$ and hence that $\|T\| = M_T$. From the definition of the number M_T there is a sequence $\{x_n\} \subseteq H$ such that $\|x_n\| = 1$ for all n and

$$M_T = \lim \langle Tx_n, x_n \rangle$$

Hence we may write

$$\|Tx_n - M_T x_n\|^2 = \|Tx_n\|^2 - 2M_T \langle Tx_n, x_n \rangle + M_T^2 \leq 2M_T^2 - 2M_T \langle Tx_n, x_n \rangle.$$

Now since the right-hand side of this inequality tends to zero as n increases we see that M_T is a generalized eigenvalue of T. Thus $M_T \in \sigma(T)$ and hence this set is non-empty. We could give a similar argument to show that m_T is in the spectrum of T. However, we may also argue as follows: Set $S = -T$ so that $-m_T = M_S$, and $M_S \in \sigma(S)$ by what we have just proved. But then $(T - m_T I) = -(S - M_S I)$ does not have a bounded inverse showing that $m_T \in \sigma(T)$.

Finally, suppose that the number λ is *not* a generalized eigenvalue of T. Then there is an $\varepsilon > 0$ such that

$$\varepsilon \|y\| \leq \|(T - \lambda I)y\|$$

for all $y \in H$ (Section 2.4, Definition 1 and Exercises 2.4 problem 3). Again using Theorem 2 we see that such a λ could not be in $\sigma(T)$. Thus every point of this set must be a generalized eigenvalue of T.

Exercises 1

* 1. Let T be a bounded, self-adjoint operator on H, let λ, μ be two distinct eigenvalues of T and let x, $y \in H$ be eigenvectors corresponding, respectively, to these eigenvalues. Show that $\langle x, y \rangle = 0$ and that $M_\lambda = [(T - \lambda I)(H)]^\perp$.

  2. Let T be a bounded, self-adjoint operator on H and suppose that $\langle Tx, x \rangle$ is non-negative for every $x \in H$. Show that $\sigma(T)$ is a subset of $\{\lambda \in \mathbf{R} | \lambda \geq 0\}$.

  3. Let S be a bounded, linear operator on H. We saw above that if S is self-adjoint then $\langle Sx, x \rangle$ is real for all x in H. Show that the converse is

true (this was already done in Exercises 4.3, problem 2(e)) as follows: For $\alpha, \beta \in \mathbb{C}$ we have

$$\langle S(\alpha x + \beta y), \alpha x + \beta y\rangle = \alpha\overline{\beta}\langle Sx, y\rangle + \overline{\alpha}\beta\langle Sy, x\rangle$$

is real. Now:

   a. Set $\alpha = \beta = 1$ and conclude that the imaginary part of $\langle Sx, y\rangle$ is minus the imaginary part of $\langle Sy, x\rangle$.

   b. Set $\alpha = 1$, $\beta = i$ and conclude that $\langle Sx, y\rangle$ and $\langle Sy, x\rangle$ have the same real part.

   c. Combine (a) and (b) to get $\langle Sx, y\rangle = \langle S^*x, y\rangle$ and so $\langle (S - S^*)x, y\rangle = 0$ for all x, y in H. Conclude that $S = S^*$.

4. Let S be a bounded, linear operator on H and define

$$\sigma(x, y) = \langle S^*Sx, y\rangle, \quad \chi(x, y) = \langle SS^*x, y\rangle$$

for all x, y in H.

   a. Show that σ and χ are bilinear forms on H (Section 4.3, Definition 1).

   b. Compute $\hat{\sigma}(x)$ and $\hat{\chi}(x)$. These are the quadratic forms associated with σ and χ respectively (Exercises 4.3, problem 2). Conclude (Exercises 4.3, problem 3) that $\sigma = \chi$.

   c. Use (b) to show that S is normal iff $\|Sx\| = \|S^*x\|$ for all $x \in H$.

   d. Assume that S is a normal operator. Use (c) to show that λ is an eigenvalue of S iff $\overline{\lambda}$ is an eigenvalue of S^*.

5. Recall the multiplication operator P on $L^2[a, b]$ (see Section 2.4(b) and Exercises 4.2 problem 4). Find the spectral bounds for P. Do this problem for the operator $A(\{z_n\}) = \{\frac{z_n}{n}\}$ on ℓ^2.

6.2 An Important Function Space

In Section 5.1, we asked for a "reasonably large" class of functions $\mathcal{F}$ and a "reasonably large" class of linear operators $\mathcal{D}$ (on a finite dimensional space) such that $f(T)$ has meaning for all $f \in \mathcal{F}$ and all $T \in \mathcal{D}$. We eventually took $\mathcal{F}$ to be the class of functions which are analytic in some neighborhood of zero and we chose $\mathcal{D}$ to be the class of diagonalizable operators. Even then we could only define $f(T)$ when all of the eigenvalues of T were less than the radius of convergence of the Maclaurin's series for f. Our subsequent work in

Chapter 5 led us to restrict our attention to normal operators and, at first, even to only self-adjoint operators. So we know what our $\mathcal{D}$ is going to be. We shall work with self-adjoint operators throughout most of this chapter and only later will try to extend our results to the class of normal operators.

Now what about the class $\mathcal{F}$? It turns out that the class of analytic functions is to restrictive for our purposes. We must be able to define $f(T)$ for functions f which may even be discontinuous at some points. The purpose of this section is to describe a class of functions L which contains all continuous functions and others as well, and has the properties we shall need. The class L, it is a function space as we shall see, is defined on an interval $[m, M]$. As we mentioned above we could not define $f(T)$ for any $f \in \mathcal{F}$ and any $T \in \mathcal{D}$. We had to know that the spectrum of T (i.e., its eigenvalues in this case) were contained in the circle of convergence of $f(z)$. Similarly, we cannot define $f(T)$ for every $f \in L[m, M]$ and every self-adjoint operator T. We have to know that the spectrum of T is contained in $[m, M]$.

In this section we shall construct our function space L. Later we shall bring in operators and discuss the question of defining functions of these operators.

Let m, M be fixed real numbers with $m < M$. The set of all polynomials $p(t)$ such that: (i) $p(t)$ has real coefficients; (ii) $p(t) \geq 0$ for all $t \in [m, M]$, will be denoted by $P^+[m, M]$. So $p(t) = 8t^4 + 7t^2$ is in this set and so also is $p(t) = M - t$.

Definition 1

The set of all real-valued functions f defined on $[m, M]$ for which there is a sequence $\{p_n\} \subseteq P^+[m, M]$ such that: (a) $0 \leq p_{n+1}(t) \leq P_n(t)$ for all $t \in [m, M]$ and each n; (b) $\lim_n p_n(t) = f(t)$ for each $t \in [m, M]$, will be denoted by $L^+[m, M]$.

Clearly $P^+[m, M] \subseteq L^+[m, M]$ for, given any $p \in P^+[m, M]$ we may set $p_n(t) = p(t)$ for $n = 1, 2, \ldots$ and observe that the sequence $\{p_n\}$ so obtained satisfies (a) and (b) of Definition 1. Also notice that if $f \in L^+[m, M]$ and $\{p_n\}$ is the sequence in $P^+[m, M]$ satisfying (a) and (b) then $f(t) \leq p_1(t)$ for all $t \in [m, M]$. Since $p_1(t)$ is bounded on $[m, M]$ so also is f, thus every function in L^+ is bounded on $[m, M]$.

It is clear, and easy to prove from the definition, that the sum of two functions in L^+ is a function in L^+ and that a positive, even non-negative, scalar multiple of such a function is also in L^+. This set is not a vector space however since the product of an element of L^+ with a negative scalar is not, in general, in L^+. However, it is easy to construct a vector space from our set.

Definition 2

The set of all real-valued functions h defined on $[m, M]$ for which there are functions f, g in $L^+[m, M]$ such that $h = f - g$, will be denoted by $L[m, M]$.

Note that any polynomial with real coefficients is in $L[m, M]$. To see this let such a polynomial $p(t)$ be given and choose a positive real number α such that $p(t) \leq \alpha$ for all $t \in [m, M]$. Then $f(t) \equiv p(t) + \alpha \in P^+[m, M] \subseteq L^+[m, M]$ and $g(t) \equiv \alpha \in P^+[m, M]$ also. Thus $p(t) = f(t) - g(t)$ is in $L[m, M]$.

We would like to show that every real-valued, continuous function on $[m, M]$ is in $L[m, M]$. In order to do this we shall use a construction which will be useful in other arguments as well. For this reason it is convenient to explicitly state it as a lemma.

Lemma 1

Given any real-valued, continuous function $f(t)$ on $[m, M]$ and any integer n we can find a polynomial $p_n(t)$ with real-coefficients such that

$$f(t) + \frac{1}{n+1} < p_n(t) < f(t) + \frac{1}{n} \quad \text{for all } t \in [m, M].$$

Proof. Given any $\varepsilon > 0$ the Weierstrass approximation theorem (Section 1.7, Corollary 3 to Theorem 1) tells us that we can choose a polynomial $p_\varepsilon(t)$, with real coefficients since f is real-valued, such that

$$|f(t) - p_\varepsilon(t)| < \varepsilon \text{ for all } t \in [m, M]$$

Thus $f(t) - \varepsilon < p_\varepsilon(t) < f(t) + \varepsilon$ or, setting $q_\varepsilon(t) = p_\varepsilon(t) + \varepsilon$, $f(t) < q_\varepsilon(t) < f(t) + 2\varepsilon$ for all $t \in [m, M]$. Thus, from this last chain of inequalities we see that for any fixed integer n

$$f(t) + \frac{1}{n+1} < q_\varepsilon(t) + \frac{1}{n+1} < f(t) + 2\varepsilon + \frac{1}{n+1}$$

Now let us take ε, which was arbitrary, to be $\frac{1}{2n(n+1)}$ so that $2\varepsilon + \frac{1}{n+1} = \frac{1}{n}$ and let us set $p_n(t) = q_\varepsilon(t) + \frac{1}{n+1}$. Then

$$f(t) + \frac{1}{n+1} < p_n(t) < f(t) + \frac{1}{n} \text{ for all } t \in [m, M]$$

as was to be shown.

Theorem 1

Every real-valued, continuous function on $[m, M]$ is in the vector space $L[m, M]$.

Proof. Given any such function, call it f, we can set

$$f^+(t) = \max\{f(t), 0\}, \quad f^-(t) = \max\{-f(t), 0\}$$

for all t, and note that both f^+ and f^- are non-negative, real-valued functions which are continuous on $[m, M]$. Furthermore,

$$f = f^+ - f^-$$

It follows from this, see Definition 2, that $f \in L$ if we can show that f^+ and f^- are both in L^+. Thus our theorem will be proved if we can show that any non-negative, real-valued function which is continuous on $[m, M]$ is in $L^+[m, M]$. Assume now that f has these properties and for each $n = 1, 2, \ldots$ choose a polynomial $p_n(t)$ with real-coefficients such that

$$f(t) + \frac{1}{n+1} < p_n(t) < f(t) + \frac{1}{n} \text{ for all } t \in [m, M]$$

We are, of course, using Lemma 1. Then

$$0 < \frac{1}{n+2} + f(t) < p_{n+1}(t) < f(t) + \frac{1}{n+1} < p_n(t)$$

showing that

$$0 \leq p_{n+1}(t) \leq p_n(t) \text{ for all } t \in [m, M] \text{ and each } n.$$

Also, $\lim p_n(t) = f(t)$ for each t, so our sequence $\{p_n(t)\}$ has properties (a) and (b) of Definition 1. It follows that $f \in L^+[m, M]$.

We shall now define a set of functions $e_s(t)$, one for each fixed real number s, which will be very useful in our subsequent discussions. All of these functions are, as we shall show, in the space $L[m, M]$.

If $s < m$ we shall let $e_s(t) = 0$ for all t, and if $M \leq s$ then we shall let $e_s(t) = 1$ for all t. Suppose that $m \leq s < M$. Then we set $e_s(t) = 0$ for all $t > s$ and we set $e_s(t) = 1$ for each $t \leq s$. Observe that each of these functions is non-negative and that $[e_s(t)]^2 = e_s(t)$ for all t and each fixed s. This last fact will enable us to show that when we replace t by an operator T we will obtain a projection operator.

Lemma 2

For each fixed real number s the function $e_s(t)$ is in the set $L^+[m, M]$.

Proof. The result is trivial when $s < m$ or $M \leq s$. Assume that $m \leq s < M$ and let N be the first integer such that $s + \frac{1}{N} \leq M$. For each $n \geq N$ we define a function $f_n(t)$ as follows:

$$\begin{aligned} f_n(t) &= 1 \text{ for } t \leq s; \\ f_n(t) &= -nt + ns + 1 \text{ for } s < t < s + \frac{1}{n}; \\ f_n(t) &= 0 \text{ for } s + \frac{1}{n} \leq t. \end{aligned}$$

So $f_n(t)$ is identically one until $t = s$, then it falls linearly to zero in the interval from s to $s + \frac{1}{n}$, and after that it remains zero. Clearly each f_n is continuous and non-negative. We also have $f_{n+1}(t) \leq f_n(t)$ for all t and each $n \geq N$, and $\lim_n f_n(t) = e_s(t)$ for each t. These facts are, perhaps, best seen by graphing $f_n(t)$ for several values of n. It is easy to prove them analytically from the definitions.

It follows from Lemma 1 that we may, for each $n \geq N$, find a polynomial $p_n(t)$, with real-coefficients, such that $f_n(t) + \frac{1}{n+1} < p_n(t) < f_n(t) + \frac{1}{n}$ for all $t \in [m, M]$. Then $\{p_n(t)\} \subseteq P^+[m, M]$, because $f_n(t) > 0$, $p_{n+1}(t) \leq p_n(t)$ for all t in $[m, M]$ and each n, from our inequalities, and $\lim p_n(t) = \lim f_n(t) = e_s(t)$ for each t. Thus the function e_s is in $L^+[m, M]$.

As we have said we want to define $f(T)$ for T a self-adjoint operator and $f \in L[m, M]$ where m, M are the spectral bounds of T. We shall do that in the next section. Perhaps it has occurred to the reader that for any polynomial $p(t)$ and any bounded, linear operator A on a Hilbert space H, the operator $p(A)$ can be defined in a natural way. We end this section by presenting this definition and discussing some of its consequences. Let

$$p(t) = a_0 + a_1 t + a_2 t^2 + \cdots + a_n t^n \qquad (1)$$

be a polynomial with real or complex coefficients. Then we set

$$p(A) = a_0 I + a_1 A + a_2 A^2 + \cdots + a_n A^n \qquad (2)$$

and we note that $p(A)$ is certainly a bounded, linear operator on H. The question we want to answer is this: What is the spectrum of the operator $p(A)$? Our next theorem will tell us, but first note that our definitions imply

$$p(A)q(A) = q(A)p(A) \qquad (3)$$

for any two polynomials p, q.

Theorem 2

For any bounded, linear operator A on H and any polynomial $p(t)$ the spectrum of the operator $p(A)$ is the set of all values assumed by $p(t)$ on the set $\sigma(A)$; i.e.,

$$\sigma[p(A)] = \{p(\lambda) | \lambda \in \sigma(A)\}.$$

Proof. The result is trivial if $p(t)$ is a constant and so we shall assume that $p(t)$ is given by (1) with $n \geq 1$ and $a_n \neq 0$.

For any complex number λ we see, trivially, that λ is a root of the polynomial $p(t) - p(\lambda) = 0$. Thus $t - \lambda$ is a factor of this polynomial. In other words, there is a polynomial $q(t)$ such that

$$p(t) - p(\lambda) = (t - \lambda)q(t) \qquad (4)$$

Thus, from our definitions and (3),

$$p(A) - p(\lambda)I = (A - \lambda I)q(A) = q(A)(A - \lambda I). \tag{5}$$

Suppose that $p(\lambda)$ is a regular value for the operator $p(A)$. Then $p(A) - p(\lambda)I$ has a bounded, linear inverse. Hence by (4) both $(A-\lambda I)q(A)$ and $q(A)(A-\lambda I)$ are invertible (they are the same operator but they look different) and they have the same inverse, namely $\{p(A) - p(\lambda)I\}^{-1}$. So we may write:

$$I = (A - \lambda I)q(A)\{p(A) - p(\lambda)I\}^{-1} = \{p(A) - p(\lambda)\}^{-1}q(A)(A - \lambda I).$$

These two equations, the middle operator equal to I and the last operator equal to I, almost say that $(A - \lambda I)$ is invertible. More explicitly we mean the equations

$$(A - \lambda I)[q(A)\{p(A) - p(\lambda)I\}^{-1}] = I, \quad [\{p(A) - p(\lambda)I\}^{-1}q(A)](A - \lambda I) = I$$

The trouble here is that we have two slightly different operators in the square brackets. Comparing these we see that we will have shown that $(A - \lambda I)$ is invertible once we show that

$$q(A)\{p(A) - p(\lambda)I\}^{-1} = \{p(A) - p(\lambda)I\}^{-1}q(A).$$

This equation can be obtained by multiplying both sides of

$$\{p(A) - p(\lambda)I\}q(A) = q(A)\{p(A) - p(\lambda)I\} \tag{6}$$

by $\{p(A) - p(\lambda)I\}^{-1}$ first on the left and then on the right. But (6) is clearly true by (3) above.

We have shown that for any complex number λ the following holds: If $p(\lambda)$ is a regular value for $p(A)$ then λ is a regular value for A. Hence we have shown: If $\lambda \in \sigma(A)$ then $p(\lambda) \in \sigma[p(A)]$.

Now suppose that $\mu \in \sigma[p(A)]$ is given. We shall find a $\lambda \in \sigma(A)$ such that $p(\lambda) = \mu$. The first thing we do is factor the polynomial $p(t) - \mu$. It has n complex roots $\lambda_1, \lambda_2, \ldots, \lambda_n$ and we may write

$$p(t) - \mu = a_n(t - \lambda_1)(t - \lambda_2)\ldots(t - \lambda_n), \quad a_n \neq 0 \tag{7}$$

Then (7) gives us the operator equation

$$p(A) - \mu I = a_n(A - \lambda_1 I)(A - \lambda_2 I)\ldots(A - \lambda_n I) \tag{8}$$

If every λ_j, $1 \leq j \leq n$, is a regular value for A then each of the operators $(A - \lambda_j I)$ has a bounded, linear inverse. Clearly then, from (8), $p(A) - \mu I$ would be invertible contradicting the choice of μ. We conclude that some λ_s, $1 \leq s \leq n$, is in $\sigma(A)$. But then (7) gives

$$p(\lambda_s) - \mu = a_n(\lambda_s - \lambda_1)(\lambda_s - \lambda_2)\ldots(\lambda_s - \lambda_n) = 0.$$

Hence $p(\lambda_s) = \mu$ and we are done.

Corollary 1

Let T be a bounded, self-adjoint operator on H and let $p(t)$ be a polynomial with real coefficients. Then $p(T)$ is a self-adjoint operator and

$$\|p(T)\| = \max\{|p(\lambda)|\,|\,\lambda \in \sigma(T)\}.$$

Proof. We refer to Exercises 4.3 problems 1(a) and 1(b) to see that, since T is self-adjoint and the coefficients of $p(t)$ are real, $p(T)$ is a self-adjoint operator. Our theorem tells us that

$$\sigma[p(T)] = \{p(\lambda)\,|\,\lambda \in \sigma(T)\}.$$

Now recall that for any self-adjoint operator S with spectral bounds m_s and M_s we have

$$\sigma(S) \subseteq [m_s, M_s], \quad m_s, M_s \text{ are in } \sigma(S), \quad \|S\| = \max\{|m_s|, \|M_s\|\}$$

by Theorem 1.3 and the discussion just before it. Since $p(T)$ is self-adjoint and we know its spectrum we see that

$$\|p(T)\| = \max\{|p(\lambda)|\,|\,\lambda \in \sigma(T)\}.$$

as claimed.

Exercises 2

* 1. Show that $L[m, M]$ is a vector space over $\mathbf{R}$. This amounts to showing that for $f, g \in L[m, M]$ the function $f + g$ is in this set and that, for any real number α, the function αf is in $L[m, M]$ since once we have these facts the other axioms for a vector space are obviously satisfied.

2. a. For each $n = 1, 2, \ldots$ let $p_n(t) = t^n$. Show that each of these functions is in $P^+[0, 1]$.

   b. Referring to part (a) show that $0 \le p_{n+1}(t) \le p_n(t)$ for all $t \in [0, 1]$ and each n.

   c. Show that $\lim_n p_n(t)$ exists for each $t \in [0, 1]$ and thus the limit function, call it $f(t)$, is in $L^+[0, 1]$. Graph $f(t)$ and show that $\{p_n(t)\}$ could *not* converge uniformly to $f(t)$ over $[0, 1]$.

3. Graph each of the functions $e_s(t)$ for various values of s.

4. If f, g are in $L^+[m, M]$ (respectively, $L[m, M]$) show that the function fg is in $L^+[m, M]$ (respectively, $L[m, M]$).

* 5. Prove Dini's theorem: Let $\{g_n(t)\}$ be a sequence of real-valued, continuous functions on $[m, M]$ such that: (i) $g_{n+1}(t) \leq g_n(t)$ for all $t \in [m, M]$ and each n; (ii) $\lim_n g_n(t)$ exists, call it $g(t)$, for each $t \in [m, M]$. Suppose further, and this is crucial (see problem 2(c) above), that $g(t)$ is continuous on $[m, M]$. Then $\{g_n(t)\}$ converges to $g(t)$ uniformly over $[m, M]$.

6. a. If $\alpha \neq 0$ is a scalar what is the inverse of the operator αI?

   b. Suppose that $A_1, A_2, \ldots, A_n$ is a finite set of invertible operators on H. Find the inverse of their product, $A_1 \cdot A_2 \cdots A_n$, in terms of the inverses of $A_1, A_2, \ldots, A_n$.

7. Let A be a bounded, linear operator on H and let $p(t)$ be a polynomial. Then:

   a. Show that the operator $p(A)$ is invertible iff $p(\lambda) \neq 0$ for all $\lambda \in \sigma(A)$.

   b. Suppose that the operator $p(A)$ is invertible with inverse $p(A)^{-1}$. Show that $\sigma[p(A)^{-1}] = \{\frac{1}{p(\lambda)} | \lambda \in \sigma(A)\}$. Hint: For any $\mu \neq 0$, $p(A)^{-1} - \mu I = -\mu p(A)^{-1}[p(A) - \mu^{-1} I]$.

6.3 Functions of a Bounded Self-adjoint Operator

Let H be a fixed Hilbert space over $\mathbb{C}$ and let $\mathcal{S}$ be the set of all bounded, self-adjoint operators on H. We choose, and fix once and for all, an element $T \in \mathcal{S}$ and we let

$$m = \inf\{\langle Tx, x\rangle | \|x\| = 1\}, \quad M = \sup\{\langle Tx, x\rangle | \|x\| = 1\}$$

be the spectral bounds of T. Each of these numbers is in $\sigma(T)$ and this latter set is contained in $[m, M]$. Here we shall show that for any function $f \in L[m, M]$ the operator $f(T)$ is defined and is an element of $\mathcal{S}$. Note again that our function space is defined on $[m, M]$ and that these numbers are obtained from the given self-adjoint operator T.

The process for defining $f(T)$ a little bit complicated and so we shall go through it once leaving out the proofs of certain assertions. The proofs will be given at the end of the section. We will also introduce some new concepts without much comment about their properties. These will be investigated later in the section and in the exercises.

We have seen how to define $p(T)$ when $p(t)$ is any polynomial. Note that when $p(t) \in P^+[m, M]$ then operator $p(T) \in \mathcal{S}$ (see Section 2 just before Definition 1). Actually more is true as we shall see.

Definition 1

An operator $S \in \mathcal{S}$ is said to be a non-negative (respectively positive) operator if $\langle Sx, x \rangle \geq 0$ (respectively > 0) for every non-zero vector $x \in H$. The set of all non-negative operators will be denoted by $\mathcal{S}^+$.

Lemma 1

If $p(t) \in P^+[m, M]$, then $p(T) \in \mathcal{S}^+$.

Now given $f \in L^+[m, M]$ we know that there is a sequence $\{p_n\} \subseteq P^+[m, M]$ such that: (a) $0 \leq p_{n+1}(t) \leq p_n(t)$ for all $t \in [m, M]$ and each n; (b) $\lim_n p_n(t) = f(t)$ for each $t \in [m, M]$ (see Section 2, Definition 1). Thus to define $f(T)$ we first look at the sequence $\{p_n(T)\}$ which, by Lemma 1, is in $\mathcal{S}^+$. We shall see that there is, because of the properties of the sequence $\{p_n(t)\}$, an operator $S \in \mathcal{S}^+$ such that

$$\lim \|\{S - p_n(T)\}x\| = 0 \quad \text{for every } x \in H.$$

Of course, this must be proved. There is also another problem here that must be dealt with. The problem is this: Suppose that we can find another sequence $\{q_n(t)\} \subseteq P^+[m, M]$ such that $q_{n+1}(t) \leq q_n(t)$ for all $t \in [m, M]$ and each n, and $\lim q_n(t) = f(t)$ for all $t \in [m, M]$. Then, as we claimed above for the sequence $\{p_n(t)\}$, there must be an operator $S_q \in \mathcal{S}^+$ such that

$$\lim \|\{S_q - q_n(T)\}x\| = 0 \text{ for all } x \in H.$$

Now if we are going to define $f(T)$ to be S, as seems natural, we must be able to show that $S = S_q$. Otherwise our definition is ambiguous.

Finally, given $h \in L[m, M]$ we know that $h = f - g$ for $f, g \in L^+[m, M]$ and so we would naturally set $h(T)$ equal to $f(T) - g(T)$. Again, we have the problem that h may equal $f_1 - g_1$, where f_1, g_1 are in L^+. So we have to be able to show that

$$f(T) - g(T) = f_1(T) - g_1(T)$$

So the process for defining $h(T)$ for any $h \in L[m, M]$ is rather natural and straightforward given the way this function space was defined. All we need to do now is prove all of the facts asserted in our discussion.

Now let $S, T \in \mathcal{S}$. Referring to Definition 1 we set $S \leq T$ if, and only if, $(T - S) \in \mathcal{S}^+$. This means

$$0 \leq \langle (T - S)x, x \rangle$$

for all $x \in H$ and clearly this is equivalent to saying that

$$\langle Sx, x \rangle \leq \langle Tx, x \rangle$$

for all $x \in H$.

Lemma 2

Let S_1, S_2, S_3 belong to $\mathcal{S}$. Then we have:

  i. $S_1 \leq S_1$;
  ii. $S_1 \leq S_2$ and $S_2 \leq S_3$ imply $S_1 \leq S_3$;
  iii. $S_1 \leq S_2$ and $S_2 \leq S_1$ imply $S_1 = S_2$.

Furthermore, if α, β are any real numbers then: (iv) $S_1 \leq S_2$ and $0 \leq \alpha$ imply $\alpha S_1 \leq \alpha S_2$; (v) $S_1 \in \mathcal{S}^+$ and $\alpha \leq \beta$ imply $\alpha S_1 \leq \beta S_1$.

Finally, (vi) $S_1 \leq S_2$ if, and only if, $-S_2 \leq -S_1$; (vii) $S_1 \leq S_2$ implies $S_1 + S_3 \leq S_2 + S_3$.

Proof. We shall prove (iii) only and leave the other assertions to the exercises. Define $\sigma(x,y) = \langle S_1 x, y\rangle$, $\chi(x,y) = \langle S_2 x, y\rangle$ for all x,y in H. These are bounded, bilinear forms on H (Section 4.3, Definition 1). Because of our assumptions in (iii) the quadratic forms associated with σ and χ are equal for every $x \in H$ (Exercises 4.3, problems 2 and 3) and so $\sigma(x,y) = \chi(x,y)$ for all $x, y \in H$. Thus

$$\langle (S_1 - S_2)x, y\rangle = 0$$

for all x, y in H giving us $S_1 = S_2$ as was to be shown.

Let us now prove Lemma 1. We have $p(t) \in P^+[m, M]$ and so the spectrum of $p(T)$ is contained in $\{\lambda \in \mathbb{R} | \lambda \geq 0\}$ by Theorem 2.2 and the way $P^+[m, M]$ is defined (Section 2, just before Definition 1). Now $p(t)$ has real coefficients and $T \in \mathcal{S}$ so $p(T)$ certainly belongs to $\mathcal{S}$. However, the lower spectral bound for this operator is non-negative and so

$$0 \leq \inf\{\langle p(T)x, x\rangle | \|x\| = 1\} \leq \langle p(T)x, x\rangle$$

for all $x \in H$, $\|x\| = 1$. Clearly then

$$0 \leq \langle p(T)\frac{y}{\|y\|}, \frac{y}{\|y\|}\rangle = \langle p(T)y, y\rangle \|y\|^{-2}$$

shows that $p(T) \in \mathcal{S}^+$.

Lemma 3

Let $T \in \mathcal{S}^+$. Then

$$|\langle Tx, y\rangle|^2 \leq \langle Tx, x\rangle \langle Ty, y\rangle$$

for all x, y in H. This is a generalized C.S.B. inequality.

Proof. We define $\sigma(x,y)$ to be $\langle Tx, y \rangle$ for all x, y in H. Then σ has all of the properties of an inner product (Section 1.4, Definition 1) except that $\sigma(x,x)$ may be zero for $x \neq 0$. However we can get around this problem. In the proof of the C.S.B. inequality (Section 1.4, Theorem 1) we used the fact that $\langle u, v \rangle$ not zero implies neither u, nor v is the zero vector. We needed this fact because we then set $\alpha = \langle u, v \rangle / \langle v, v \rangle$. Using this α we proved the theorem by direct calculation. Let us examine this matter here just a little further.

We have $\langle u, 0 \rangle = \langle 0, u \rangle = 0$ by Exercises 1.4, problem 1(a). Thus $\langle u, v \rangle \neq 0$ implies $u \neq 0 \neq v$. In the case of an inner product we could then assert that $\langle u, u \rangle = \|u\|^2 \neq 0$ and $\langle v, v \rangle = \|v\|^2 \neq 0$. The rest of our proof then follows. Here now we are working with $\sigma(x,y)$ which behaves like an inner product except that $\sigma(x,x)$ can be zero for $x \neq 0$. Suppose we are given that $\sigma(x,y) \neq 0$. Then certainly we can say that neither x nor y is the zero vector. But this is not enough to insure that $\sigma(y,y) \neq 0$. We need more argument. What we do is this: For any scalar β we have, since $T \in \mathcal{S}^+$,

$$0 \leq \sigma(x - \beta y, x - \beta y) = \sigma(x,x) - \beta \sigma(y,x) - \overline{\beta} \sigma(x,y) + |\beta|^2 \sigma(y,y)$$

Now if $\sigma(y,y) = 0$ we could take $\beta = [\sigma(x,x) + 1]/2\sigma(y,x)$ and putting this in our inequality gives $0 \leq -1$ an obvious contradiction.

The proof now follows as in the case of $\langle , \rangle$.

Lemma 4

Let $T \in \mathcal{S}^+$. Then

$$\|Tx\|^2 \leq \|T\| \langle Tx, x \rangle$$

for every $x \in H$.

Proof. Let us set $y = Tx$ and apply Lemma 3; again $\sigma(x,y) = \langle Tx, y \rangle$. We have

$$\begin{aligned} \|Tx\|^4 &= [\langle Tx, Tx \rangle]^2 = [\langle Tx, y \rangle]^2 \leq \langle Tx, x \rangle \langle Ty, y \rangle \\ &\leq \langle Tx, x \rangle \|Ty\| \|y\| \leq \langle Tx, x \rangle \|T\| \|y\|^2 = \langle Tx, x \rangle \|T\| \|Tx\|^2 \end{aligned}$$

where in going from the first line to the one below it we used the ordinary C.S.B. inequality. The result now follows by cancelling the common positive factor $\|Tx\|^2$; if $\|Tx\| = 0$ the result is trivially true.

Theorem 1

Let $\{T_n\} \subseteq \mathcal{S}$, $S \in \mathcal{S}$ and suppose that

$$T_1 \leq T_2 \leq \ldots \leq T_n \leq \ldots \leq S.$$

then there exists an operator $T \in \mathcal{S}$ such that
$$T_n \leq T \leq S \text{ for all } n \text{ and } \lim T_n x = Tx$$
for each $x \in H$.

Proof. By Lemma 2, we have
$$0 \leq T_m - T_n \leq S - T_1 \text{ for } 1 \leq n \leq m$$
and, furthermore,
$$\|T_m - T_n\| = \sup\langle (T_m - T_n)x, x\rangle \leq \sup\langle (S - T_1)x, x\rangle = \|S - T_1\|$$
by Theorem 1.1, here the suprema are over all x with $\|x\| = 1$. For any fixed $x \in H$ we have
$$\langle Tx, x\rangle \leq \langle T_2 x, x\rangle \leq \ldots \leq \langle T_n x, x\rangle \leq \ldots \leq \langle Sx, x\rangle$$
and
$$\lim_{m,n\to\infty} \langle (T_m - T_n)x, x\rangle = 0$$
because an increasing sequence of real numbers which is bounded above is convergent. Now by Lemma 4
$$\|(T_m - T_n)x\|^2 \leq \|T_m - T_n\|\langle (T_m - T_n)x, x\rangle \leq \|S - T_1\|\langle (T_m - T_n)x, x\rangle$$
Thus we see that
$$\lim_{m,n\to\infty} \|(T_m - T_n)x\| = 0$$
which tells us that the sequence $\{T_n x\}$ is a Cauchy sequence in H. It must converge and, anticipating what we are going to prove, we denote the limit of this sequence by Tx. Thus
$$\lim T_n x = Tx$$
for each x and so
$$\begin{aligned} T(\alpha x + \beta x) &= \lim T_n(\alpha x + \beta y) = \lim(\alpha T_n x + \beta T_n y) \\ &= \alpha \lim T_n x + \beta \lim T_n y = \alpha Tx + \beta Ty \end{aligned}$$
showing that the map $x \to Tx$ is a linear mapping on H. Also
$$\langle Tx, y\rangle = \lim\langle T_n x, y\rangle = \lim\langle x, T_n y\rangle = \langle x, Ty\rangle$$
and so T is self-adjoint. Finally,
$$\|Tx\| = \lim\|T_n x\| = \lim\|T_n\|\|x\| \leq (\sup\{\|T_n\| | n = 1, 2, \ldots\})\|x\|$$

and the supremum is finite by our assumption that $T_n \leq S$ for each n and Theorem 1.1. Thus T is a bounded, self-adjoint operator on H.

Finally,
$$\langle T_n x, x \rangle \leq \lim_m \langle T_m x, x \rangle = \langle Tx, x \rangle \leq \langle Sx, x \rangle$$
for every $x \in H$.

Corollary 1

Suppose that $\{T_n\} \subseteq \mathcal{S}$, $S \in \mathcal{S}$ and that
$$S \leq \ldots \leq T_n \leq \ldots \leq T_2 \leq T_1.$$
then there exists an operator $T \in \mathcal{S}$ such that
$$S \leq T \leq T_n \text{ for all } n \text{ and } Tx = \lim T_n x$$
for each $x \in H$.

Proof. We set $T'_n = -T_n$ for each n and $S' = -S$ and apply the theorem.

Let us return now to the problem of defining a function of an operator. Recall that we chose and fixed $T \in \mathcal{S}$ with spectral bounds m, M respectively. Then for any $p(t) \in P^+[m, M]$ we saw, Lemma 1, that $p(T) \in \mathcal{S}^+$. Now given $f \in L^+[m, M]$ we find $\{p_n(t)\} \subseteq P^+[m, M]$ such that: (a) $0 \leq p_{n+1}(t) \leq p_n(t)$ for all $t \in [m, M]$ and each n; (b) $\lim p_n(t) = f(t)$ for each $t \in [m, M]$. Since $\{p_n(t) - p_{n+1}(t)\} \in P^+[m, M]$ by (a) we see from Lemma 1 that
$$0 \leq p_{n+1}(T) \leq p_n(T)$$
for each n. Thus the corollary to Theorem 1 tells us that there is an operator $S \in \mathcal{S}$ such that
$$0 \leq S \leq p_n(T)$$
and
$$\lim p_n(T)x = Sx$$
for each $x \in H$. Thus $S \in \mathcal{S}^+$ and we shall set $f(T) = S$.

Suppose now that $\{q_n\} \subseteq P^+[m, M]$ also satisfies: (a) $0 \leq q_{n+1}(t) \leq q_n(t)$ for all $t \in [m, M]$ and each n; (b) $\lim q_n(t) = f(t)$ for each $t \in [m, M]$. Then $\{q_n(T)\}$ converges to an operator $S_q \in \mathcal{S}^+$. By Dini's theorem (Exercises 2, problem 5) the sequence $\{q_n\}$ and the sequence $\{p_n\}$ both converge uniformly to f over $[m, M]$. Hence for any given integer r we can, since $f(t) \leq q_r(t) \leq q_i(t) + \frac{1}{r}$ and $f(t) \leq p_r(t) \leq p_r(t) + \frac{1}{r}$, find an integer s such that both
$$p_s(t) \leq q_r(t) + \frac{1}{r} \text{ and } q_s(t) \leq p_r(t) + \frac{1}{r}$$

Thus we have both

$$p_s(T) \leq q_r(T) + \frac{1}{r}I \text{ and } q_s(T) \leq p_r(T) + \frac{1}{r}I$$

and letting $s \to \infty$ and then $r \to \infty$ we get both

$$S \leq S_q \text{ and } S_q \leq S$$

Hence $f(T)$ is well-defined.

It is clear that, for $f, g \in L^+[m, M]$, $(f + g)(T) = f(T) + g(T)$. This shows that our definition of $h(T)$ for every $h \in L[m, M]$ is unique. For if $h = f - g = f_1 - g_1$ then $f + g_1 = f_1 + g$ and so $(f + g_1)T = (f_1 + g)T$, which is true since f, g_1, f_1, g are in $L^+[m, M]$, gives us $f(T) - g(T) = f_1(T) - g_1(T)$.

Theorem 2

Let B be a bounded, linear operator on H such that $BT = TB$. Then for every $f \in L[m, M]$, $Bf(T) = f(T)B$.

Proof. It suffices to show that B commutes with $f(T)$ whenever $f \in L^+[m, M]$. Given such an f we choose a sequence of polynomials $\{p_n(t)\} \subseteq P^+[m, M]$ such that

$$\lim \|\{f(T) - p_n(T)\}x\| = 0$$

for each $x \in H$. Then we can write

$$\|\{Bf(T) - Bp_n(T)\}x\| \leq \|B\|\|\{f(T) - p_n(T)\}x\|$$

and

$$\|\{f(T)B - p_n(T)B\}x\| = \|\{f(T) - p_n(T)\}y\|$$

where $y = Bx$. It follows that

$$\lim \|\{Bf(T) - Bp_n(T)\}x\| = 0$$

and

$$\lim \|\{f(T)B - p_n(T)B\}x\| = 0$$

for all $x \in H$. However, since $TB = BT$ and $p_n(T)$ is a polynomial in T, $p_n(T)B = Bp_n(T)$ for each n. It follows that

$$\|\{f(T)B - Bf(T)\}x\| = 0$$

for every $x \in H$. Thus $f(T)B = Bf(T)$.

Definition 2

Let $\{A_n\}_{n=1}^\infty$ be a sequence of bounded, linear operators on H. We shall say that this sequence is strongly convergent to the bounded, linear operator A_0 on H if
$$\lim \|(A_0 - A_n)x\| = 0$$
for each $x \in H$.

Since $\|(A_0 - A_n)x\| \leq \|A_0 - A_n\|\|x\|$ we see that any sequence which converges for the operator norm is also strongly convergent and to the same limit. The converse is false however, as we shall now show.

Observe first that if P is any orthogonal projection on H (Exercises 4.2, problem 2) then $R(P) \equiv \{Py | y \in H\} = \{x \in H | Px = x\} = \{x \in H | \|x\| = \|Px\|\}$. Since $P^2 = P$ it is clear that these last two sets contain $R(P)$. Suppose that $x \notin R(P)$; then $x = Px + P^\perp x$ and $P^\perp x \neq 0$, where $P^\perp$ is the orthogonal projection of H onto $\mathrm{Ker}\, P$ (Section 4.2, Theorem 2). Hence $\|x\|^2 = \|Px\|^2 + \|P^\perp\|^2 > \|Px\|^2$ showing that x is not in either of the sets defined above.

Theorem 3

Let P, Q be orthogonal projections on H with ranges $R(P)$ and $R(Q)$ respectively. Then these are equivalent: (a) $R(P) \subseteq R(Q)$; (b) $QP = P$; (c) $PQ = P$; (d) $\|Px\| \leq \|Qx\|$ for all $x \in H$; (e) $P \leq Q$ (meaning $Q - P \in \mathcal{S}^+$ which means $0 \leq \langle (Q-P)x, x \rangle$ for all $x \in H$); (f) $Q - P$ is an orthogonal projection.

Proof. The equivalence of (a), (b) and (c) was proved before (Exercises 4.2, problem 2(b)). We shall repeat the proof here since it is very short. Assume (a). Since $Px \in R(P) \subseteq R(Q)$, $Q(Px) = Px$ for all $x \in H$ and this is (b). Assuming that (b) is true we get $PQ = P^*Q^* = (QP)^* = P^* = P$ which is (c). Now assume (c). Then $(Q - P)^* = Q^* - P^* = Q - P$ and $(Q - P)^2 = Q^2 - QP - PQ + P^2 = Q - (PQ)^* - P + P = Q - P$. This shows that $(Q - P)$ is an orthogonal projection on H; i.e., (c) implies (f). Now suppose that (f) is true. Then

$$\langle (Q-P)x, x \rangle = \langle (Q-P)^2 x, x \rangle = \langle (Q-P)x, (Q-P)x \rangle = \|(Q-P)x\|^2$$

giving us (e). Assuming (e) we have

$$\|Qx\|^2 - \|Px\|^2 = \langle Qx, x \rangle - \langle Px, x \rangle = \langle (Q-P)x, x \rangle \geq 0$$

for all x, showing that $\|Px\| \leq \|Qx\|$ for all x. Thus (e) gives (d). Finally, assume (d). For any $x \in R(P)$ we have

$$\|x\| = \|Px\| \leq \|Qx\| \leq \|x\|$$

which says $\|x\| = \|Qx\|$. Our observation just before this theorem shows that $x \in R(Q)$. Hence (d) gives us (a).

Now let H be a separable Hilbert space with countably infinite orthonormal basis $\{e_n\}_{n=1}^{\infty}$ (Section 1.9, Definition 1 and Section 1.6, Definition 4). For each n let M_n be the closed, linear span of $\{e_1, \ldots, e_n\}$, $n = 1, 2, 3, \ldots$. Clearly $\{0\} \subsetneq M_1 \subsetneq M_2 \subsetneq \ldots \subsetneq M_n \subsetneq M_{n+1} \subsetneq \ldots \subsetneq H$. For each n let P_n be the orthogonal projection of H onto M_n. By parts (a) and (e) of Theorem 3 we have
$$0 \le P_1 \le P_2 \le \ldots \le P_n \le \ldots \le I,$$
here 0 is the projection of H onto $\{0\}$. Also, $P_n x = \sum_{j=1}^n \langle x, e_j \rangle e_j$, so
$$\lim_n P_n x = \lim_n \sum_{j=1}^n \langle x, e_j \rangle e_j = \sum_{j=1}^\infty \langle x, e_j \rangle e_j = x$$
because $\{e_j\}$ is a basis for H. Thus $\{P_n\}_{n=1}^\infty$ is strongly convergent to I. However, $I - P_n$ is the orthogonal projection of H onto $M_n^\perp$ and so $\|I - P_n\| = 1$ for each n showing that $\{P_n\}$ does *not* converge to I for the operator norm.

Exercises 3

1. Let $T \in \mathcal{S}$ be fixed and let m, M be its respective spectral bounds. Define a map $\Omega : L[m, M] \to \mathcal{S}$ as follows: For each $f \in L[m, M]$ we set $\Omega(f) = f(T)$.

    a. Show that $\Omega(\alpha f + \beta g) = \alpha \Omega(f) + \beta \Omega(g)$ for all scalars α, β and all f, g.

    b. Show that Ω is a "multiplicative map," i.e., show that $\Omega(fg) = \Omega(f)\Omega(g)$ for all $f, g \in L[m, M]$. The first thing to do is show that for $f, g \in L^+$, $fg \in L^+$ and then prove this for functions in L.

    c. For $f, g \in L$ let us write $f \le g$ to mean $f(t) \le g(t)$ for all t. Show that when $f \le g$ then $\Omega(f) \le \Omega(g)$; i.e., $\Omega(g) - \Omega(f) \in \mathcal{S}^+$.

2. Suppose that $T \in \mathcal{S}$ and has spectral bounds m, M and that $\sigma(f)$ is a real-valued function which is analytic at zero. So $\sigma(t) = \sum_{n=0}^\infty a_n t^n$ where the series converges for $|t| < R$ and $R > 0$. If $[m, M] \subseteq (-R, R)$ then $f \in L$ (why?) and so $f(T)$ is defined. Consider $\{\sum_{n=0}^k a_n T^n\}_{k=1}^\infty$. Show that this sequence converges for the operator norm to an operator S which is self-adjoint; i.e., $S \in \mathcal{S}$. Next show that $S = f(T)$.

* 3. Suppose that $\{A_n\}_{n=1}^{\infty}$ is a sequence of bounded, linear operators on H which converges strongly to the operator B. Show that

$$\lim_n \langle A_n x, x \rangle = \langle Bx, x \rangle$$

for every $x \in H$.

4. a. For any operator $C \in \mathcal{S}$ show that $C^2 \in \mathcal{S}^+$.

   b. For any operator $B \in \mathcal{S}^+$ show that $B^n \in \mathcal{S}^+$ for each $n = 1, 2, \ldots$.
   Hint: Consider the case of n even and the case of n odd separately.

6.4 The Resolution of the Identity

Once again we shall work with a fixed self-adjoint operator T on our Hilbert space H with spectral bounds m, M respectively. We have seen how one can define $f(T)$ for every $f \in L[m, M]$ and that $f(T)$ is also a self-adjoint operator. Now recall the functions $\{e_s(t) | s \in \mathbb{R}\}$ defined in Section 2. When $s < m$, $e_s(t) = 0$ for all t and when $M \leq s$, $e_s(t) = 1$ for all t. If $s \in [m, M)$ then $e_s(t) = 0$ whenever $s < t$ and $e_s(t) = 1$ whenever $t \leq s$. Each e_s is non-negative and $e_s^2 = e_s$ (i.e., $0 \leq e_s(t)$ for all t and any fixed s, and $e_s(t)^2 = e_s(t)$ for all t and each fixed s). By Lemma 2.2, we have $\{e_s | s \in \mathbb{R}\} \subseteq L^+[m, M]$. It follows that $e_s(T) \in \mathcal{S}$ for each fixed s and that $0 \leq e_s(T)$, $e_s(T)^2 = e_s(T)$ (Exercises 3 problem 1(c) and (b)). Thus $e_s(T)$ is a projection on H, because it is equal to its own square, and since it is self-adjoint it is an orthogonal projection (Section 4.1, Definition 1 and Section 4.2, Theorem 2). Let us set $E(s)$ equal to $e_s(T)$. Then:

Theorem 1

For each s the operator $E(s)$ is an orthogonal projection on H. Furthermore: (a) $E(s) \leq E(t)$ for $s \leq t$; (b) $E(t) = 0$ for $t < m$; (c) $E(t) = I$ for $M \leq t$; (d) $\lim_{\varepsilon \to 0+} \|[E(s + \varepsilon) - E(s)]x\| = 0$ for each $x \in H$ and each fixed s.

Proof. Let us define, as we did in Exercises 3, $\Omega(f) = f(T)$ for all $f \in L$. Then $\Omega(e_s) = e_s(T) = E(s)$ for each s. Now Ω is linear, it is multiplicative and it is order preserving (Exercises 3, problem 1(a), (b), (c)). Since $e_s(u) \leq e_t(u)$ for all u when $s \leq t$, $\Omega(e_s) \leq \Omega(e_t)$ or, in another notation, $E(s) \leq E(t)$. This proves (a). Part (b) is easy because e_t is the zero function when $t < m$ and the zero function is the zero vector in L so any linear map takes this to the zero vector in $\mathcal{S}$. To prove (c) we let $f \in L$ be given and note that $1 \in L$. Thus, since $f(t) = 1 f(t)$, $\Omega(f) = \Omega(1 \cdot f) = \Omega(1)\Omega(f) = \Omega(f \cdot 1) = \Omega(f)\Omega(1)$. Clearly $\Omega(1) = I$.

The proof of part (d) is a little more difficult. We have $e_s(t) \le e_{s+\frac{1}{n}}(t)$ for every t and each n. Also, $e_{s+\frac{1}{n+1}}(t) \le e_{s+\frac{1}{n}}(t)$ for all t and each n. Furthermore, all of these funtions are in L^+. Hence we may choose a polynomial $p_n \in P^+[m, M]$ such that $e_{s+\frac{1}{n}}(t) \le p_n(t)$ for all t and each n, such that $p_{n+1}(t) \le p_n(t)$ for all t and each n, and such that $\lim p_n(t) = e_s(t)$ for each t. It follows that

$$\Omega(e_s) \le \Omega(e_{s+1/n}) \le \Omega(p_n) \text{ for each } n,$$

or

$$E(s) \le E(s+\frac{1}{n}) \le p_n(T) \text{ for all } n.$$

However, $\{p_n(T)\}$ converges strongly to $E(s)$ hence the same must be true of the sequence $\{E(s+\frac{1}{n})\}$. Thus by Exercises 3, problem 3, $\lim \langle E(s+\frac{1}{n})x, x \rangle = \langle E(s)x, x \rangle$ for all $x \in H$. Combining the last fact with the fact that, by (a) and Theorem 3.3,

$$E(s+\frac{1}{n}) - E(s)$$

is a projection, we have:

$$\begin{aligned}
\|[E(s+\frac{1}{n}) - E(s)]x\|^4 &= \|[E(s+\frac{1}{n}) - E(s)]x\|^2 \|(E(s+\frac{1}{n}) - E(s))x\|^2 \\
&\le \|E(s+\frac{1}{n}) - E(s)\|^2 \|x\|^2 \langle [E(s+\frac{1}{n}) - E(s)]x, \\
&\quad [E(s+\frac{1}{n}) - E(s)]x \rangle \\
&\le \|x\|^2 \langle [E(s+\frac{1}{n}) - E(s)]x, x \rangle
\end{aligned}$$

It follows that

$$\lim_n \|[E(s+\frac{1}{n}) - E(s)]x\| = 0$$

From this and (a) we have (d).

Corollary 1

If an operator S on H commutes with T then it commutes with every $E(s)$, s real, i.e., $E(s)T = TE(s)$ for all s.

Proof. This follows from Theorem 3.2.

Let us look now at an example. Recall the operator P on $L^2[0, 1]$ defined by

$$P[g(x)] = xg(x)$$

for every $g \in L^2[0,1]$ (Section 2.4 (b)). This operator is self-adjoint (Exercises 4.2, problem 4) and its spectrum coincides with the set $[0,1]$ (Section 2.4, Theorem 4). Thus the spectral bounds for P are $m=0$, $M=1$. Observe that $P^2[g(x)] = P[P[g(x)]] = P[xg(x)] = x^2g(x)$ and in fact $p(P)[g(x)] = p(x)g(x)$ for any polynomial $p(x)$.

Suppose now that $f \in L^+[0,1]$ and choose a sequence $\{p_n(t)\} \subseteq P^+[0,1]$ which converges pointwise to $f(t)$ and satisfies

$$0 \le p_{n+1}(t) \le p_n(t)$$

for all t and each n. Then we know that the sequence $\{p_n(P)\}$ is contained in $\mathcal{S}^+$ and that it is strongly convergent to an operator $S \in \mathcal{S}^+$ (Corollary 1 to Theorem 3.1). Now for each n

$$p_n(P)[g(x)] = p_n(x)g(x)$$

and $\lim_n p_n(x) = f(x)$ for each $x \in [0,1]$ so we would expect that

$$S(P)[g(x)] = f(x)g(x)$$

for every $g \in L^2[0,1]$. In order to prove this we shall do two things: First we shall show that the formula

$$S[g(x)] = f(x)g(x)$$

for $g \in L^2[0,1]$ defines a bounded, linear operator on our Hilbert space and, secondly, that the sequence $\{p_n(P)\}$ converges strongly to this operator; i.e., that

$$\lim \|(S - p_n(P))[g(x)]\| = 0$$

for every $g \in L^2[0,1]$.

Our first task is quite easy. Since $f \in L^+[0,1]$ there is a constant M such that $|f(x)| \le M$ for all $x \in [0,1]$ (Section 6.2, just after Definition 1). Thus

$$\int_0^1 |f(x)g(x)|^2 dx \le M\|g\|_2^2$$

which shows that S is a bounded, linear operator on $L^2[0,1]$ with $\|S\| \le M$. It is easy to see, by direct calculation, that $S \in \mathcal{S}^+$. Now

$$\|(S - p_n(P))g(x)\|^2 = \int_0^1 |S[g(x)] - p_n(P)[g(x)]|^2 dx$$
$$= \int_0^1 |f(x) - p_n(x)|^2 |g(x)|^2 dx$$

Note that the sequence $\{|f(x) - p_n(x)|\}_{n=1}^\infty$ is dominated by the function $2|p_1(x)|$, and converges pointwise to zero. By the Lebesgue dominated convergence theorem (Section 1.8, Theorem 6) our last integral above must tend to zero as $n \to \infty$. Thus $\{p_n(P)\}$ converges strongly to S as was to be proved.

Finally, if $h \in L[0,1]$ then $h = f - g$ where $f, g \in L^+[0,1]$ and

$$\begin{aligned} h(P)[\theta(x)] &= (f(P) - g(P))[\theta(x)] = f(P)[\theta(x)] - g(P)[\theta(x)] \\ &= f(x)\theta(x) - g(x)\theta(x) = (f(x) - g(x))\theta(x) = h(x)\theta(x) \end{aligned}$$

for every $\theta \in L^2[0,1]$.

Our discussion of the operator P shows that, in particular, the resolution of the identity corresponding to this operator is as follows:

$$E(s) = e_s(P) \text{ where } E(s)[g(x)] = e_s(P)[g(x)] = e_s(x)g(x)$$

for every $g \in L^2[0,1]$. A direct calculation shows that $E(s)$ is an orthogonal projection. The properties listed in Theorem 1 are also obvious for this particular family because of the way the functions $e_s(t)$ were defined.

Exercises 4

1. Recall the operator A on ℓ^2 defined by $A(\{a_n\}) = \{a_n/n\}$. See Exercises 3.3, problem 1, and Exercises 5.4, problem 1. This is a compact operator whose spectrum is the set $\{0\} \cup \{1/n | n = 1, 2, \ldots\}$.

   a. For any $f \in L^+[0,1]$ show that the operator $f(A)$ is just:

   $$f(A)[\{a_n\}] = \{f(1/n)a_n\}$$

   for every $\{a_n\} \in \ell^2$. Prove this for $f \in L[0,1]$ as well.

   b. Show that $f(A)$ is a compact self-adjoint operator for any $f \in L[0,1]$, and find the spectrum of $f(A)$.

   c. Compute the spectral decomposition of the operator $f(A)$.

2. Let P be the multiplication operator on $L^2[0,1]$; so $P[g(x)] = xg(x)$. Suppose that $f \in L^+[0,1]$ is continuous. Compute $\sigma[f(P)]$.

6.5 The Riemann-Stieltjes Integral

We have already said that the spectral representation of a self-adjoint operator involves a process of integration rather than a sum. The process we shall use is very similar to the classical Riemann-Stieltjes integral, so let us review this integral here. This generalization of the familiar Riemann integral is rather straightforward. We shall avoid proofs for the most part and concentrate on the properties of the integral and the important classes of functions that arise in its study. Besides its use in the spectral theorem this theory also enables us to prove a Riesz representation theorem for the Banach space $(C[a,b], \|\cdot\|_\infty)$. We shall state this result at the end of this section.

Let a, b be two fixed, real numbers with $a < b$. A partition P of $[a, b]$ is a finite set $\{x_0, x_1, \ldots, x_n\}$ such that

$$a = x_0 < x_1 < \ldots < x_n = b$$

The maximum of the numbers $x_j - x_{j-1}$, $1 \leq j \leq n$, is called the mesh of the partition. Now let f, g be two real-valued functions on $[a, b]$. The $R - S$ sums of f with respect to g for the partition P are

$$\sum_{j=1}^{n} f(x_j^*)[g(x_j) - g(x_{j-1})]$$

where, for each fixed j, $x_j^* \in [x_{j-1}, x_j]$. If there is a number A such that for any $\varepsilon > 0$ we can find a $\delta > 0$ for which

$$\left| A - \sum_{j=1}^{n} f(x_j^*)[g(x_j) - g(x_{j-1})] \right| < \varepsilon$$

whenever $P = \{x_0, \ldots, x_n\}$ is any partition of $[a, b]$ having mesh less than δ (and $x_j^* \in [x_{j-1}, x_j]$ is arbitrary for $j = 1, 2, \ldots, n$), then we say that f is integrable over $[a, b]$ with respect to g. In this case we shall write

$$A = \int_a^b f(x) dg(x) \tag{1}$$

Observe that when $g(x) = x$ the $R - S$ sums reduce to the familiar Riemann sums and the integral of f over $[a, b]$ with respect to this g is just the Riemann integral of f over $[a, b]$. Furthermore, if g is continuous on $[a, b]$ and has a continuous derivative on (a, b) then one can show that

$$\int_a^b f(x) dg(x) = \int_a^b f(x) g'(x) dx$$

However, the functions g which we must deal with are often discontinuous, so this "reduction" will not concern us.

The function f in (1) is called, as usual, the integrand and the function g will be called the integrator. Our integral in linear in both of these; i.e., for any constants α, β

$$\int_a^b (\alpha f_1 + \beta f_2) dg = \alpha \int_a^b f_1 dg + \beta \int_a^b f_2 dg \qquad (2)$$

$$\int_a^b f d(\alpha g_1 + \beta g_2) = \alpha \int_a^b f dg_1 + \beta \int_a^b f dg_2 \qquad (3)$$

provided that all of these integrals exist.

The theory of the Riemann-Stieltjes integral most closely parallels that of the Riemann integral when the terms $g(x_j) - g(x_{j-1})$ appearing in the $R-S$ sums are non-negative. That is, for integrators which are of the following type.

Definition 1

A real-valued function g on $[a,b]$ is said to be monotonically increasing on this interval (respectively monotonically decreasing) if $g(x) \leq g(y)$ whenever $x \leq y$ (respectively $g(x) \geq g(y)$ whenever $x \leq y$). A function of either type is said to be monotonic on $[a,b]$.

A function which is monotonic on $[a,b]$ is clearly bounded on this interval; experience impells us to point out that such a function is never ∞ since it was assumed real-valued. Such functions can only have jump discontinuities and, since they are bounded, only countably many of these (see problem 1(c) below).

Theorem 1

Suppose that g is monotonically increasing on $[a,b]$ and that f is continuous on this interval. Then the integral of f with respect to g over $[a,b]$ exists and

$$|\int_a^b f(x) dg(x)| \leq \|f\|_\infty [g(b) - g(a)].$$

Proof. We shall only prove the inequality. For any $R-S$ sum we have

$$|\sum_{j=1}^n f(x_j^*)[g(x_j) - g(x_{j-1})]| \leq [\max_{a \leq x \leq b} |f(x)|] \sum_{j=1}^n [g(x_j) - g(x_{j-1})]$$
$$= \|f\|_\infty [g(b) - g(a)].$$

Since the integral is linear in the integrator we would expect that the first statement in Theorem 1, the existence of the integral for any continuous f, to be true for all g in the vector space generated by the monotonic functions. More explicitly: If g_1, g_2 are monotonically increasing on $[a,b]$ then so also is

THE RIEMANN-STIELTJES INTEGRAL   227

$g_1 + g_2$ and αg for $\alpha \geq 0$. Then (Section 2 just before Definition 2 and Exercises 2, problem 1) the set of all functions h such that there are monotonically increasing functions g_1, g_2 with $h = g_1 - g_2$, is a vector space and we would expect that

$$\int f\,dh = \int f\,dg_1 - \int f\,dg_2$$

exists for every f which is continuous on $[a, b]$. This is the case and the inequality in Theorem 1 has a nice analogue in this, more general, setting. Before formally stating the result let us look more closely at the vector space just mentioned.

Definition 2

Let h be a real-valued function on $[a, b]$. For any partition $P = \{x_0, x_1, \ldots, x_n\}$ of this interval we set

$$V(h, P) = \sum_{j=1}^{n} |h(x_j) - h(x_{j-1})|.$$

This is called the variation of h on P. The supremum of the set $\{V(h, P) | P$ any partition of $[a, b]\}$ is called the total variation of h on $[a, b]$ and is denoted by $V_a^b(h)$, or just $V(h)$. We shall say that h is of bounded variation on $[a, b]$ if $V_a^b(h) < \infty$.

The set of all functions which are of bounded variation on $[a, b]$ will be denoted by $BV[a, b]$, or just BV.

Any function g which is monotonically increasing on $[a, b]$ is of bounded variation on this interval and $V_a^b(g) = g(b) - g(a)$. The class of functions of bounded variation also arises naturally in discussions of the arc length of parametrically defined curves.

Theorem 2

A real-valued function h on $[a, b]$ is of bounded variation on this interval if, and only if, it can be written as the difference of two monotonically increasing functions (i.e., if, and only if, there exist g_1, g_2 both monotonically increasing on $[a, b]$ such that $h(x) = g_1(x) - g_2(x)$ for all x).

Proof. We take $x \in [a, b]$ and consider the total variations of h on $[a, x]$; i.e., $V_a^x(h)$. It is easy to see that this function is monotonically increasing on $[a, b]$. However, it turns out that

$$V_a^x(h) - h(x)$$

is also monotonically increasing on this interval. Thus

$$h(x) = V_a^x(h) - [V_a^x(h) - h(x)].$$

Corollary 1

Suppose that h is of bounded variation on $[a, b]$ and that f is continuous on this interval. Then the integral of f with respect to h over $[a, b]$ exists and

$$\left| \int_a^b f(x) dh(x) \right| \leq \|f\|_\infty V_a^b(h).$$

Proof. Again we prove only the inequality. For any $R - S$ sum we have

$$\left| \sum_{j=1}^n f(x_j^*)[h(x_j) - h(x_{j-1})] \right| \leq \max_{a \leq x \leq b} |f(x)| \sum_{j=1}^n |h(x_j) - h(x_{j-1})| \leq \|f\|_\infty V_a^b(h).$$

In Theorem 3 (below) we shall have to consider the question of whether two different functions can, when used as integrators, give us the same integral for every continuous function. In this connection we have:

a. Given any $h \in BV[a, b]$ we can find a function $h_1 \in BV[a, b]$ which is continuous from the right (i.e., $\lim_{\varepsilon \to 0+} h_1(x + \varepsilon) = h_1(x)$ for each $x \in [a, b)$) and satisfies

$$(*) \qquad \int_a^b f(x) dh(x) = \int_a^b f(x) dh_1(x)$$

for every function f which is continuous on $[a, b]$;

b. $h, h_1 \in BV[a, b]$, if both of these functions is continuous from the right and if $(*)$ holds for these functions, then $h(x) - h_1(x)$ is a constant on $[a, b]$.

All of the results of this section are valid for complex-valued integrands. One simply works with the real and imaginary parts separately. There are even instances when one must work with complex-valued integrators. We shall discuss this further in another section. For now we simply state that remarks (a) and (b) apply in this case also.

The Riemann-Stieltjes integral can be used to obtain a Riesz representation theorem for the space $(C[a, b], \|\cdot\|_\infty)$. Recall that a linear map from this space into the underlying field ($\mathbb{R}$ or $\mathbb{C}$) is called a linear functional on $C[a, b]$. We are interested in characterizing all continuous, linear functionals on this space. Note that, for $h \in BV[a, b]$, we may define

$$\sigma_h(f) = \int_a^b f(x) dh(x)$$

for every $f \in C[a,b]$. By (2) above, σ_h is a linear functional and, by Corollary 1 Theorem 2 we have
$$|\sigma_n(f)| \leq \|f\|_\infty V_a^b(h)$$
Thus σ_h is a continuous, linear functional on our space. The question we now ask is this: Given any continuous, linear functional σ on $C[a,b]$ does there exist on $h \in BV[a,b]$ such that
$$\sigma(f) = \sigma_h(f)$$
for all f? One obvious problem is the fact that two different functions of bounded variation can give us the same integral for all continuous functions. However, using (a) and (b) above we may state:

Theorem 3

Let σ be any continuous, linear functional on the Banach space $(C[a,b], \|\cdot\|_\infty)$. Then there is a unique function h on $[a,b]$ such that

i. h is of bounded variation on $[a,b]$, h is continuous from the right and $h(a) = 0$;

ii. $\sigma(f) = \int_a^b f(x) dh(x)$ for all $f \in C[a,b]$;

iii. $\|\sigma\| = V_a^b(h)$.

Proof. We shall only discuss how (a) and (b) enter into this result. Given σ one shows, never mind how, that there is an h on $[a,b]$, of bounded variation on this interval, such that
$$\int_a^b f(x) dh(x) = \sigma(f) \text{ for all } f \in C[a,b].$$

By (a) we can replace h by another $h_1 \in BV$ which is continuous from the right. But there are lots of these: $h_1(x) + c$ will also do. We fix the constant c by requiring that $h_1(x) + c$ equal zero when $x = a$. This gives us a unique function satisfying (i) and (ii). One then proves that (iii) holds as well.

Exercises 5

* 1. Let g be a monotonically increasing function on $[a,b]$.

   a. Show that $\lim_{x \to a^+} g(x)$ and $\lim_{x \to b^-} g(x)$ both exist.

CHAPTER 6 THE SPECTRAL THEOREM, PART II

b. For any $c \in (a,b)$ show that $\lim_{x \to c^-} g(x)$ and $\lim_{x \to c^+} g(x)$ both exist and that the first of these is less than or equal to the second.

c. Show that the set of points of $[a,b]$ at which g is not continuous is a countable set. Hint: Work on (a,b) and, for each fixed n, let $D_n = \{y \in (a,b) \mid \lim_{x \to y^+} g(x) - \lim_{x \to y^-} g(x) \geq \frac{1}{n}\}$. Show that each D_n is a finite set.

d. Extend the results in (a), (b), and (c) to the class BV.

2. a. Show that any vector space of functions on $[a,b]$ which contains the monotonic functions also contain $BV[a,b]$.

b. Show that any function $h \in BV[a,b]$ can be written as the difference of two monotonically decreasing functions on $[a,b]$.

c. If h is continuous on $[a,b]$ and has a bounded derivative on (a,b), then $h \in BV[a,b]$.

d. Let $f(x) = \sin\left(\frac{1}{x}\right)$ for $0 \leq x \leq 1$, $f(0) = 0$. Show that f is not of bounded variation on $[a,b]$.

3. Let $h, h_1 \in BV[a,b]$ and let α, β be any constants.

a. Show that $(\alpha h + \beta h_1) \in BV[a,b]$ and that $V(\alpha h + \beta h_1) \leq |\alpha|V(h) + |\beta|V(h_1)$.

b. Show that $V(h) = 0$ if, and only if, h is constant on $[a,b]$.

c. If h_1 is continuous at the point $x_0 \in (a,b)$, show that $\lim_{n \to \infty} V_{x_0}^{x_0 + \frac{1}{n}}(h_1) = 0$.

4. Choose and fix a partition $\{z_j\}_{j=1}^n$ of $[a,b]$ and let $a_0, a_1, \ldots, a_n$ be positive real numbers. Define $g(a) = 0$, $g(x) = a_0$ for $x \in (z_0, z_1)$, $g(x) = a_0 + a_1$ for $x \in (z_1, z_2)$, ..., $g(x) = a_0 + a_1 + \ldots + a_{k-1}$ for $x \in (z_{k-1}, z_k)$ and, finally, $g(b) = a_0 + a_1 + \ldots + a_n$. The function g can be defined at $z_1, z_2, \ldots, z_{n-1}$ is any way. Show that for any continuous function f on $[a,b]$

$$\int_a^b f(x) dg(x) = \sum_{k=0}^n a_k f(z_k).$$

5. Let f_1, f_2 be continuous on $[a,b]$ and let g be monotonically increasing on this interval.

a. For any $c \in (a,b)$ we have $\int_a^b f_1(x) dg(x) = \int_a^c f_1(x) dg(x) + \int_c^b f_1(x) dg(x)$.

b. Suppose that $f_1(x) \leq f_2(x)$ for all $x \in [a,b]$. Show that $\int_a^b f_1(x) dg(x) \leq \int_a^b f_2(x) dg(x)$.

6.6 The Spectral Theorem for Self-Adjoint Operators

We come at last to the major result of this chapter. Let T be a fixed, bounded, self-adjoint operator on a Hilbert space H, and let it have spectral bounds m, M (Definition 1.1). In Section 4 we associated with any such T a family of orthogonal projection $\{E(s)|s \text{ real }\}$ such that (Theorem 4.1):

a. $E(s) \leq E(t)$ when $s \leq t$;

b. $E(t) = 0$ for $t < m$;

c. $E(t) = I$ for $M \leq t$;

d. $\lim_{\varepsilon \to 0+} \|\{E(s+\varepsilon) - E(s)\}x\| = 0$

where, in (d), the limit is zero for each s and every $x \in H$. We may think of the function $s \to E(s)$ as a projection-valued function of the real variable s which increases monotonically from 0 to I. The spectral theorem shows that the operator T is the "Riemann-Stieltjes" integral of the function $f(t) = t$ with respect to the monotonic function $s \to E(s)$. A corollary of this result shows that, for any continuous function $f(t)$, the operator $f(T)$, which has meaning by Section 6.3, is the "Riemann-Stieltjes" integral of $f(t)$ with respect to $s \to E(s)$.

Definition 1

Let a, b be two real numbers with $a < m$ and $M \leq b$. A real-valued function f on $[a, b]$ is said to be an E-integrable function if there is an operator $S \in \mathcal{S}$ (class of all self-adjoint operators on H) such that: For any $\varepsilon > 0$ there is a $\delta > 0$ for which

$$\|S - \sum_{j=1}^{n} f(t_j^*)[E(t_j) - E(t_{j-1})]\| < \varepsilon$$

whenever, $a = t_0 < t_1 < \ldots < t_n = b$ is any partition of $[a, b]$ with mesh (Section 5) less than δ (as usual, t^*, denotes any arbitrary element of $[t_{j-1}, t_j]$, $j = 1, 2, \ldots, n$). We call S the E-integral of f.

Observe that since $t_{j-1} < t_j$ the difference $E(t_j) - E(t_{j-1})$ is an orthogonal projection on H (Theorem 3.3). Also note that the operator S, if it exists, is unique. Let us sketch the proof of this. Suppose that S and S_1 were two operators such that, for the same function f, both S and S_1 are E-integrals of f. Then for any $\varepsilon > 0$ we could choose two partitions P, P_1 such that $\|S - \sum(P)\| < \varepsilon$ and $\|S_1 - \sum(P_1)\| < \varepsilon$; here $\sum(P)$ is a short-hand for the

232   CHAPTER 6   THE SPECTRAL THEOREM, PART II

sum in Definition 1. Now $P_1 \cup P$ is also a partition of $[a,b]$ and its mesh is no larger than the larger of the meshes of P and P_1. Hence we must also have

$$\|S - \sum(P \cup P_1)\| < \varepsilon \text{ and } \|S_1 - \sum(P \cup P_1)\| < \varepsilon$$

But then

$$\|S - S_1\| \leq \|S - \sum(P \cup P_1)\| + \|\sum(P \cup P_1) - S_1\| < 2\varepsilon$$

Since $\varepsilon > 0$ was arbitrary it follows that $S = S_1$.

We may now, without ambiguity, write

$$S = \int_a^b f(t)dE(t)$$

when f and S satisfy Definition 1.

Now which functions are E-integrable? Well clearly any constant function is. In fact, if $f(t) = \alpha$ for all t, then $\int_a^b f(t)dE(t) = \int_a^b \alpha dE(t) = \alpha I$. Also:

Theorem 1

(Spectral Theorem). Let a, b be any two real numbers with $a < m$, $M \leq b$. Then the function $f(t) = t$ from $[a,b]$ into $\mathbb{R}$ is E-integrable. Furthermore,

$$T = \int_a^b t\, dE(t)$$

Proof. Let s, u be fixed elements of $[m, M)$ and suppose that $s < u$. Recall that $e_u(t) = 0$ when $u < t$, and it equals one when $t \leq u$. The function $e_s(t)$ is defined similarly. Observe now that

$$\begin{aligned} e_u(t) - e_s(t) &= 1 \text{ for } t \in (s,u) \cap [m,M] \\ &= 0 \text{ for all other } t. \end{aligned}$$

Thus for all t we have the chain of inequalities:

$$s[e_u(t) - e_s(t)] \leq t[e_u(t) - e_s(t)] \leq u[e_u(t) - e_s(t)]$$

Now let us use once again the map Ω defined in Exercises 3. Recall $\Omega(f) = f(T)$ for all $f \in L[m,M]$ and, see the proof of Theorem 4.1, $\Omega(e_s) = e_s(T) = E(s)$ for each s. Also, we have seen (Exercises 3, problem 1(a), (b), and (c)) that Ω is linear, multiplicative and order preserving (so $\Omega(fg) = \Omega(f)\Omega(g)$, $f \leq g$ implies $\Omega(g) \leq \Omega(g)$). We now apply Ω to our chain of inequalities:

$$\Omega(s)[\Omega(e_u(t)) - \Omega(e_s(t))] \leq \Omega(t)[\Omega(e_u(t)) - \Omega(e_s(t))] \leq \Omega(u)[\Omega(e_u(t)) - \Omega(e_s(t))]$$

THE SPECTRAL THEOREM FOR SELF-ADJOINT OPERATORS 233

or, since $\Omega(t) = T$, $\Omega(s) = sI$ and $\Omega(u) = uI$,

$$sI[E(u) - E(s)] \leq T[E(u) - E(s)] \leq uI[E(u) - E(s)] \qquad (1)$$

Now given $\varepsilon > 0$ choose a partition $P = \{s_0, s_1, \ldots, s_n\}$ of $[a, b]$ with mesh less than ε; so $a = s_0 < s_1 < \ldots < s_n = b$ and $\max_{1 \leq j \leq n}(s_j - s_{j-1}) < \varepsilon$. Using (1) repeatedly we obtain:

$$\sum_{k=1}^{n} s_{k-1} I[E(s_k) - E(s_{k-1})] \leq T \sum_{k=1}^{n} [E(s_k) - E(s_{k-1})] \leq \sum_{k=1}^{n} s_k I[E(s_k) - E(s_{k-1})]$$

However, $a < m < M \leq b$ and so the middle sum is just

$$E(b) - E(a) = I - 0 = I$$

Thus our last chain of inequalities is, dropping the I,

$$\sum_{k=1}^{n} s_{k-1}[E(s_k) - E(s_{k-1})] \leq T \leq \sum_{k=1}^{n} s_k[E(s_k) - E(s_{k-1})] \qquad (2)$$

Next we choose, for $k = 1, 2, \ldots, n$, arbitrary numbers $t_k \in [s_{k-1}, s_k]$ and subtract $\sum_{k=1}^{n} t_k[E(s_k) - E(s_{k-1})]$ from each term in (2) to get:

$$\sum_{k=1}^{n}(s_{k-1} - t_k)[E(s_k) - E(s_{k-1})] \leq T - \sum_{k=1}^{n} t_k[E(s_k) - E(s_{k-1})] \qquad (3)$$

$$\leq \sum_{k=1}^{n}(s_k - t_k)[E(s_k) - E(s_{k-1})]$$

We need two more inequalities. Since $s_{k-1} < s_k$, $E(s_{k-1}) < E(s_k)$ so $E(s_k) - E(s_{k-1}) \geq 0$. Also, since our partition has mesh less than ε, $-\varepsilon < s_{k-1} - t_k$ and $s_k - t_k < \varepsilon$. Using these and Lemma 3.2 part (v) we may write:

$$-\varepsilon[E(s_k) - E(s_{k-1})] \leq (s_{k-1} - t_k)[E(s_k) - E(s_{k-1})] \qquad (4)$$

$$(s_k - t_k)[E(s_k) - E(s_{k-1})] \leq \varepsilon[E(s_k) - E(s_{k-1})] \qquad (5)$$

Finally, by combining (4) and (5) with (3) we obtain

$$-\varepsilon I \leq T - \sum_{k=1}^{n} t_k[E(s_k) - E(s_{k-1})] \leq \varepsilon I \qquad (6)$$

The definition of $\leq$ for self-adjoint operators (Definition 3.1) tells us that, from (6), the middle operator has spectral bounds which are between $-\varepsilon$ and ε (Definition 1.1). Thus, by Theorem 1.1, our chain of inequalities (6) is equivalent to

$$\|T - \sum_{k=1}^{n} t_k[E(s_k) - E(s_{k-1})]\| < \varepsilon \qquad (7)$$

and this proves our theorem.

Corollary 1

Any real-valued function f which is continuous on $[a, b]$ is E-integrable and

$$f(T) = \int_a^b f(t) dE(t)$$

Proof. We will first prove the result for a polynomial $p(t)$. In order to do this it suffices to show that for any integer $r \geq 0$,

$$T^r = \int_a^b t^r dE(t)$$

By property (a) of a spectral family and Theorem 3.3. we see that $\{[E(s_k) - E(s_{k-1})]\}_{k=1}^n$ is a family of orthogonal projections. Also

$$[E(s_k) - E(s_{k-1})][E(s_\ell) - E(s_{\ell-1})] = 0$$

when $\ell \neq k$; just recall that $E(s) = e_s(T)$ and use the properties of the functions $e_s(t)$. Thus

$$\{\sum_{k=1}^n t_k [E(s_k) - E(s_{k-1})]\}^r = \sum_{k=1}^n t_k^r [E(s_k) - E(s_{k-1})]$$

and our result follows from this.

Now given any continuous, real-valued function f on $[a, b]$ and any $\varepsilon > 0$ we choose a polynomial $p(t)$ such that

$$|f(t) - p(t)| < \varepsilon/3$$

for each $t \in [a, b]$ (Section 1.7, Corollary 3 to Theorem 1). Then

$$-\frac{\varepsilon}{3} I \leq f(T) - p(T) \leq \frac{\varepsilon}{3} I$$

by Exercises 3, problem 1(c). It follows from Theorem 1.1 that

$$\|f(T) - p(T)\| < \varepsilon/3$$

Now let $P = \{s_1, s_1, \ldots, s_n\}$ be any partition of $[a, b]$ and write

$$S_f = \sum_{k=1}^n f(s_k^*)[E(s_k) - E(s_{k-1})], \quad S_p = \sum_{k=1}^n p(s_k^*)[E(s_k) - E(s_{k-1})]$$

where $s_k^* \in [s_{k-1}, s_k]$ for each $k = 1, 2, \ldots, n$. Observe that

$$-\frac{\varepsilon}{3} I \leq S_f - S_p \leq \frac{\varepsilon}{3} I$$

THE SPECTRAL THEOREM FOR SELF-ADJOINT OPERATORS

and hence, as above, we have
$$\|S_f - S_p\| < \frac{\varepsilon}{3}$$

Finally, choose $\delta > 0$ so that $\|p(T) - S_p\| < \frac{\varepsilon}{3}$ for any partition having mesh less than δ; we can do this because $p(t)$ is a polynomial and so we may use the first part of this proof. Thus for any partition having mesh less than δ we can combine our observations above to write

$$\|f(T) - S_f\| \leq \|f(T) - p(T)\| + \|p(T) - S_p\| + \|S_p - S_f\| < \varepsilon$$

and this completes the proof.

Our results above, the theorem and its corollary, have a number of useful variants which we discuss now in a series of remarks.

1. The corollary is valid for any continuous, complex-valued function on $[a, b]$ for given such a function we may work with its real and imaginary parts separately.

2. Let $x \in H$. Then line (7) in the proof of the spectral theorem can be used to write
$$\|Tx - \sum_{k=1}^{n} t_k[E(s_k)x - E(s_{k-1})x]\| \leq \varepsilon\|x\|$$
for any partition having mesh less than ε. More generally, see the last inequality in the proof of Corollary 1,
$$\|f(T)x - \sum_{k=1}^{n} f(t_k)[E(s_k)x - E(s_{k-1})x]\| \leq \varepsilon\|x\|$$
for any fixed real- (or complex-) valued function f which is continuous on $[a, b]$ and any partition having mesh less than ε. We express these results symbolically as follows:
$$Tx = \int_a^b t\,d(E(t)x) \quad f(T)x = \int_a^b f(t)d(E(t)x)$$

3. For any two vectors $x, y \in H$ we have, again by line (7) of the proof of the spectral theorem,
$$|\langle Tx, y\rangle - \sum_{k=1}^{n} t_k[\langle E(s_k)x, y\rangle - \langle E(s_{k-1})x, y\rangle]| \leq \varepsilon\|x\|\|y\|$$
for all partitions having mesh less than ε. Thus
$$\langle Tx, y\rangle = \int_a^b t\,d\langle E(t)x, y\rangle, \quad \langle f(T)x, y\rangle = \int_a^b f(t)d\langle E(t)x, y\rangle$$

where f is as in Remark 2. Here we really have a "Riemann-Stieltjes" type integral even though our integrator, $\langle E(t)x, y\rangle$ as a function of t with x and y fixed, is complex-valued. What one may do is use the polar identity (Exercises 1.4, problem 1(d)) to write $\langle E(t)x, y\rangle$ as a linear combination of four real-valued, monotonically increasing functions (see also Exercises 4.3, problem 2(c)).

4. For any $x \in H$ and any complex-valued, continuous functions f on $[a, b]$ we have (see Theorem 3.3)

$$\Big| \|f(T)x\|^2 - \sum_{k=1}^{n} |f(t_k)|^2 [\|E(s_k)x\|^2 - \|E(s_{k-1})x\|^2] \Big| \leq \varepsilon \|x\|^2$$

for any partition having norm less than ε. Thus we write

$$\|f(T)x\|^2 = \int_a^b |f(t)|^2 d(\|E(t)x\|^2)$$

for each $x \in H$.

5. Recall now that m, M are the spectral bounds of our operator T and that we chose a, b to be arbitrary real numbers such that $a < m$ and $M \leq b$. The reader may wonder why we did this? Why not simply work with $[m, M]$ and state our results for this interval? The trouble here is that the family $\{E(s)\}$, which we have seen is continuous from the right (i.e., $\lim_{\varepsilon \to 0+} \|[E(s+\varepsilon) - E(s)]x\| = 0$ for each $x \in H$), need not be continuous from the left. In particular,

$$\lim_{s \to m^-} \|[E(s) - E(m)]x\|$$

need *not* be zero (see Theorem 2 below). We could force this limit to be zero by defining $E(m) = 0$, but this would destroy the right-hand continuity of the family $\{E(s)\}$ and so we shall not do that. Some writers define

$$\int_{m-0}^{M} t dE(t) = \lim_{a \to m^-} \int_a^M t dE(t)$$

Using this notation we have

$$T = \int_{m-0}^{M} t dE(t), \quad f(T) = \int_{m-0}^{M} f(t) dE(t)$$

etc.

6. We have associated with our self-adjoint operator T, having spectral bounds m and M, a family of orthogonal projections $\{E(s)|s \in \mathbb{R}\}$ such that

   a. $E(s) \leq E(t)$ when $s \leq t$;
   b. $E(t) = 0$ for $t < m$;
   c. $E(t) = I$ for $M \leq t$;
   d. $\lim_{\varepsilon \to 0^+} \|[E(s+\varepsilon) - E(s)]x\| = 0$

for each s and every $x \in H$. This is called the spectral family of T on $[m, M]$. Here we shall justify this terminology by showing that the family $\{E(s)\}$ is uniquely determined by T.

Suppose that $\{F(s)|s \in \mathbb{R}\}$ is another family of orthogonal projections which satisfies (a)–(d). Then for any $x, y \in H$ and all real-valued, continuous functions $f(t)$ on $[a, b]$ we have:

$$\int_a^b f(t)d\langle F(t)x, y\rangle = \int_a^b f(t)d\langle E(t)x, y\rangle$$

where a, b are arbitrary but $a < m$, $M \leq b$ (see Remark 3). It follows that

$$\langle E(t)x, y\rangle - \langle F(t)x, y\rangle = \langle [E(t) - F(t)]x, y\rangle$$

is a constant function of t (Section 5, Remark (b)). However, for $M \leq t$ we have both $F(t) = I$ and $E(t) = I$, hence this constant must be zero. Thus

$$\langle [E(t) - F(t)]x, y\rangle = 0$$

for all t and each fixed $x, y \in H$. But since y is arbitrary our equation can only hold if

$$[E(t) - F(t)]x = 0$$

for all $x \in H$ and this implies that $E(t) = F(t)$ for all t.

Corollary 2

Let T be a bounded, self-adjoint operator on H with spectral bounds m, M and let $\{E(s)|s \in \mathbb{R}\}$ be the spectral family of T. A bounded, linear operator S on H commutes with T if, and only if, it commutes with every $E(s)$, $s \in \mathbb{R}$.

Proof. We have already seen that if $ST = TS$ then $SE(s) = E(s)S$ for every $s \in \mathbb{R}$, (Theorem 4.1, and its corollary). Suppose now that $SE(s) =$

$E(s)S$ for every s. We refer once again to line (7) in the proof of the spectral theorem. For any partition with mesh less than ε we have

$$\|ST - TS\| \leq \|ST - S\sum_{k=1}^{n} t_k[E(s_k) - E(s_{k-1})]\| +$$

$$\|\{\sum_{k=1}^{n} t_k[E(s_k) - E(s_{k-1})]\}S - TS\| < 2\varepsilon$$

because

$$S\sum_{k=1}^{n} t_k[E(s_k) - E(s_{k-1})] = \{\sum_{k=1}^{n} t_k[E(s_k) - E(s_{k-1})]\}S$$

Since $\varepsilon > 0$ is arbitrary $ST = TS$.

Using the notation of Corollary 2 we know that

$$\lim_{\varepsilon \to 0+} \|[E(s+\varepsilon) - E(s)]x\| = 0$$

for each s and each $x \in H$. We have already remarked that

$$\lim_{\varepsilon \to 0+} \|[E(s) - E(s-\varepsilon)]x\|$$

need not be zero (Remark 5). Let us define

$$E(s-0) = \lim_{\varepsilon \to 0+} E(s-\varepsilon)$$

where this is a strong limit (see Exercises 5 problem 1(b)). Then we have:

Theorem 2

For each fixed $s_0 \in \mathbb{R}$ the operator $E(s_0) - E(s_0 - 0)$ is the orthogonal projection of H onto the subspace $G(s_0) = \{x \in H | Tx = s_0 x\}$. Consequently s_0 is an eigenvalue of T if, and only if, $E(s_0 - 0) < E(s_0)$.

Proof. By Theorem 3.3 we know that, since $E(s_0 - 0) \leq E(s_0)$,

$$E(s_0) - E(s_0 - 0)$$

is an orthogonal projection on some subspace G of H. For $x \in G$, (i) $E(s)x = 0$ for $s < s_0$; (ii) $E(s)x = x$ for $s_0 \leq s$; because $E(s) \leq E(s_0)$ when $s \leq s_0$ and $E(s_0) \leq E(s)$ when $s_0 \leq s$ (see Theorem 3.3 again). Now let $a < m$ $M \leq b$ and consider any Riemann-Stieltjes sum approximating the integral

$$\int_a^b t \, d(E(t)x)$$

(Remark 2). By adjoining the point s_0 to any partition of $[a,b]$ we do not increase its mesh, and so we may assume that all the approximating sums we work with have s_0 as a point of this underlying partitions of $[a,b]$. But from (i) and (ii) any such sum reduces to one single term $s_0(E(s_0)x) = s_0 x$ (Exercises 5 problem 4). Thus

$$Tx = \int_a^b t\,d(E(t)x) = s_0 x$$

showing that $G \subseteq G(s_0)$.

Next suppose that $(T - s_0 I)x = 0$. Then by Remark 4 (iii) $\|(T - s_0 I)x\|^2 = \int_a^b (t - s_0)^2 d\|E(t)x\|^2 = 0$ where $t \to \|E(t)x\|^2$ is a monotonically increasing function on $\mathbf{R}$. Suppose that $\|E(s')x\|^2 < \|x\|^2$ for some $s' > s_0$. Then we have

$$\int_a^b (t - s_0)^2 d\|E(t)x\|^2 \geq \int_{s'}^b (t - \mu)^2 d\|E(t)x\|^2 \geq (s' - s_0)^2 \|E(s')x\|^2 > 0$$

which contradicts (iii). Thus $\|E(s')x\|^2 = \|x\|^2$ for $s' \geq s_0$. Finally, suppose $0 < \|E(s')x\|^2$ for some $s' < s_0$. Then

$$\int_a^b (t - s_0)^2 d\|E(t)x\|^2 \geq \int_a^{s'} (t - s_0)^2 d\|E(t)x\|^2 \geq (s' - s_0)^2 \|E(s')x\|^2 > 0$$

which again contradicts (iii). Thus $\|E(s')x\|^2 = 0$ for $s' < s_0$. We conclude that $[E(s_0) - E(s_0 - 0)]x = x$ and hence $G(s_0) \supseteq G$.

Theorem 2 tells us that the function $s \to E(s)$, which is a monotonically increasing projection-valued function of the real variable s, has a jump discontinuity at every eigenvalue of T; and, conversely, any point where this function has a jump discontinuity is an eigenvalue of this operator. Since the only discontinuities of a monotonic function are of this kind all points of the spectrum of T which are not eigenvalues of this operator and all regular points of T are points of continuity of our function. The next theorem, which we will not prove, shows that every regular point of T has a neighborhood on which our function is constant. Thus, this function does all of its increasing on the set $\sigma(T)$.

Theorem 3

Let $\{E(s)|s \in \mathbf{R}\}$ be the spectral family of the bounded, self-adjoint operator T. A number $s_0 \in \mathbf{R}$ is a regular value for T if, and only if, there is an $\varepsilon > 0$ such that $E(s_0 - \varepsilon) = E(s_0 + \varepsilon)$. Consequently, $s_0 \in \sigma(T)$ if, and only if, $E(s_0 - \varepsilon) < E(s_0 + \varepsilon)$ for all $\varepsilon > 0$.

In Section 5.5, we extended the spectral theorem for compact, self-adjoint operators to the class of compact, normal operators by making use of the fact that any normal operator A can be written in the form

$$A = S + iT$$

where S, T are self-adjoint operators and $ST = TS$. Not surprisingly, we can extend the results of this section to the class of normal operators as well. There is an interesting analogy here. The class of normal operators "corresponds" to $\mathbb{C}$, and the class of self-adjoint operators "corresponds" to $\mathbb{R}$. We can push this analogy a little bit further. There is a class of operators that corresponds to the unit circle in $\mathbb{C}$ (i.e., the set $\{z \in \mathbb{C} | |z| = 1\}$).

Definition 2

Let U be a bounded, linear operator on H. We shall say that U is a unitary operator if U is invertible and $U^{-1} = U^*$.

This class of operators is very important and arises naturally in many investigations (see Chapter 7). For now let us simply mention that there is a spectral theorem for unitary operators also. We shall give a precise statement of the spectral theorems for normal and unitary operators in Appendix III (see also section 4 of Chapter 7).

Exercises 6

1. Recall once again the multiplication operator on $L^2[a, b]$, $P[g(x)] = xg(x)$ for all g in this space. Let $\{E(s)\}$ be the spectral family defined by P (see the example in Section 4) and consider the map $s \to E(s)$. Explain why this function must be continuous.

2. Show that the spectral theorem for a compact, self-adjoint operator is a special case of Theorem 1 above as follows: Work with the case of a compact, self-adjoint operator A which is *not* of finite rank and suppose that all of its eigenvalues $\{\lambda_j\}_{j=1}^{\infty}$ are negative. By Section 5.4 Theorem 2 we have $A = \sum_{j=1}^{\infty} \lambda_j P_j$ where P_j is the orthogonal projection of H onto the eigenspace M_j corresponding to the eigenvalue λ_j. We have assumed that $\lambda_1 < \lambda_2 < \ldots < 0$, and $\lim_{j \to \infty} \lambda_j = 0$ as we know. Suppose $\lambda_0 = 0$ and $M_0 = \{x \in H | Ax = 0\}$. Then $\sigma(A) = \{\lambda_j\}_{j=0}^{\infty}$.

    a. Compute the spectral bounds for A.

    b. For each $\lambda \in \mathbb{R}$ let $P(\lambda)$ be the orthogonal projection of H onto the subspace $\{\sum \oplus M_j | \lambda_j \leq \lambda\}$. Show that the family $\{P(\lambda) | \lambda \in \mathbb{R}\}$

satisfies (a), (b), (c) and (d) listed above, and so this is the spectral family of A.

c. Show that, if m, M are the spectral bounds for A and if $a < m$, $M \leq b$, $\int_a^b \lambda dP(\lambda) = \sum_{j=1}^{\infty} \lambda_j P_j$ (see Exercises 5, problem 4).

d. Suppose now that f is a real-valued, continuous function on $[a, b]$ such that $f(0) = 0$ (see problem 3). Show that

$$f(A) = \int_a^b f(\lambda)dP(\lambda) = \sum_{j=1}^{\infty} f(\lambda_j)P_j$$

3. Let A be a compact, self-adjoint operator with spectral bounds m, M.

   a. Explain why $0 \in [m, M]$.

   b. For any real-valued, continuous function f on $[m, M]$ the operator $f(A)$ has meaning (Section 3 just after Corollary 1) and is a self-adjoint operator. In general, however, it will *not* be compact. For example $p(t) = t + 3$ gives $p(A) = A + 3I$ which is not a compact operator (unless H is finite dimensional). However, if $p(t)$ is polynomial with real coefficients (to insure self-adjointness) and zero constant term then $p(A)$ is a compact operator (Exercises 2.5, problem 2(a) and 6(b)). Show that for any real-valued, continuous function f on $[m, M]$ such that $f(0) = 0$, the operator $f(A)$ is self-adjoint (that's easy) and compact.

4. Let U be a unitary operator on H.

   a. Show that $\langle Ux, Uy \rangle = \langle x, y \rangle$ for all $x, y \in H$. Observe that $\|U\| = 1$ and that U^* is also a unitary operator.

   b. Show that $\sigma(U) \subseteq \{z \in \mathbb{C} | |z| = 1\}$. Hint: Use Corollary 2 to Theorem 2 (Section 2.3) and then use Corollary 3 to this theorem together with the fact that zero is a regular value of U.

7

UNBOUNDED OPERATORS

Many of the important operators which arise in the applications of operator theory are unbounded. Operators of this kind also arise quite naturally in the study of the resolvent of a bounded, linear operator A (i.e., the study of $(A - \lambda I)^{-1}$ for various λ). We shall present the details of this below. The properties of unbounded operators are somewhat more subtle than those of bounded, linear operators and so we shall begin our discussion with some simple examples.

7.1 Closed Operators

We shall begin our discussion of unbounded operators by looking at a well-known type of second-order differential equation. Suppose that ℓ is a positive real number, that λ_0 is a complex number, and that $f(x)$ is a given function in the space $L^2[0, \ell]$. We want to solve the equation

$$-\frac{d^2u}{dx^2} - \lambda_0 u = f(x), \quad 0 \leq x \leq \ell \qquad (1)$$

together with the boundary conditions $u(0) = 0 = u(\ell)$. The first thing we do is replace equation (1) by the associated "eigenvalue problem"

$$-\frac{d^2\varphi}{dx^2} = \lambda\varphi, \quad 0 \le x \le \ell, \quad \varphi(0) = 0 = \varphi(\ell) \tag{2}$$

where here we regard λ as an unknown constant. It is easy to see that (2) has a non-trivial (i.e., not identically zero) solution only for certain values of the constant λ (see problem 4 below). Specifically, only for $\lambda = \lambda_n$ where

$$\lambda_n = \frac{n^2\pi^2}{\ell^2}, \quad n = 1, 2, 3, \ldots \tag{3}$$

For each n the solution to (2) when $\lambda = \lambda_n$ is $\sin n\pi x/\ell$ which is in the space $L^2[0, \ell]$ and hence can be normalized. Thus corresponding to the sequence (3) we have a sequence of normal eigenvectors

$$\varphi_n(x) = \left(\frac{2}{\ell}\right)^{1/2} \sin\frac{n\pi x}{\ell}, \quad n = 1, 2, 3, \ldots \tag{4}$$

We can solve equation (1) for any given f in $L^2[0, \ell]$ because the eigenfunctions (4) are an orthonormal basis for this space (Section 1.6 Definition 3, Exercises 1.7 problem 3, Section 1.9 just after Definition 1). We can write

$$f(x) = \sum_{n=1}^{\infty} f_n\varphi_n(x), \quad u(x) = \sum_{n=1}^{\infty} u_n\varphi_n(x) \tag{5}$$

where

$$f_n = \langle f, \varphi_n \rangle = \left(\frac{2}{\ell}\right)^{1/2} \int_0^\ell f(x) \sin\frac{n\pi x}{\ell} dx, \quad n = 1, 2, 3, \ldots \tag{6}$$

$$u_n = \langle u, \varphi_n \rangle = \left(\frac{2}{\ell}\right)^{1/2} \int_0^\ell u(x) \sin\frac{n\pi x}{\ell} dx, \quad n = 1, 2, 3, \ldots$$

Remember that here $f(x)$ is known and hence the numbers f_n are known. What we want to find are the numbers u_n so that we can then get $u(x)$ from the second series in (5). We may calculate these numbers as follows: Multiply equation (1) by $\varphi_n(x)$ and integrate from 0 to ℓ to get

$$-\left(\frac{2}{\ell}\right)^{1/2} \int_0^\ell u''(x) \sin\frac{n\pi x}{\ell} dx - \lambda_0 u_n = f_n$$

Integrating by parts and using the boundary conditions on $u(x)$ we obtain

$$\left(\frac{n^2\pi^2}{\ell^2} - \lambda_0\right) u_n = f_n, \quad n = 1, 2, 3, \ldots \tag{7}$$

Now if the given constant λ_0 is *not* one of the eigenvalues then we can solve for each u_n and get

$$u(x) = \sum_{n=1}^{\infty} \left(\frac{f_n}{\frac{n^2\pi^2}{\ell^2} - \lambda_0} \right) \varphi_n(x)$$

$$= \sum_{n=1}^{\infty} \left(\frac{2/\ell}{\frac{n^2\pi^2}{\ell^2} - \lambda_0} \right) \left(\int_0^\ell f(x) \sin \frac{n\pi x}{\ell} dx \right) \sin \frac{n\pi x}{\ell} \quad (8)$$

One can show that (8) is the unique solution to (1) with the given boundary conditions. The solution to this problem when λ_0 is an eigenvalue is discussed in problem 4 below.

Let us now look at the discussion above from the point of view of operator theory. The first thing we did was to regard $(-d^2/dx^2)$ as a linear operator on $L^2[0, \ell]$ and to then find its eigenfunctions. Now this really isn't an operator on $L^2[0, \ell]$ since we cannot differentiate every L^2-function. What we have here is mapping from some subset D of $L^2[0, \ell]$ into this space. The set D must be specified or we have not really defined $(-d^2/dx^2)$ as an operator in L^2. In this case it is natural to take D to be the set of all functions $u(x)$ in $L^2[0, \ell]$ which have second derivatives in this space and which satisfy $u(0) = 0 = u(\ell)$. It is easy to see that D is a linear manifold and that, since it contains each of the functions (4), it is dense in $L^2[0, \ell]$ (Section 1.9 Definition 1). Finally, we observe that the linear mapping $(-d^2/dx^2)$ from D into $L^2[0, \ell]$ is unbounded; for each φ_n has norm one while equations (2) and (3) show that the norm of $(-d^2\varphi_n/dx^2)$ is $n^2\pi/\ell^2$ which tends to infinity with n.

Our solution to problem (1) and the discussion just given shows the importance of unbounded, linear operators which are defined in some dense, linear manifold of a Hilbert space. We turn now to a more systematic discussion of this kind of operator. Let us begin with a very simple, but illuminating, example. Recall the Banach space $(C[0,1], \|\cdot\|_\infty)$; here $C[0,1]$ is the space of all continuous functions on $[0,1]$ and, for each f in this space,

$$\|f\|_\infty = \max_{0 \le t \le 1} |f(t)|$$

Now, of course, some of the functions in this space have derivatives and some do not. So if we want to consider $\frac{d}{dx}$ as an operator on $C[0,1]$ we have to realize that it cannot operate on every function in this space and, once we realize that, we have to specify just what functions it is to operate on.

Suppose we say that $\frac{d}{dx}$ is to operate on all f in $C[0,1]$ such that $f'(t)$ exists at each $t \in (0,1)$. There are two immediate difficulties. We want $\frac{d}{dx}$ to map a subset of $C[0,1]$ into $C[0,1]$. Hence our first problem is: How are $f'(0)$ and $f'(1)$ to be defined — they must be defined if $\frac{df}{dx}$ is to be in $C[0,1]$. Our second problem is that f' may exist at each point of $(0,1)$ without the function f' being continuous on this interval. Now there is an easy way of avoiding both problems. We let $D = \{f \in C[0,1] | f'$ exists and is uniformly continuous on $(0,1)$}. Recall that, by Exercises 3.2 problem 4(a), a function which is uniformly continuous on $(0,1)$ can be extended in one, and only one, way to a function which is (uniformly) continuous on $[0,1]$. Thus for each $f \in D$ the function f' can be extended to a unique function, which we shall call f' also, which is continuous on $[0,1]$. With this understanding we may define an operator in (not on) $C[0,1]$ as follows: For each $f \in D$,

$$\frac{df}{dx} = f'$$

Observe that

$$\frac{d}{dx} : D \to C[0,1]$$

that D is a linear manifold in $C[0,1]$ and that $\frac{d}{dx}$ is a linear map.

The linear manifold D contains every polynomial function and so (Section 1.9 just after Definition 1) D is dense in $(C[0,1], \|\cdot\|_\infty)$. Let us show that our operator is unbounded: Each of the functions $x, x^2, \ldots, x^n, \ldots$ is in D and each of them has norm one (remember that we are using the sup norm here), however

$$\|\frac{d}{dx} x^n\|_\infty = n, \quad n = 1, 2, \ldots$$

Definition 1

Let $(E, \|\cdot\|)$ be a normed space and let D be a linear manifold in E. A linear map T,

$$T : D \to E$$

will be called a linear operator in E. We shall call D the domain of T and often denote it by $D(T)$.

Many of the examples we shall give will be of operators in E whose domains are dense in E. These operators are especially important.

There is another way to see that the operator $\frac{d}{dx}$ defined in $(C[0,1], \|\cdot\|_\infty)$, as discussed above, is not bounded. Choose $f \in C[0,1]$ such that f has no derivative at any point of $(0,1)$; we know that there are lots of functions like this. By the Weierstrass approximation theorem (Section 1.7, Corollary 3 to Theorem 1) we can find a sequence of polynomials $p_1, p_2, \ldots$ such that $\lim p_n(t) = f(t)$ uniformly over $[0,1]$; i.e.,

$$\lim_{n \to \infty} \|f - p_n\|_\infty = 0$$

Now, for each n, the function $\frac{d}{dx} p_n(t)$ is certainly in $C[0,1]$. But the sequence $\{p'_n(t)\}$ could not converge for $\|\cdot\|_\infty$; for, if it did, its limit, say $g(t)$, would be in $C[0,1]$ (because $(C[0,1], \|\cdot\|_\infty)$ is a Banach space) and, furthermore, $\frac{d}{dx} f(t) = g(t)$ for all $t \in (0,1)$ contradicting the fact that f is not differentiable. Here we are using a standard theorem from advanced calculus: If $\{p_n\}$ converges uniformly to f and if $\{p'_n\}$ converges uniformly to g, then f is differentiable and $f' = g$ (see [12; p. 602]). We can restate this theorem as follows: If $\{f_n\} \subseteq D$ and if we have both (i) $\lim f_n = f$ and (ii) $\lim \frac{d}{dx} f_n = g$ for $\|\cdot\|_\infty$, then $f \in D$ and $\frac{df}{dx} = g$.

This important property is shared by many unbounded operators.

Definition 2

Let $(E, \|\cdot\|)$ be a normed space, let D be a linear manifold in E and let $T : D \to E$ be a linear operator in E with domain D. We shall say that T is a closed linear operator if for any sequence $\{x_n\} \subseteq D$ such that both $\{x_n\}$ and $\{Tx_n\}$ converge in E, i.e.

$$\lim x_n = y \quad \text{and} \quad \lim Tx_n = z$$

we have $y \in D$ and $Ty = z$.

An operator which is not closed may have a closed "extension." We shall see examples of this below. For now we just define this concept.

Definition 3

Let $(E, \|\cdot\|)$ be a normed space and let S, T be two linear operators in E with domains $D(S)$ and $D(T)$ respectively. We shall say that T is an extension of S, and we shall write $S \subseteq T$ if: (a) $D(S) \subseteq D(T)$; (b) $T|D(S) = S$ (i.e., $Tx = Sx$ for each $x \in D(S)$). We shall say that S and T are equal, and we shall write $S = T$, if we have both $S \subseteq T$ and $T \subseteq S$.

Given a linear operator T in E with domain $D(T)$ we may regard T as a linear mapping from the space $D(T)$ into the space E. Now if E has a norm then we can give $D(T)$ this norm and regard T as a linear mapping between two normed spaces. Whenever we say that an operator T in $(E, \|\cdot\|)$ is bounded we shall mean that it is a bounded, linear map from the normed space $(D(T), \|\cdot\|)$ to the normed space $(E, \|\cdot\|)$ (Section 2.1, Definition 1).

Exercises 1

1. Let $(E, \|\cdot\|)$ be a normed space, let D be a linear manifold in E and let T be a linear operator in E with domain D.

    a. Suppose that T is bounded. Show that T is a closed operator if, and only if, D is a closed, linear manifold; i.e., if, and only if, D is a subspace of E.

b. Suppose that D is not closed, but again suppose that T is bounded. Show that T has a closed extension. Hint: The extension has domain $\overline{D}$.

2. Let $(H, \langle,\rangle)$ be a Hilbert space, let D be a linear manifold in H and let T be a linear operator in H with domain D. Suppose that T is bounded, and suppose (see problem 1(b)) that D is a subspace of H. Let $D^{\perp}$ be the orthogonal complement of D (Section 4.2, Definition 1). Define $\overline{T}$ as follows: $\overline{T}x = Tx$ for all $x \in D$, $\overline{T}y = 0$ for all $y \in D^{\perp}$ and $\overline{T}$ is linear. Show that $\overline{T}$ is a well-defined, bounded, linear operator on H.

3. Let A be a closed, linear operator in the normed space E with domain $D(A)$ and range $R(A)$.

   a. Suppose that A is one-to-one so that we may define a linear operator A^{-1} in E by $A^{-1}y = x$ for all $y \in R(A)$. Show that A^{-1} is a closed, linear operator in E.

   b. Let λ be a complex number. We shall say that λ is an eigenvalue of A if $Ax = \lambda x$ for some non-zero vector in $D(A)$. Show that $(A - \lambda I)^{-1}$ exists if, and only if, λ is *not* an eigenvalue of A.

   c. Suppose that λ is not an eigenvalue of A so that $(A - \lambda I)^{-1}$ exists (by part (b)). Show that this operator is closed. Hint: By problem 3 it suffices to show that $A - \lambda I$ is a closed operator.

4. a. Consider the differential equation $-d^2\varphi/dx^2 = \lambda\varphi$ and its general solution $\varphi(x) = A \sin \nu x + B \cos \nu x$, where here A, B are arbitrary constants and $\nu^2 = \lambda$. If we restrict ourselves to $[0, \ell]$ and require that $\varphi(0) = 0 = \varphi(\ell)$, then B must be zero and ν must be $n\pi/\ell$, $n = 1, 2, 3, \ldots$.

   b. Now consider the problem $-d^2u/dx^2 - \lambda_0 u = f(x)$ on $[0, \ell]$, $u(0) = 0 = u(\ell)$ where $f(x)$ is a given function in $L^2[0, \ell]$. Suppose that $\lambda_0 = k^2\pi^2/\ell^2$ for some fixed integer k.

   i. Show that this problem has no solution unless
   $$\int_0^\ell f(x) \sin \frac{k\pi x}{\ell} dx = 0$$

   ii. When the condition stated in (i) is satisfied the problem has general solution
   $$u(x) = A \sin \frac{k\pi x}{\ell} + \sum_{n=1, n\neq k}^{\infty} \left(\frac{2/\ell}{\frac{n^2\pi^2}{\ell^2} - \frac{k^2\pi^2}{\ell^2}} \right) \sin \frac{n\pi x}{\ell} \int_0^\ell f(x) \sin \frac{n\pi x}{\ell} dx$$
   where A is an arbitrary constant.

7.2 The Adjoint and the Graph of an Operator

In this section, and in the remainder of the chapter, we shall work only with linear operators defined in a Hilbert space $(H, \langle,\rangle)$ over $\mathbb{C}$. Suppose that T is such an operator and that for some $y \in H$ there is a $z \in H$ such that

$$\langle Tx, y \rangle = \langle x, z \rangle \tag{1}$$

for all $x \in D(T)$. We would like to define $T^*y = z$, but the vector z need not be unique. We can get around this problem when $D(T)$ is dense in H. In that case we argue as follows: Suppose z_1, $z_2 \in H$ and both satisfy (1). Then $\sigma(x) = \langle x, z_1 - z_2 \rangle$ for all $x \in D(T)$ is a continuous, linear functional on this linear manifold (Section 4.3, Theorem 2) and, furthermore, $\sigma(x) = 0$ for all $x \in D(T)$. Thus by Exercises 3.2, problem 4(a), $\sigma(x) = 0$ for all $x \in \overline{D(T)} = H$ and this shows $z_1 = z_2$.

Definition 1

Let T be a linear operator in the Hilbert space H and suppose that $D(T)$ is dense in this space. Let $D(T^*) = \{y \in H |$ there is a $z \in H$ with $\langle Tx, y \rangle = \langle x, z \rangle$ for all $x \in D(T)\}$. Then for each $y \in D(T^*)$ we define T^*y to be the corresponding vector z.

It is clear that, for any T, $0 \in D(T^*)$ and $T^*0 = 0$. However, 0 may be the only element of $D(T^*)$. Fortunately there are interesting cases in which we can show that this last linear manifold is dense in H.

There is a very useful and illuminating way of looking at unbounded operators which was suggested by J. von Neumann. We begin by introducing a new space as follows: Let $\mathcal{H} = H \times H = \{(x, y) | x$ and y are vectors in $H\}$ and define the vector space operations in $\mathcal{H}$ in the natural way; i.e.,

$$(x_1, y_1) + (x_2, y_2) \equiv (x_1 + x_2, y_1 + y_2), \quad \lambda(x, y) \equiv (\lambda x, \lambda y).$$

We define an inner product on $\mathcal{H}$ in a similar way:

$$\langle (x_1, y_1), (x_2, y_2) \rangle = \langle x_1, x_2 \rangle + \langle y_1, y_2 \rangle$$

It is easy to show that, with these definitions, $\mathcal{H}$ is a Hilbert space over $\mathbb{C}$ (see problem 1 below). Any operator T in H has a "graph" $\mathcal{G}(T) \subseteq \mathcal{H}$ which is the set

$$\{(x, Tx) | x \in D(T)\}$$

Referring to Definition 3 of section 1 we see immediately that $S \subseteq T$ if, and only if, $\mathcal{G}(S) \subseteq \mathcal{G}(T)$, and that $S = T$ if, and only if, $\mathcal{G}(S) = \mathcal{G}(T)$.

Furthermore:

Lemma 1

Let T be a linear operator in H. Then the graph of T, $\mathcal{G}(T)$, is a linear manifold in $\mathcal{H}$. Moreover $\mathcal{G}(T)$ is a subspace of $\mathcal{H}$ (i.e., it is a closed set) if, and only if, T is a closed operator.

Proof. The fact that $\mathcal{G}(T)$ is a linear manifold follows immediately from the definitions and the fact that T is linear.

Suppose now that $\mathcal{G}(T)$ is a subspace and let $\{x_n\} \subseteq D(T)$ be a sequence such that both $\{x_n\}$ and $\{Tx_n\}$ converge to, say, y and z respectively. Then $\{(x_n, Tx_n)\}$ is a sequence in $\mathcal{G}(T)$ which converges for the norm of $\mathcal{H}$ to (y, z). But $\mathcal{G}(T)$ is a closed set and so (y, z) must belong to $\mathcal{G}(T)$. Thus $Ty = z$ showing that T is a closed operator.

Now suppose that T is a closed operator and let $\{(x_n, y_n)\}$ be a sequence in $\mathcal{G}(T)$ which converges for the norm of $\mathcal{H}$ to (x_0, y_0). Then, from the definition of this norm, $\{x_n\}$ converges to x_0 and $\{y_n\}$ converges to y_0. But $(x_n, y_n) \in \mathcal{G}(T)$ tells us that $Tx_n = y_n$ and so we have

$$\lim x_n = x_0 \quad \text{and} \quad \lim Tx_n = y_0$$

Since T is a closed operator we can conclude that $x_0 \in D(T)$ and that $Tx_0 = y_0$. Thus $(x_0, y_0) \in \mathcal{G}(T)$ and this shows that the latter is a subspace (Section 2.4, Theorem 1).

Let us now define two linear operators on $\mathcal{H}$ as follows:

$$\mathcal{U}(x, y) = (y, x) \quad \text{and} \quad \mathcal{V}(x, y) = (y, -x)$$

for all (x, y) in $\mathcal{H}$. These are unitary operators (i.e., invertible bounded linear operators whose inverses coincide with their adjoints, see Section 6.6, Definition 2). If we let $\mathcal{I}$ denote the identity operator on $\mathcal{H}$, then:

$$\mathcal{U}\mathcal{V} = -\mathcal{V}\mathcal{U} \quad \text{and} \quad \mathcal{U}^2 = -\mathcal{V}^2 = \mathcal{I}$$

Now suppose that T is a linear operator in H with dense domain $D(T)$ and recall (Definition 1) that $T^*y = y^*$ if, and only if, $\langle Tx, y \rangle = \langle x, y^* \rangle$ for all $x \in D(T)$. Thus

$$\langle \mathcal{V}(x, Tx), (y, y^*) \rangle = \langle (Tx, -x), (y, y^*) \rangle = \langle Tx, y \rangle - \langle x, y^* \rangle = 0$$

for all $x \in D(T)$. Now $(y, y^*) \in \mathcal{G}(T^*)$ and $\mathcal{V}(x, Tx)$ is in $\mathcal{V}[\mathcal{G}(T)]$ so this last equation shows that $\mathcal{G}(T^*)$ is the orthogonal complement of the linear manifold $\mathcal{V}[\mathcal{G}(T)]$; equivalently, $\mathcal{G}(T^*)$ is the orthogonal complement of the subspace $\overline{\mathcal{V}[\mathcal{G}(T)]}$ (Section 4.2, Definition 1 and Exercises 4.2, problems 1(a) and (b)). But the orthogonal complement of a linear manifold is always a subspace and so we have proved: If T is a linear operator in H which has a dense domain, then T^* is a closed operator.

We have just observed that $\mathcal{G}(T^*)$ is the orthogonal complement of $\overline{\mathcal{V}[\mathcal{G}(T)]}$. However, from the definition of $\mathcal{V}$ we can easily see that

$$\overline{\mathcal{V}[\mathcal{G}(T)]} = \mathcal{V}(\overline{\mathcal{G}(T)})$$

and hence

$$\mathcal{G}(T^*) = \mathcal{H} \ominus \mathcal{V}(\overline{\mathcal{G}(T)}) \qquad (2)$$

where the "circled minus sign," $\ominus$, means that $\mathcal{G}(T^*) \oplus \mathcal{V}(\overline{\mathcal{G}(T)}) = \mathcal{H}$.

Lemma 2

If T is a linear operator on H with dense domain, then the existence of T^{-1}, T^* and $(T^{-1})^*$ implies the existence of $(T^*)^{-1}$ and, moreover, $(T^*)^{-1} = (T^{-1})^*$.

Proof. Observe that $\mathcal{G}(T^{-1}) = \mathcal{U}(\mathcal{G}(T))$ and so, by (2),

$$\begin{aligned}\mathcal{G}((T^{-1})^*) &= \mathcal{H} \ominus \mathcal{V}(\overline{\mathcal{G}(T^{-1})}) = \mathcal{H} \ominus \mathcal{V}\mathcal{U}(\overline{\mathcal{G}(T)}) \\ &= \mathcal{U}[\mathcal{U}\mathcal{H} \ominus \mathcal{V}(\overline{\mathcal{G}(T)})] = \mathcal{U}[\mathcal{H} \ominus \mathcal{V}(\overline{\mathcal{G}(T)})] \\ &= \mathcal{U}(\mathcal{G}(T^*)).\end{aligned}$$

Theorem 1

Let T be a linear operator in H which is closed and has a dense domain. Then the domain of T^* is also dense in H and, furthermore, $T^{**} = T$.

Proof. The existence of T^{**} will follow once we show that the domain of T^* is dense in H; see the discussion just before Definition 1.

Let us denote the domain of T^* by D_* and suppose $z \in H$ is orthogonal to D_* (Section 4.2, Corollary 1 to Theorem 1). It follows that the element $(0, z) \in \mathcal{H}$ is orthogonal to $\{(T^*y, -y)|y \in D_*\}$. Thus this element is in the orthogonal complement of $\mathcal{V}[\mathcal{G}(T^*)]$. Now the orthogonal complement of $\mathcal{G}(T^*)$ is $\mathcal{V}[\overline{\mathcal{G}(T)}]$, as we saw above. Recall here that $\mathcal{V}$ is a unitary operator and so (Exercises 6.6 problem 4(a))

$$\langle \mathcal{V}(x_1, y_1), \mathcal{V}(x_2, y_2)\rangle = \langle (x_1, y_1), (x_2, y_2)\rangle.$$

Thus the orthogonal complement of $\mathcal{V}[\mathcal{G}(T^*)]$ is

$$\mathcal{V}[\overline{\mathcal{G}(T)}] = \mathcal{V}^2[\overline{\mathcal{G}(T)}] = -\mathcal{I}[\overline{\mathcal{G}(T)}] = \overline{\mathcal{G}(T)} = \mathcal{G}(T)$$

the last equality comes from the fact that T is a closed operator and hence $\mathcal{G}(T)$ is a closed set. It follows that $(0,z) \in \mathcal{G}(T)$ which tells us that $T0 = z$, or $z = 0$. We conclude that $D_*^\perp$ contains only the zero vector and hence that D_* is dense in H (Section 4.2, Lemma 1).

We now know that T^{**} exists. Since $\mathcal{G}(T^{**})$ is the orthogonal complement of $\mathcal{V}[\mathcal{G}(T^*)]$, we just proved this in the paragraph above, we must have

$$\mathcal{G}(T^{**}) = \mathcal{G}(T)$$

and so $T^{**} = T$.

Corollary 1

Let T be a linear operator in H with dense domain. Then the domain of T^* is dense in H if, and only if, T has a closed, linear extension. Furthermore, every closed, linear extension of T is also an extension of T^{**}.

Proof. Suppose first that T^* has a dense domain. Then T^{**} exists and, clearly, $T \subseteq T^{**}$. Since $T^{**} = (T^*)^*$ and an adjoint operator is necessarily a closed operator, we see that T has a closed, linear extension in this case.

Now suppose that T has a closed, linear extension, say, S. Then $D(T) \subseteq D(S)$ and since $D(T)$ is dense in H so also is $D(S)$. Thus S^* has a dense domain and $S = S^{**}$ by Theorem 1. However, $T \subseteq S$ implies $S^* \subseteq T^*$ which shows that T^* has a dense domain.

Finally, $T \subseteq T^{**} \subseteq S^{**} = S$ shows that S is an extension of T^{**}.

Theorem 2

A closed, linear operator in H which is defined at every point of H (so that it is actually an operator on H) is a bounded, linear operator.

Proof. Let T be a closed, linear operator with $D(T) = H$. Then $D(T)$ is certainly dense in H and so T^* is defined in H and its domain, D_*, is, by Theorem 1, dense in H. We shall show that T^* is bounded on D_*. Suppose that it is not. Then we must have a sequence $\{y_n\} \subseteq D_*$ such that $\|y_n\| = 1$ for every n, and $\sup\|T^*y_n\| = \infty$. For each n we define a linear functional F_n on H as follows:

$$F_n(x) = \langle Tx, y_n \rangle = \langle x, T^*y_n \rangle$$

Let us emphasize that, since T is defined on all of H, so also is each F_n. Furthermore, since

$$\|F_n\| \leq \|T^*y_n\|$$

for every n, each of these linear functionals is bounded on H (Section 4.3, Definition 1). Even more is true. For each fixed $x \in H$ the sequence $\{F_n(x)\} \subseteq$

$\mathbb{C}$ is bounded because

$$|F_n(x)| \leq |\langle Tx, y_n \rangle| \leq \|Tx\|\|y_n\| \leq \|Tx\|$$

We conclude, from the Banach-Steinhaus theorem (Section 4.4, Theorem 1) that there is a constant M such that

$$\|F_n\| \leq M$$

for all n. However, this is a contradiction because

$$|F_n(T^*y_n)| = \langle T^*y_n, T^*y_n \rangle = \|T^*y_n\|^2 \to \infty$$

Hence T^* is bounded on its domain D_*.

Now T is a closed operator and so D_* must be dense in H. Since T^* is bounded on D_* it has a unique bounded, linear extension to $\overline{D_*} = H$. We shall denote this extension by T^* also since the extension clearly satisfies

$$\langle Tx, y \rangle = \langle x, T^*y \rangle$$

for all x, y in H. Thus we have proved: If T is a closed, linear operator on H then T^* is a bounded, linear operator on H. Applying this to T^* we see that T^{**} is a bounded, linear operator on H. But, since T is given to be closed, $T = T^{**}$ and we have proved that T is a bounded, linear operator on H.

Exercises 2

1. a. Show that the space $\mathcal{H}$ defined just before Lemma 1 is a Hilbert space over $\mathbb{C}$.

   b. Show that the operators $\mathcal{U}, \mathcal{V}$ defined just after the proof of Lemma 1 are unitary and that $\mathcal{UV} = -\mathcal{VU}, \mathcal{U}^2 = -\mathcal{V}^2 = \mathcal{I}$.

   c. Show that $\overline{\mathcal{V}[\mathcal{G}(T)]} = \mathcal{V}[\overline{\mathcal{G}(T)}]$ as claimed in the text.

2. Let S be a bounded, self-adjoint operator on H and recall (Section 6.1, Lemma 1) that the complex number λ is an eigenvalue of S if, and only if, $\overline{[(S - \lambda I)(H)]} \neq H$; i.e., if, and only if, $(S - \lambda I)(H)$ is *not* dense in H. Suppose that $\lambda \in \sigma(S)$, but λ is not an eigenvalue of this operator. Show that $(S - \lambda I)^{-1}$ exists, and that it is an unbounded, linear operator in H with dense domain.

7.3 Differentiation in the Space $L^2[a,b]$

One of the most important applications of operator theory to physics is to provide a mathematical framework for quantum mechanics. In this application various differential operators on L^2 play an important role. As in Section 1 we have to be careful to specify the domain of such an operator. Here we shall present the real variable theory which will enable us to do that. Let a, b be two fixed, real numbers with $a < b$. In Section 1.8 we defined two measurable functions on $[a, b]$ to be equivalent if they were equal almost everywhere in this interval (see Definition 2 and the discussion just after it, and Definition 3 also). We defined $L^2[a, b]$ to be the set of all (equivalence classes of) measurable functions f on $[a, b]$ such that

$$\int_a^b |f(t)|^2 dt < \infty$$

We then shows that this set is a vector space, that the definition

$$\langle f, g \rangle = \int_a^b f(t)\overline{g(t)} dt$$

defines an inner product on this vector space, and that $(L^2[a, b], \langle, \rangle)$ is a Hilbert space. We might also recall that the polynomials are dense in this Hilbert space (Section 1.9 just after Definition 1). Consider now the following:

Definition 1

The set $L^1[a, b]$ consists of all (equivalence classes of) measurable functions f on $[a, b]$ such that

$$\int_a^b |f(t)| dt < \infty$$

For each f in this set we shall take $\|f\|_1$ to be the number

$$\int_a^b |f(t)| dt$$

It turns out that $L^1[a, b]$ is a vector space, that $\|\cdot\|_1$ is a norm on this vector space, and that $(L^1[a, b], \|\cdot\|_1)$ is a Banach space (Section 1.3, Definition 2) which is not a Hilbert space (see the exercises below).

Now suppose that we take a function $f \in L^1[a, b]$ and consider its indefinite integral; i.e., the function

$$F(x) = \int_a^x f(t) dt$$

We would expect $F(x)$ to be differentiable, and we would expect $F'(x)$ to be $f(x)$. If f were a continuous function, this would be true. Since f is only an L^1-function our expectation is only "almost" true. The next few remarks will clarify this strange statement.

Definition 2

A function g on $[a,b]$ is said to be absolutely continuous on this interval if for any given $\varepsilon > 0$ these is a $\delta > 0$ such that

$$\sum_{j=1}^{\infty} |g(b_j) - g(a_j)| < \varepsilon$$

whenever $\{(a_j, b_j)\}_{j=1}^{\infty}$ is any countable family of non-overlapping subintervals of $[a,b]$ for which

$$\sum_{j=1}^{\infty} (b_j - a_j) < \delta.$$

It is easy to see that a function which is absolutely continuous on $[a,b]$ is continuous on this interval. We leave the proof of this to the exercises. Furthermore, if g is continuous on $[a,b]$, if g' exists at each point of (a,b) and if there is a constant M such that

$$|g'(t)| \leq M$$

for all $t \in (a,b)$, then g is absolutely continuous on $[a,b]$; because

$$|g(b_j) - g(a_j)| = |g'(t_0)|(b_j - a_j)$$

for some $t_0 \in (a_j, b_j)$. So all polynomials are absolutely continuous. It turns out that the indefinite integral of any $f \in L^1[a,b]$, the function F defined above, is absolutely continuous and we also have:

Theorem 1

If F is absolutely continuous on $[a,b]$, then $F'(x)$ exists almost everywhere in $[a,b]$, this derivative is in the space $L^1[a,b]$ and, moreover,

$$F(x) = F(a) + \int_a^x F'(t) dt$$

for all x in $[a,b]$.

We even have an "integration by parts" formula for these functions: If F, G are absolutely continuous on $[a,b]$, then both $F'G$ and $G'F$ are in $L^1[a,b]$ and

$$\int_a^b F(t)G'(t)dt + \int_a^b F'(t)G(t)dt = F(t)G(t) \mid_a^b = F(b)G(b) - F(a)G(a)$$

Finally, we observe that, from Theorem 1, a continuous nowhere differentiable function on $[a,b]$ could not be absolutely continuous on $[a,b]$.

Exercises 3

1. a. Show that $L^1[a,b]$ is a vector space and that $\|\cdot\|_1$ is a norm on this space.

   b. Readers who know some measure theory may want to show that $L^2[a,b] \subseteq L^1[a,b]$. Explain why the fact that $[a,b]$ has finite measure is crucial here. See Exercises 8.1 problem 5 for a proof that avoids measure theory.

2. Let f be absolutely continuous on $[a,b]$.

   a. Show that f is continuous on $[a,b]$.

   b. Show that any constant multiple of f is absolutely continuous on $[a,b]$.

   c. Suppose that g is absolutely continuous on $[a,b]$. Show that $f+g$ is absolutely continuous on this interval.

* 3. Let $D = \{f \in L^2[0, 2\pi] | f$ is absolutely continuous on $[0, 2\pi]$, $f(0) = 0 = f(2\pi)$, and $f' \in L^2[0, 2\pi]\}$. Show that D is a linear manifold in $L^2[0, 2\pi]$ which is dense in this space. Hint: See Section 1.8, Theorem 5 or Section 1.9, just after Definition 1.

7.4 Some Differential Operators

The operators we shall present here illustrate the importance of, and the interrelationships between, the ideas presented in Section 1, 2, and 3. We begin with: $D(T) = \{f \in L^2[0, 2\pi] | f$ is absolutely continuous on $[0, 2\pi]$, $f(0) = 0 = f(2\pi)$ and $f' \in L^2[0, 2\pi]\}$. We have seen, Exercises 3 problem 3, that $D(T)$ is a linear manifold in $L^2[0, 2\pi]$ which is dense in this space. For each $f \in D(T)$ we set

$$Tf = if'$$

where $i = \sqrt{-1}$. Clearly T is a linear operator in $L^2[0, 2\pi]$.

(a) The operator T is a closed operator.

Choose a sequence $\{f_n\} \subseteq D(T)$ and suppose that both $\{f_n\}$ and $\{f'_n\}$ converge, for the L^2-norm, to functions f, h respectively. Then, by the C.S.B. inequality (Section 1.4, Theorem 1)

$$|f_n(x) - f_m(x)| = |\int_a^x \{f'_n(t) - f'_m(t)\} \cdot 1 dt| \leq \|f'_n - f'_m\|_2 \sqrt{2\pi}$$

It follows that $\{f_n\}$ converges to f uniformly over $[0, 2\pi]$. A similar calculation gives us $|f_n(x) - \int_0^x h(t)dt| \leq \|f'_n - h\|_2 \sqrt{2\pi}$ and since $\|f'_n - h\|_2 \to 0$, $f(x) = \int_0^x h(t)dt$. It follows that f is absolutely continuous on $[0, 2\pi]$ and that $f' = h \in L^2[0, 2\pi]$. Furthermore, since $f_n(0) = 0 = f_n(2\pi)$ for every n and $\{f_n\}$ converges uniformly to f, we must have $f(0) = 0 = f(2\pi)$. Thus $f \in D(T)$ and

$$Tf = if' = ih$$

which shows that T is closed.

(b) The adjoint of T, T^*, has a dense domain; in fact, $D(T^*) = \{g \in L^2[0, 2\pi] | g$ is absolutely continuous and

$$g' \in L^2[0, 2\pi]\}$$

and, furthermore,

$$T^*g = ig'$$

for each g in $D(T^*)$.

The fact that T^* has a dense domain follows from the fact that T is a closed operator with dense domain (Section 2, Theorem 1). These properties of T also imply that $T = T^{**}$. Here we are identifying the domain of T^* and showing that this operator is a proper, closed extension of T.

Suppose that g, g^* are in $L^2[0, 2\pi]$ and that

$$\langle Tf, g \rangle = \langle f, g^* \rangle$$

for all $f \in D(T)$. Then, recalling the definition of T, we have

$$\begin{aligned} \langle f, g^* \rangle &= \langle Tf, g \rangle = \langle if', g \rangle = i \int_0^{2\pi} f'(t) \overline{g(t)} dt \\ &= f(t) \overline{g^{**}(t)} \Big|_0^{2\pi} - \int_0^{2\pi} f'(t) \overline{g^{**}(t)} dt \end{aligned}$$

where we have set

$$u = f(t),\ du = f'(t);\ dv = \overline{g^*(t)}dt,\ v = \overline{g^{**}(t)}$$

and used our integration by parts formula (Section 3). Now since $f \in D(T)$, $f(0) = 0 = f(2\pi)$ showing that the first turn on the right-hand side of our equation is zero. Transposing we can write:

$$i \int f'(t)\{\overline{g(t)} + \overline{ig^{**}(t)}\} dt = 0$$

This last equation implies $g(t) + ig^{**}(t) = c$, a constant, almost everywhere. We shall prove this below, for now let us use it. The function $g^* \in L^2[0, 2\pi] \subseteq$

$L^1[0,2\pi]$ (Exercises 3, problem 1(b)) and so g^{**}, since it is the indefinite integral of an L^1-function, is absolutely continuous on $[0,2\pi]$. It follows then that g is absolutely continuous on this interval; for $g(t) = c - ig^{**}$. Thus we may write
$$g'(t) + ig^*(t) = 0 \text{ a.e.}$$
It follows from this that $g' \in L^2[0,2\pi]$.

To sum up: If g, g^* are in $L^2[0,2\pi]$ and satisfy
$$\langle Tf, g \rangle = \langle f, g^* \rangle$$
for all $f \in D(T)$, then g is absolutely continuous and $g' \in L^2[0,2\pi]$. Furthermore,
$$T^*g = g^* = \frac{-1}{i}g'(t) = ig(t)$$
Except for the claim made above, which we have yet to prove, this completes the proof of (b).

Claim: $i\int_0^{2\pi} f'(t)\{\overline{g(t)} + \overline{ig**(t)}\}dt = 0$ for every $f \in D(T)$ if, and only if, $g(t) + ig^{**}(t)$ is constant almost everywhere.

Proof. Let us set $h(t) = g(t) + ig^{**}(t)$ and suppose first that $h(t) = c$, a constant, a.e. Then
$$i\int_0^{2\pi} f'(t)\overline{h(t)}dt = i\bar{c}\int_0^{2\pi} f'(t)dt = i\bar{c}f(t)\,|_0^{2\pi} = i\bar{c}(f(2\pi) - f(0)) = 0$$
This proves the easy half of our claim. Let us now suppose that
$$i\int_0^{2\pi} f'(t)\overline{h(t)}dt = 0$$
Then, since g, g^{**} are in L^2, so is the function h. So, again by Exercises 3 problem 1(b), $h \in L^1[0,2\pi]$. Let us set
$$f'(x) = h(x) - c$$
where c is a constant as yet undetermined. Then $f' \in L^2$ and
$$f(x) = \int_0^x h(t)dt - \int_0^x c\,dt$$
shows that f is absolutely continuous on $[0,2\pi]$; it also shows that $f(0) = 0$. We now choose the constant c so that $f(2\pi) = 0$. Thus $f \in D(T)$. Now
$$\int_0^{2\pi} |h(t) - c|^2 dt = \int_0^{2\pi} \{h(t) - c\}\{\overline{h(t) - c}\}dt$$
$$= \int_0^{2\pi} \{h(t) - c\}\overline{h(t)}dt - \bar{c}\int_0^{2\pi} h(t)dt + c\bar{c}\,2\pi$$
$$= \int_0^{2\pi} f'(t)\overline{h(t)}dt$$

because of the way we chose c. This last integral is zero since, as we have just seen, $f \in D(T)$. Hence
$$\int_0^{2\pi} |h(t) - c|^2 dt = 0$$
and from this we can conclude (Section 1.8 just before Definition 2) that
$$|h(t) - c| = 0 \text{ a.e.}$$
recalling our definition of h,
$$g(t) + i\, g^{**}(t) = c \text{ a.e.}$$

We began this section by defining an operator T as follows: First $D(T) = \{f \in L^2[0, 2\pi] | f$ is absolutely continuous on $[0, 2\pi]$, $f(0) = 0 = f(2\pi)$, and $f' \in L^2[0, 2\pi]\}$. Then we set $Tf = if'$ for every f in $D(T)$. We have seen that $D(T)$ is dense in $L^2[0, 2\pi]$ and that T is a closed operator.

Next we calculated T^*. We found: $D(T^*) = \{g \in L^2[0, 2\pi] | g$ is absolutely continuous on $[0, 2\pi]$ and $g' \in L^2[0, 2\pi]\}$, and $T^*g = ig'$ for every $g \in D(T^*)$. We see immediately that $T \subseteq T^*$, $T \neq T^*$.

Definition 1

Let T be a linear operator in H. We shall say that T is a symmetric operator if $T \subseteq T^*$ (meaning $D(T) \subseteq D(T^*)$ and $T^*|D(T) = T$) and we shall say that T is a self-adjoint operator if $T = T^*$.

The differential operator considered in this section is a symmetric operator, but it is not self-adjoint. Let us finish by showing that our operator has a self-adjoint extension. As usual, we start by defining a dense, linear manifold in $L^2[0, 2\pi]$. Set $D(T_1) = \{f \in L^2[0, 2\pi] | f$ is absolutely continuous on $[0, 2\pi]$, $f(0) = f(2\pi)$, and $f' \in L^2[0, 2\pi]\}$. Recall that $f \in D(T)$ implies, among other things, that $f(0) = 0 = f(2\pi)$, so $D(T) \subseteq D(T_1)$ but these linear manifolds are distinct; of course this inclusion shows that $D(T_1)$ is dense in our space. We now define
$$T_1 f = if'$$
for every $f \in D(T_1)$. Then $T \subseteq T_1$, $T \neq T_1$ and we leave it to the reader (see the exercises below) to prove that T_1 is a closed operator. We shall show, now, that $T_1^* = T_1$.

We know that T_1^* exists, that it has a dense domain, $D(T_1^*)$, and that it is a closed operator (Section 2, Theorem 1). The same argument that we used to prove that $T \subseteq T^*$ can be used here to show that $T_1 \subseteq T_1^*$. Let $g \in D(T_1^*)$ and let $g^* = T_1^* g$. Then we have
$$\langle if', g \rangle = \langle f, g^* \rangle$$

for all $f \in D(T_1)$. However, the function $f(t) = 1$ for all $t \in [0, 2\pi]$, is in $D(T_1)$ and so
$$0 = \langle f, g^* \rangle = \int_0^{2\pi} g^*(t) dt$$
We see from this that if we set
$$g^{**}(x) = \int_0^x g^*(t) dt$$
then $g^{**}(2\pi) = 0$, and clearly $g^{**}(0) = 0$ also. As in our discussion of the operator T we see here again that
$$g(t) + ig^{**}(t) = c, \text{ a.e.}$$
We draw two conclusions from this. First, $g(0) = g(2\pi)$. Next, $ig'(t) = g^*(t)$. It follows that $g \in D(T_1)$ and we have shown that $D(T_1) = D(T_1^*)$. Clearly then $T_1 = T_1^*$.

Exercises 4

1. Find a function which is in $D(T^*)$ but not in $D(T)$. Find a function in $D(T_1)$ which is not in $D(T)$.

2. Prove that the operator T_1 defined above (just after Definition 1) is a closed operator.

3. Let T_1 be a linear operator in H which has dense domain and let T_2 be a linear operator in H with $T_1 \subseteq T_2$.

    a. Show that the domain of T_2 is dense in H and that $T_1^* \supseteq T_2^*$.

    b. Suppose now that T_1 is a self-adjoint operator and that T_2 is a symmetric operator. Show that $T_1 = T_2$.

    c. Suppose that T_1 is a symmetric operator and that T_2 is also a symmetric operator. Show that $T_2 \subseteq T_1^*$.

4. Let T_1 be a linear operator in H which has a self-adjoint extension T_2. Show that T_1 must be a symmetric operator.

7.5 Cayley Transform, Deficiency Indices

In the last section we looked at a symmetric operator which was not self-adjoint but did have a self-adjoint extension. Here we shall characterize those symmetric operators which have self-adjoint extensions. More generally, we shall characterize all symmetric extensions of a given symmetric operator.

Let S be a given, symmetric operator in H. Then

$$\begin{aligned}\|(S \mp iI)x\|^2 &= \langle Sx, Sx\rangle \mp i\langle x, Sx\rangle \pm i\langle Sx, x\rangle + \langle x, x\rangle \\ &= \|Sx\|^2 + \|x\|^2\end{aligned} \quad (1)$$

It follows from this that both of the linear operators $S + iI$ and $S - iI$ have zero null spaces, and that

$$\|(S - iI)x\| = \|(S + iI)x\| \quad (2)$$

for all $x \in D(S)$. Hence it makes sense to talk about $(S + iI)^{-1}$.

Definition 1

Let S be a symmetric operator in H. Then the linear operator

$$V = (S - iI)(S + iI)^{-1}$$

is called the Cayley transform of S.

We observe that:

a. The Cayley transform is always an isometric operator; i.e., $\|Vx\| = \|x\|$ for all $x \in D(V)$. To see this suppose that $y = (S + iI)^{-1}x$. Then $(S - iI)y = Vx$ so $\|Vx\| = \|(S - iI)y\|$. Now by (2) we have $\|(S - iI)y\| = \|(S + iI)y\| = \|x\|$.
b. $D(V) = \{(S + iI)x | x \in D(S)\}$.
c. $R(V)$, the range of V, is $\{(S - iI)x | x \in D(S)\}$.
d. $S = i(I + V)(I - V)^{-1}$.

We leave the proofs of (b), (c), (d) to the exercises.

Theorem 1

The Cayley transform of a self-adjoint operator is a unitary operator and, conversely, if the Cayley transform of a symmetric operator is unitary then the operator is self-adjoint.

Proof. First assume that S is a self-adjoint operator. From (1) we have

$$\|(S + iI)x\| \geq \|x\|$$

and so by taking $x = (S+iI)^{-1}y$ we get

$$\|y\| \geq \|(S+iI)^{-1}y\|$$

showing that $(S+iI)^{-1}$ is a continuous linear operator. Similarly, $(S-iI)^{-1}$ is continuous. Let us show that the domains of these operators are dense in H. Suppose, to be specific, that the domain of $(S-iI)^{-1}$ is not dense. Then there is a $z \neq 0$ such that

$$\langle (S-iI)x, z \rangle = 0$$

for every $x \in D(S)$. But then z is in the domain of $(S-iI)^*$ which is equal to $S+iI$ because S is self-adjoint. So we have

$$\langle x, (S+iI)z \rangle = 0$$

for all $x \in D(S)$. Since $D(S)$ is dense in H we see that

$$(S+iI)z = 0$$

However, $S+iI$ is a one-to-one mapping and so $z = 0$. It follows that $(S-iI)^{-1}$, $(S+iI)^{-1}$ and hence V, are bounded, linear operators defined everywhere in H.

Let us show that V is an onto mapping. Recall that $D(V) = \{(S+iI)x | x \in D(S)\}$. Given $y \in H$ we have

$$(S-iI)^{-1}y \in D(S)$$

because $S-iI$ of this element exists and equals y of course showing that S of this element is defined. But

$$V\{(S+iI)[(S-iI)^{-1}y]\}$$

then has meaning and clearly, from the definition of V, it is just y.

Finally, we recall that a bounded, linear operator U was defined to be a unitary operator if $\langle Ux, y \rangle = \langle x, U^{-1}y \rangle$ for all x, y in H; i.e., $U^* = U^{-1}$. The operator V is isometric

$$\langle Vx, Vy \rangle = \langle x, y \rangle$$

for all x, y in H and so

$$\langle Vx, y \rangle = \langle Vx, VV^{-1}y \rangle = \langle x, V^{-1}y \rangle$$

Thus V is unitary.

Now assume that the Cayley transform of the given symmetric operator S is a unitary operator; so we suppose that V is unitary. Let $y \in D(S^*)$ and set $y^* = S^*y$. Then we have

$$\langle Sx, y \rangle = \langle x, y^* \rangle$$

for all $x \in D(S)$ and since these x are of the form

$$(I - V)z$$

for all $z \in D(V) = H$ (recall that $S = i(I + V)(I - V)^{-1}$) we have

$$\langle i(I + V)z, y \rangle = \langle (I - V)z, y^* \rangle$$

or

$$i\langle z, y \rangle + i\langle Vz, y \rangle = \langle z, y^* \rangle - \langle Vz, y^* \rangle$$

for all $z \in H$. Since V is unitary we can replace $\langle z, y \rangle$ by $\langle Vz, Vy \rangle$ and $\langle z, y^* \rangle$ by $\langle Vz, Vy^* \rangle$ to get

$$\langle Vz, -iVy - iy - Vy^* + y^* \rangle = 0$$

Now as z runs through H, Vz also runs through H and so

$$-iVy - iy - Vy^* + y^* = 0$$

We may solve this in two different ways to get:

$$y = (I - V)(\frac{y - iy^*}{2}), \quad y^* = i(I + V)(\frac{y - iy^*}{2})$$

Thus we can write

$$\begin{aligned} y^* &= S^*y = i(I + V)\left(\frac{y - iy^*}{2}\right) = i(I + V)(I - V)^{-1}(I - V)\left(\frac{y - iy^*}{2}\right) \\ &= i(I + V)(I - V)^{-1}y = Sy \end{aligned}$$

This shows that S is a self-adjoint operator.

In the case of an arbitrary closed, symmetric operator S the closed, isometric operator V is not a unitary operator. Hence $D(V)$ and the range of V, $R(V)$, are proper (closed) subspaces of H. The spaces $H \ominus D(V)$ and $H \ominus R(V)$ are called the deficiency subspaces, and their dimensions are called the deficiency indices, of S (or of V). Theorem 2 shows that S is self-adjoint if, and only if, both of its deficiency indices are zero.

If S' is a closed, symmetric extension of the given closed, symmetric operator S then the definition of the Cayley transform shows that the Cayley transform of S', call it V', is an isometric extension of V; the Cayley transform of S. Note that $D(V)$ will be a subspace of $D(V')$ and that $D(V') \ominus D(V)$ will be mapped by V' into $R(V') \ominus R(V)$. Thus the deficiency indices of S' are obtained from those of S by diminishing each of them by the same number (finite or infinite).

Theorem 2

Every isometric extension U of the Cayley transform V of the given operator S, determines a symmetric extension S' of S whose Cayley transform, call it V', is equal to U.

Proof. We first show that $I - U$ has zero kernel. So suppose that $(I - U)z = 0$ and for $y \in D(U)$ let $x = (I - U)y$ and write

$$\begin{aligned}\langle z, x \rangle &= \langle z, y \rangle - \langle z, Uy \rangle = \langle Uz, Uy \rangle - \langle z, Uy \rangle \\ &= -\langle (I - U)z, Uy \rangle = -\langle 0, Uy \rangle = 0\end{aligned}$$

Thus z is orthogonal to the image of $I - U$ and hence to the image of $I - V$. In particular z is orthogonal to the domain of S. However, the latter is dense in H and so $z = 0$.

We now may define an operator S' as follows:

$$S' = i(I + U)(I - U)^{-1}$$

Clearly S' is an extension of S because U is an extension of V. Thus S' has a dense domain. Let x, y be in this domain. Then $x = (I - U)x'$, $y = (I - U)y'$ and so

$$S'x = i(I + U)x', \quad S'y = i(I + U)y',$$

$$\begin{aligned}\langle Sx, y \rangle &= \langle i(I + U)x', (I - U)y' \rangle \\ &= i\{\langle Ux', y' \rangle - \langle x', Uy' \rangle\}\end{aligned}$$

because $\langle x', y' \rangle = \langle Ux', Uy' \rangle$. Continuing our calculation:

$$= \langle (I - U)x', i(I + U)y' \rangle = \langle x, S'y \rangle$$

and so S' is a symmetric operator.

Finally, let us show that the Cayley transform, V', of S' is the given operator U. For $x = (I - U)x'$ we have

$$S'x = i(I + U)x', (S' + iI)x = 2ix', (S' - iI)x = 2iUx'$$

Thus $D(V') = \{2ix' | x' \in D(U)\}$ and $V'(2ix) = 2iUx' = U(2ix')$ showing that $V' = U$.

We now see that the problem of finding all symmetric extensions of a closed, symmetric operator S is equivalent to the problem of finding all isometric extensions of its Cayley transform V. In order to extend V we have only to map the deficiency subspace $H \ominus D(V)$, or a subspace of it, isometrically into $H \ominus R(V)$. It is thus possible to exhaust the deficiency subspace of smallest dimension. The resulting extension will be a "maximal" symmetric operator, i.e., a symmetric operator which has no proper symmetric extension. If the two deficiency subspaces are of the same dimension we obtain a unitary extension of V and hence a self-adjoint extension of S. Thus we may state: (i) S is self-adjoint if, and only if, its deficiency indices are both zero; (ii) S is maximal if, and only if, one if its deficiency indices is zero; (iii) S admits a self-adjoint extension if, and only if, its deficiency indices are equal.

Let us end this second by computing the deficiency indices of the operator discussed above. Recall $H = L^2[0, 2\pi]$, $D(T) = \{f \in L^2[0, 2\pi] | f$ is absolutely continuous, $f' \in L^2$, $f(0) = f(2\pi) = 0\}$. $Tf = if'$ for all $f \in D(T)$. If V is the Cayley transform of T then

$$D(V) = \{(T + iI)f | f \in D(T)\}$$

and

$$R(V) = \{(T - iI)f | f \in D(T)\}$$

The set $\{e^{inx} - 1 | -\infty < n < \infty\} \subseteq D(T)$ and so

$$\{ine^{inx} - i(e^{inx} - 1)\} \subseteq D(V)$$

Thus all functions of the form $i(n+1)e^{inx} - i$ are in $D(V)$; n is an integer here. Suppose $\sigma \in L^2[0, 2\pi]$ is given. Then

$$\sigma(x) = \sum c_n e^{inx}$$

where we know that

$$\sum |c_n|^2 < \infty$$

Now consider

$$\frac{c_n}{i(n+1)}[i(n+1)e^{inx} - i] = c_n e^{inx} - \frac{c_n}{n+1}$$

Since we have both

$$\sum |c_n|^2 < \infty \quad \text{and} \quad \sum \left(\frac{1}{n+1}\right)^2 < \infty$$

we see that

$$\sum c_n e^{inx} - \sum \frac{c_n}{n+1} = \sigma(x) - \alpha$$

is in the domain of V. But σ was an arbitrary L^2-function. It follows that $H \ominus D(V) = \{\alpha \cdot 1 | \alpha$ complex $\}$ showing that $\dim[H \ominus D(V)] = 1$. Similarly, one can show that $\dim[H \ominus R(V)] = 1$. Thus our deficiency indices are $1, 1$.

Exercises 5

1. Let S be a linear operator in H and let V be its Cayley transform. Show that

    i. $D(V) = \{(S + iI)x | x \in D(S)\}$;
    ii. $R(V) = \{(S - iI)x | x \in D(S)\}$;
    iii. $S = i(I + V)(I - V)^{-1}$

2. Let S be a self-adjoint operator in H and let V be its Cayley transform. Show that $I - V$ is a one-to-one mapping from H onto $D(S)$. Conclude that 1 is not an eigenvalue for V.

7.6 The Spectral Theorem for Unitary Operators

We shall apply our knowledge of unbounded linear operators to study the spectrum of a bounded, unitary operator on our Hilbert space H. Unfortunately, the reasoning in much of the first part of this section is rather convoluted. We begin by recalling some terminology.

Let S be a bounded, linear operator on H and let $\lambda \in \mathbb{C}$. Then

$$M_\lambda = \{x \in H | Sx = \lambda x\}$$

is equal to

$$[(S^* - \overline{\lambda}I)(H)]^\perp$$

as we saw in Section 6.1. This space, M_λ, is only interesting if λ is in the spectrum of S; otherwise $(S - \lambda I)$ has a bounded, linear inverse and so $M_\lambda = \{0\}$. Now when S is a self-adjoint operator, $S = S^*$ and λ is real (because $\sigma(S) \subseteq \mathbb{R}$ when S is self-adjoint) and so

$$M_\lambda = [(S - \lambda I)(H)]^\perp$$

in this case.

The observations above were used to show that a point $\lambda \in \sigma(S)$ is an eigenvalue of the self-adjoint operator S if, and only if,

$$\overline{(S - \lambda I)(H)} \neq H$$

i.e., if, and only if, $(S - \lambda I)(H)$ is *not* dense in H (see Section 6.1, Lemma 1 and Section 3.1, Definition 2 and Exercises 3.1, problem 4(e)). Here we shall use them to prove:

Lemma 1

Let U be a unitary operator on H and let $\lambda \in \mathbb{C}$. Then

$$M_\lambda = \{x \in H | Ux = \lambda x\} = [(U - \lambda I)(H)]^\perp$$

Consequently, λ is an eigenvalue of U if, and only if, $(U - \lambda I)(H)$ is *not* dense in H.

Proof. Suppose that for some $x \in H$, $x \neq 0$, we have $Ux = \lambda x$. Then $\lambda \in \sigma(U)$ and so $\lambda \bar{\lambda} = 1$; here, of course, the bar means complex-conjugate. Apply the operator $\bar{\lambda} U^{-1}$ to the equation $Ux = \lambda x$ to get $\bar{\lambda} x = U^{-1}x$. But then

$$\langle x, (U - \lambda I)y \rangle = \langle (U^{-1} - \bar{\lambda} I)x, y \rangle = \langle 0, y \rangle = 0$$

for all $y \in H$ showing that $x \in [(U - \lambda I)(H)]^\perp$. Since $x \in M_\lambda$, by the choice of x, we have

$$M_\lambda \subseteq [(U - \lambda I)(H)]^\perp$$

Now let x be any element of $[(U - \lambda I)(H)]^\perp$. Then

$$0 = \langle x, (U - \lambda I)y \rangle = \langle (U^{-1} - \bar{\lambda} I)x, y \rangle$$

for all y in H and this tells us that $(U^{-1} - \bar{\lambda} I)x = 0$. Since $0 \in M_\lambda$ we may suppose that the x we choose is not the zero vector. Hence $\bar{\lambda}$ is then an eigenvalue of the unitary operator U^{-1}. Again we see that $\lambda \bar{\lambda} = 1$ and we may apply the operator λU to the equation $U^{-1}x = \bar{\lambda} x$ to get $\lambda x = Ux$. Thus $x \in M_\lambda$.

We have just shown that

$$M_\lambda = [(U - \lambda I)(H)]^\perp$$

and so

$$M_\lambda^\perp = [(U - \lambda I)(H)]^{\perp\perp} = \overline{[(U - \lambda I)(H)]}$$

(Section 4.2, Lemma 1). Now the result just mentioned also contains the fact that $M_\lambda^\perp = \{0\}$ if, and only if, $\overline{M}_\lambda = H$. The number λ is an eigenvalue of U if, and only if, $M_\lambda \neq \{0\}$ which will be true if, and only if, $M_\lambda^\perp \neq \{0\}^\perp = H$.

Suppose now that λ is *not* an eigenvalue of the unitary operator U. Then $M_\lambda = \{x \in H | Ux = \lambda x\} = \{0\}$ and so the map $U - \lambda I$ is a one-to-one mapping of H onto the linear manifold $(U - \lambda I)(H)$ of H. Then $(U - \lambda I)^{-1}$ exists and it is a linear map from the linear manifold $(U - \lambda I)(H)$ into H. Now since

THE SPECTRAL THEOREM FOR UNITARY OPERATORS

we have assumed that λ is *not* an eigenvalue of U we know, from Lemma 1, that the linear manifold $(U - \lambda I)(H)$ must be dense in H. So $(U - \lambda I)^{-1}$ is a linear map which is defined on a dense linear manifold in H and takes its values in H. If this map were bounded (i.e., if we had $\|(U - \lambda I)^{-1}y\| \leq k\|y\|$ for all y in the domain and some fixed constant k), then it would be uniformly continuous on this domain (Section 2.1, Lemma 1 and Theorem 1). In that case it would have a unique extension to the closure of its domain; i.e., to all of H. Let us state now the result of this rather convoluted reasoning and present the argument again in a slightly different form:

Lemma 2

Let U be a unitary operator on a Hilbert space H and let $\lambda \in \sigma(U)$, λ not an eigenvalue of U. Then the linear map $(U - \lambda I)^{-1}$ is defined on a dense, linear manifold in H and, furthermore, this operator is unbounded on its domain.

Proof. Since λ is not an eigenvalue of U we know two things. First, $U - \lambda I$ is a one-to-one map and, second, $(U - \lambda I)(H)$ is dense in H. It follows that $(U - \lambda I)^{-1}$ exists and is defined at least on the linear manifold $(U - \lambda I)(H)$. Thus we see that $(U - \lambda I)^{-1}$ has a dense domain. Now $\lambda \in \sigma(U)$ and so $(U - \lambda I)$ does not have a bounded, linear inverse on H (Section 2.3, Definition 2). It follows from this that $(U - \lambda I)^{-1}$ could not be bounded on its domain. The reason is this: If $(U - \lambda I)^{-1}$ were bounded it would be uniformly continuous on its domain and hence it would be extendable, in one and only one way, to the closure of its domain. But its domain is dense in H so this would say that $(U - \lambda I)^{-1}$ would be extendable to all of H. But this extension would be bounded and would also be the inverse of $U - \lambda I$ on H. This contradiction shows that our operator could not be bounded.

An operator A *in* H may have a resolvent at a given complex number λ and, of course, it may not. If, for a given λ, $(A - \lambda I)^{-1}$ exists we shall call this operator the resolvent of A at λ. We have seen (Exercises 7.1, problem 3(b)) that A has a resolvent at λ if, and only if, λ is not an eigenvalue of A. Furthermore, if A is a closed operator and if it has a resolvent at λ then this resolvent is also a closed operator (Exercises 7.1, problem 3(c)).

Lemma 3

Let A be a linear operator in H. Then the two following conditions are equivalent: (a) The complex number λ is an eigenvalue of A or $(A - \lambda I)^{-1}$ exists and is an unbounded operator; (b) There exists a sequence of unit vectors $\{x_n\} \subseteq D(A)$, the domain of A, such that

$$\lim(A - \lambda I)x_n = 0$$

(Compare Section 2.4, Definition 1, and Exercises 4, problems 3 and 4).

Proof. If λ is an eigenvalue of A and $x \neq 0$ is a corresponding eigenvector, then letting $x_n = x/\|x\|$ for $n = 1, 2, \ldots$ gives us a sequence satisfying (b). The other possibility is that λ is not an eigenvalue and $(A - \lambda I)^{-1}$ is unbounded. Then we must have
$$\sup_{\|y\|=1} \|(A - \lambda I)^{-1} y\| = \infty$$
Thus there is a sequence $\{y_n\} \subseteq D[(A - \lambda I)^{-1}]$ such that $\|y_n\| = 1$ for $n = 1, 2, \ldots$ and
$$\lim \|(A - \lambda I)^{-1} y_n\| = \infty$$
Taking
$$x_n = \frac{(A - \lambda I)^{-1} y_n}{\|(A - \lambda I)^{-1} y_n\|}$$
for $n = 1, 2, \ldots$ we have $\|x_n\| = 1$ for each n and
$$\lim (A - \lambda I) x_n = \lim \frac{y_n}{\|(A - \lambda I)^{-1} y_n\|} = 0$$
Thus (a) implies (b).

Now assume (b), let $\{x_n\}$ be a sequence satisfying the conditions stated in (b) and suppose that the complex number λ is *not* an eigenvalue of A. Then by setting
$$y_n = \frac{(A - \lambda I) x_n}{\|(A - \lambda I) x_n\|}, \quad n = 1, 2, \ldots$$
we obtain a sequence of unit vectors such that
$$\lim \|(A - \lambda I)^{-1} y_n\| = \lim \frac{\|x\|}{\|(A - \lambda I) x_n\|} = \lim \frac{1}{\|(A - \lambda I) x_n\|} = \infty$$
Thus $(A - \lambda I)^{-1}$ is unbounded.

Theorem 1

The spectrum of a unitary operator consists entirely of generalized eigenvalues.

Proof. Let U be a unitary operator on H and let $\lambda \in \sigma(U)$. Now, of course, λ could be an eigenvalue of U; for any eigenvalue is also a generalized eigenvalue. Let us assume that λ is not an eigenvalue of U. Then $(U - \lambda I)^{-1}$ exists and is an unbounded operator whose domain is a dense linear manifold in H (Lemma 2). Hence, by Lemma 3, λ is a generalized eigenvalue of U.

THE SPECTRAL THEOREM FOR UNITARY OPERATORS

Let us turn now to the spectral theorem for a bounded, unitary operator U on a Hilbert space H. The results and many of the proofs are similar to those presented in Section 6.6 where we discussed the spectral theorem for self-adjoint operators. For this reason we shall omit them.

Theorem 2

(Spectral Theorem) Let U be a unitary operator on H. Then there exists a family $\{P(\lambda)|0 \leq \lambda \leq 2\pi\}$ of orthogonal projection operators with the following properties:

a. $P(\lambda) \leq P(\lambda')$ for $0 \leq \lambda \leq \lambda' \leq 2\pi$;

b. $P(0) = 0$, $P(2\pi) = I$;

c. $P(\lambda + 0) = P(\lambda)$ for $0 \leq \lambda < 2\pi$;

d. $U = \int_0^{2\pi} e^{i\lambda} dP(\lambda)$

More generally, for every continuous, complex-valued function $\sigma(x)$ on the unit circle in the complex plane we have:

$$\sigma(U) = \int_0^{2\pi} \sigma(e^{ix}) dP(\lambda)$$

Furthermore, if $\{P'(\lambda)|0 \leq \lambda \leq 2\pi\}$ is any family of orthogonal projections which satisfies (a)–(d), then $P'(\lambda) = P(\lambda)$ for all λ. Thus we may refer to $\{P(\lambda)|0 \leq \lambda \leq 2\pi\}$ as the spectral family of U.

Corollary 1

Let U be a unitary operator and let $\{P(\lambda)|0 \leq \lambda \leq 2\pi\}$ be the spectral family of U. If $\sigma(x)$ is a continuous, complex-valued function in the unit circle, then

$$\sigma(U)x = \int_0^{2\pi} \sigma(e^{ix}) dP(\lambda)x \quad \text{for all } x \in H;$$

$$\langle \sigma(U)x, y \rangle = \int_0^{2\pi} \sigma(e^{ix}) d\langle P(\lambda)x, y \rangle \quad \text{for all } x, y \in H.$$

More explicitly, for every $\varepsilon > 0$, there is a $\delta > 0$ such that $\|\sigma(U)x - \sum_{k=2}^{n} \sigma(e^{i\lambda'_k})[P(\lambda_k)x - P(\lambda_{k-1})x]\| \leq \varepsilon\|x\|$ for all $x \in H$ $|\langle \sigma(U)x, y \rangle - \sum_{k=2}^{n} \sigma(e^{i\lambda'_k})[\langle P(\lambda_0)x, y \rangle - \langle P(\lambda_{k-1})x, y \rangle]| \leq \varepsilon\|x\|\|y\|$ for all $x, y \in H$, whenever $0 = \lambda_1 < \lambda_2 < \ldots < \lambda_n = 2\pi$, $\lambda'_k \in [\lambda_{k-1}, \lambda_k]$, is any partition such that $\max_{2 \leq k \leq n}(\lambda_k - \lambda_{k-1}) < \delta$.

Corollary 2

Let U be a unitary operator on H and let $\{P(\lambda)|0 \le \lambda \le 2\pi\}$ be its spectral family. If σ is any continuous, complex-valued function on the unit circle, then $\|\sigma(U)x\|^2 = \int_0^{2\pi} |\sigma(e^{i\lambda})|^2 d\|P(\lambda)x\|^2$ for all $x \in H$. More explicitly, given $\varepsilon > 0$ there is a $\delta > 0$ such that

$$\left| \|\sigma(U)x\|^2 - \sum_{k=2}^{n} |\sigma(e^{i\lambda'_k})|^2 [\|P(\lambda_k)x\|^2 - \|P(\lambda_{k-1})x\|^2] \right| \le \varepsilon \|x\|^2$$

for all $x \in H$ whenever $0 = \lambda_1 < \ldots < \lambda_n = 2\pi$ is a partition with $\max_{2 \le k \le n}(\lambda_k - \lambda_{k-1}) < \delta$.

These results have consequences which are similar to those proved in section 6.6 for self-adjoint operators. We have:

Theorem 3

Let U be a unitary operator on H and let $\{P(\lambda)|0 \le \lambda \le 2\pi\}$ be its spectral family. Let B be a bounded, linear operator on H. Then $UB = BU$ if, and only if, $P(\lambda)B = BP(\lambda)$ for $0 \le \lambda \le 2\pi$. In particular $P(\lambda)U = UP(\lambda)$ for $0 \le \lambda \le 2\pi$.

Proof. Entirely analogous to the self-adjoint case proved above (Section 6.6).

Theorem 4

Let U be a unitary operator on H and let $\{P(\lambda)|0 \le \lambda \le 2\pi\}$ be its spectral family. The number $e^{i\mu}$, with $0 < \mu \le 2\pi$, is a regular value for U iff there exists a $\mathcal{O} > 0$ such that $P(\mu - \mathcal{O}) = P(\mu + \mathcal{O})$ (here we define $P(\lambda) = I$ for $\lambda > 2\pi$).

Corollary 3

The number $e^{i\mu}$, with $0 < \mu \le 2\pi$, belongs to $\sigma(U)$ if, and only if, $P(\mu - \mathcal{O}) < P(\mu + \mathcal{O})$ for all $\mathcal{O} > 0$ (here we define $P(\lambda) = 0$ for $\lambda < 0$ and $P(\lambda) = I$ for $\lambda > 2\pi$).

Theorem 5

Let U be a unitary operator on H and let $\{P(\lambda)|0 \le \lambda \le 2\pi\}$ be its spectral family. Let $\mu \in (0, 2\pi]$ be given. Then $P(\mu) - P(\mu - 0)$ is the orthogonal projection on the subspace $M_\mu = \{x \in H | Ux = e^{i\mu}x\}$. Consequently, $e^{i\mu}$ is an eigenvalue of U if, and only if, $P(\mu - 0) < P(\mu)$.

Exercises 6

1. We shall sketch the process by which one can define $\varphi(U)$ for any complex-valued, continuous function φ on the unit circle; here U is a fixed unitary operator on H, so $U^* = U^{-1}$.
   Let $P = \{\varphi(z) = \sum_{k=-n}^{n} \alpha_k z^k | \alpha_k \in \mathbb{C} \text{ for } -n \leq k \leq n \text{ and } n \geq 0\}$ where we regard φ as being defined on $\mathbb{C}\setminus\{0\}$. We define $\varphi(U) = \sum_{k=-n}^{n} \alpha_k U^k$, $U^0 \equiv I$, $\overline{\varphi}(U) = \sum_{k=-n}^{n} \overline{\alpha}_k U^{-k}$. Finally, let $\Gamma[\varphi(z)] = \varphi(U)$.

   a. Show that $\Gamma[\varphi_1(z)+\varphi_2(z)] = \Gamma[\varphi_1(z)]+\Gamma[\varphi_2(z)]$, $\Gamma[\alpha\varphi(z)] = \alpha\Gamma[\varphi(z)]$ and $\Gamma[\varphi_1(z)\varphi_2(z)] = \Gamma[\varphi_1(z)]\Gamma[\varphi_2(z)]$.

   b. Show that $\Gamma[\overline{\varphi(z)}] = \Gamma[\varphi(z)]^*$.

   c. Show that for any $\varphi \in P$, $\sigma[\varphi(U)] = \{\varphi(\xi) | \xi \in \sigma(U)\} = \varphi[\sigma(U)]$. Hint: $\psi(z) \equiv z^n \varphi(z) \in P$ since φ does. Note that $\eta \in \mathbb{C}$ is in $\sigma[\varphi(U)]$ iff $[\varphi(U) - \eta I] = U^{-n}[\psi(U) - \eta U^n]$ is not invertible, iff $(\psi(U) - \eta U^n)$ is not invertible, iff 0 is in the spectrum of $\psi(U) - \eta U^n$. By Section 6.2, this is the case iff $\psi(\xi) - \eta \xi^n = 0$ for some $\xi \in \sigma(U)$ iff $\eta = \psi(\xi)/\xi^n = \varphi(\xi)$ for some $\xi \in \sigma(U)$.

   d. If $\varphi \in P$ and $\varphi(\xi) \in \mathbb{R}$ for all ξ with $|\xi| = 1$ then $\varphi(U)$ is a self-adjoint operator and $\|\varphi(U)\| = \sup\{|\varphi(\xi)| | \xi \in \sigma(U)\}$. Hint: $\overline{\varphi}(\xi) = \sum_{-n}^{n} \overline{\alpha}_k \xi^{-k} = \sum_{-n}^{n} \alpha_k \xi^k = \varphi(\xi)$ for $|\xi| = 1$ so $\sum_{-n}^{n}(\alpha_k - \overline{\alpha}_{-k})\xi^k = 0$ for $|\xi| = 1$ hence $\sum_{-n}^{n}(\alpha_k - \overline{\alpha}_{-k})e^{2\pi i k\eta} = 0$ for all $\eta \in \mathbb{R}$. Now $\{e^{2\pi i k y} | -\infty < k < \infty\}$ as functions on $\mathbb{R}$ are linearly independent. Hence $\alpha_k = \overline{\alpha}_{-k}$ for $-n \leq k \leq n$. Thus $[\varphi(U)]^* = \overline{\varphi}(U) = \sum_{-n}^{n} \overline{\alpha}_k U^{-k} = \sum_{-n}^{n} \alpha_k U^k = \varphi(U)$.

2. Let $\varphi(z)$ be a continuous, complex-valued function defined on the unit circle in $\mathbb{C}$ and let U be a unitary operator. Show that there is a unique bounded, linear operator, which we shall denote by $\varphi(U)$, such that whenever $\{\varphi_n(z)\}_{n=1}^{\infty}$ is a sequence in P such that $\lim_{n\to\infty} \max\{|\varphi(z)-\varphi_n(z)| \mid z \in \sigma(U)\} = 0$, then $\varphi(U) = \lim \varphi_n(U)$.

   a. Show that $(\alpha\varphi_1 + \beta\varphi_2)(U) = \alpha\varphi_1(U) + \beta\varphi_2(U)$.

   b. Show that $[\varphi(U)]^* = \overline{\varphi}(U)$, $\|\varphi(U)\| \leq 2\max\{|\varphi(\xi)| \mid \xi \in \sigma(U)\}$.

   c. Show that the operator $\varphi(U)$ is:

      i. Self-adjoint iff $\varphi(\xi) \in \mathbb{R}$ for all $\xi \in \sigma(U)$;

      ii. Unitary iff $|\varphi(\xi)| = 1$ for all $\xi \in \sigma(U)$;

      iii. Invertible iff $\varphi(\xi) \neq 0$ for all $\xi \in \sigma(U)$.

7.7 The Spectral Theorem for Unbounded Operator

Let S be a fixed, unbounded, self-adjoint (Section 4, Definition 1) operator in a Hilbert space H. We recall that the Cayley transform, V, of S is given by

$$V = (S - iI)(S + iI)^{-1} \qquad (1)$$

and that V is, in this case, a unitary operator (Section 5, Theorem 1). If $\{Q(\mu) | 0 \le \mu \le 2\pi\}$ is the spectral family of V then

$$V = \int_0^{2\pi} e^{i\mu} dQ(\mu) \qquad (2)$$

by Theorem 2 of Section 6. Furthermore,

$$\varphi(V) = \int_0^{2\pi} \varphi(e^{i\mu}) dQ(\mu) \qquad (3)$$

for any function φ which is continuous on the unit circle in the complex plane. Now we know that

$$S = i(I + V)(I - V)^{-1} \qquad (4)$$

Exercises 5, problem 1(iii), and so we may set

$$\psi(z) = i(1 + z)(1 - z)^{-1} \qquad (5)$$

notice that $\psi(S) = V$ and try to use (3) to get a spectral representation for S. The integrand would be

$$\psi(e^{i\mu}) = i\frac{1 + e^{i\mu}}{1 - e^{i\mu}} = i\frac{e^{-\frac{i\mu}{2}} + e^{\frac{i\mu}{2}}}{e^{-\frac{i\mu}{2}} - e^{\frac{i\mu}{2}}} = -\frac{\cos\frac{1}{2}\mu}{\sin\frac{1}{2}\mu} = -\cot\left(\frac{\mu}{2}\right)$$

and we would have

$$\psi(V) = S = \int_0^{2\pi} \psi(e^{i\mu}) dQ(\mu) = -\int_0^{2\pi} \cot\left(\frac{\mu}{2}\right) dQ(\mu) \qquad (6)$$

Setting $\lambda = -\cot \mu/2$ we would obtain

$$S = \int_{-\infty}^{\infty} \lambda \, dQ(2\operatorname{arc\,cot}(-\lambda)) \qquad (7)$$

Unfortunately, the function ψ (defined by (5) is not continuous on the unit circle and so we may not use (3). It turns out that (7) is valid but we have to work quite a bit harder to prove it. Our proof is an adaptation of the one in [7; pp. 292–302]. The trick is to decompose H is a certain way.

THE SPECTRAL THEOREM FOR UNBOUNDED OPERATOR 273

Lemma 1

For each integer m let $\mu_m = 2\operatorname{arc\,cot}(-m)$ and let H_m be the subspace of H corresponding to the orthogonal projection $P_m = Q(\mu_m) - Q(\mu_{m-1})$. Then the subspaces H_m are pairwise orthogonal and $H = \sum_{m=-\infty}^{\infty} H_m$ (see Section 5.4, Lemma 1, and the discussion just before it).

Proof. For $m < n$ we have $Q(\mu_m) \leq Q(\mu_n)$ (Section 6, Theorem 2(a)) and so

$$\begin{aligned} P_m P_n &= [Q(\mu_m) - Q(\mu_{m-1})][Q(\mu_n) - Q(\mu_{n-1})] \\ &= [Q(\mu_m) - Q(\mu_{m-1})] - [Q(\mu_m) - Q(\mu_{m-1})] = 0 \end{aligned}$$

Exercises 4.2 problem 2(b). Thus H_m and H_n are orthogonal. Now 1 is not an eigenvalue of V (Exercises 5, problem 2) and so $Q(2\pi - 0) = Q(2\pi) = I$ (Section 6, Theorem 5). Thus for every $x \in H$ we have

$$\begin{aligned} \sum_{-\infty}^{\infty} P_m x &= \lim_{m \to +\infty} Q(\mu_m)x - \lim_{n \to -\infty} Q(\mu_n)x \\ &= Q(2\pi - 0)x - Q(0+0)x = x - 0 = x \end{aligned}$$

which proves our last assertion.

Note: Our decomposition gives us, for each $x \in H$, a sequence $\{x_m\}_{-\infty}^{\infty}$ of vectors such that $x_m \in H_m$ for each m and

$$\|x\|^2 = \sum_{-\infty}^{\infty} \|x_m\|^2$$

Thus we could avoid having minus infinity on the summation by writing

$$H = H_0 + \sum_{m=1}^{\infty}(H_m + H_{-m})$$

because the convergence of our sums is unaffected by such a change. We see no point in doing this however.

We would like to show that each of the subspaces H_m "reduces" the operator S. Before doing that we must define this concept for unbounded operators and prove some results analogous to those given in Section 5.3 for bounded operators.

Definition 1

Let S be an operator in a Hilbert space H and let M be a linear subspace of H. We shall say that M is invariant under S if $Sx \in M$ for all $x \in M \cap D(S)$.

The definition of a reducing subspace is somewhat more complicated than in the case of a bounded operator. We recall that a subspace M reduces the bounded, linear operator A if both M and $M^\perp = \{x \in H | \langle x, y \rangle = 0 \text{ for all } y \in M\}$ are invariant under A. So if M reduces A then so does $M^\perp$ and conversely. Our next definition preserves this symmetry.

Definition 2

Let S be a linear operator in H, let M be a linear subspace of H and let P, $P^\perp$ be the orthogonal projections of H onto M and $M^\perp$ respectively. We shall say that M reduces S if $P[D(S)] \subseteq D(S)$, $P^\perp[D(S)] \subseteq D(S)$ and if there is a linear operator S_1 in M with domain $P[D(S)] = M \cap D(S)$ and a linear operator S_2 in $M^\perp$ with domain $P^\perp[D(S)] = M^\perp \cap D(S)$ such that

$$Sx = S_1(Px) + S_2(P^\perp x) \qquad (8)$$

for all x in $D(S)$.

Lemma 2

Let S be a self-adjoint operator in H, and suppose that the linear subspace M of H reduces S. Then the operators S_1, S_2 induced by S on M and $M^\perp$ respectively (see Definition 2) are self-adjoint.

Proof. Suppose that for some y, y^* in M we have

$$(*) \quad \langle S_1 x, y \rangle = \langle x, y^* \rangle$$

for all $x \in M \cap D(S)$. Then for every $x \in D(S)$ we also have

$$\begin{aligned}
\langle Sx, y \rangle &= \langle S_1 Px + S_2 P^\perp x, y \rangle \\
&= \langle S_1 Px, y \rangle \text{ (since } S_2 P^\perp x \in M^\perp \text{ and } y \in M) \\
&= \langle Px, y^* \rangle \text{ (because } Px \in M \cap D(S)) \\
&= \langle Px + P^\perp x, y^* \rangle = \langle x, y^* \rangle
\end{aligned}$$

Since S is self-adjoint this implies that $y \in M \cap D(S) = D(S_1)$ and $y^* = Sy = S_1 y$ (see Definition 1 of Section 2 and the discussion just before it). In particular, the vector $y^* \in M$ is uniquely determined by the vector $y \in M$ and the relation (2). Thus $D(S_1) = M \cap D(S)$ is dense in M and S_1 is a self-adjoint operator in this subspace.

The proof that S_2 is self-adjoint in $M^\perp$ is similar to that just given.

Theorem 1

Let S be a linear operator in H and let P be the orthogonal projection of H onto the subspace M of H. Then M reduces S if, and only if, we have:

a. $P[D(S)] \subseteq D(S)$;

b. Both M and $M^\perp$ are invariant under S.

Proof. If M reduces S then $P[D(S)] \subseteq D(S)$ by definition. Also, from equation (1), we see that $Sx \in M$ for all $x \in M \cap D(S)$ and $Sx \in M^\perp$ for all $x \in M^\perp \cap D(S)$. Referring to Definition 1 we see that both M and $M^\perp$ are invariant under S.

Now suppose that a given operator S satisfies (a) and (b). For every $x \in D(S)$ we have $x = Px + P^\perp x$, or $P^\perp x = x - Px \in D(S)$ we see that $P^\perp[D(S)] \subseteq D(S)$. Furthermore

$$Sx = SPx + SP^\perp x$$

for all $x \in D(S)$, and this is (1) with $S_1 = SP$, $S_2 = SP^\perp$ on $M \cap D(S)$ and $M^\perp \cap D(S)$ respectively.

Theorem 2

Let S be a linear operator in H and let P be the orthogonal projection of H onto the subspace M of H. Then M reduces S if, and only if, $PS \subseteq SP$ (Section 1, Definition 3).

Proof. Suppose that M reduces S and let $x \in D(S)$ be given. Then $Px \in M \cap D(S)$, $P^\perp x \in M^\perp \cap D(S)$, $SPx \in M$, $SP^\perp x \in M^\perp$. Now for all $y \in H$ we have

$$y = Py + P^\perp y$$

and so for $x \in D(S)$

$$Sx = S[Px + P^\perp x] = SPx + SP^\perp x = PSx + P^\perp Sx$$

Since this decomposition of Sx is unique we conclude that

$$PSx = SPx$$

for all $x \in D(S) = D(PS)$. Thus $PS \subseteq SP$.

Now suppose that $PS \subseteq SP$. Since the domain of PS is just $D(S)$ this inclusion implies that for $x \in D(S)$, $Px \in D(S)$ also; i.e., $P[D(S)] \subseteq D(S)$. For every $x \in M \cap D(S)$ we have

$$Sx = SPx = PSx \in M$$

which shows M is invariant under S. For every $x \in M^\perp \cap D(S)$ we have

$$P(Sx) = S(Px) = S(0) = 0 \in M^\perp.$$

Thus $M^\perp$ is also invariant under S. Our result now follows from Theorem 1.

In Theorem 2 we considered PS and SP. Only P is bounded here. Recall that two bounded, linear operators A, B on H are said to commute if $AB = BA$.

Definition 3

Let S be a linear operator in H and let B be a bounded, linear operator on H. We shall say that S and B commute if $BS \subseteq SB$.

We shall *not* define commutability for two unbounded operators.

Theorem 3

Let V be the Cayley transform of a self-adjoint operator S in H, and let B be a bounded, linear operator on H. Then V commutes with B if, and only if, B commutes with S; i.e., $VB = BV$ if, and only if, $BS \subseteq SB$.

Proof. Suppose that B commutes with S. Then $BS \subseteq SB$ meaning $D(BS) \subseteq D(SB)$ and $SB|D(BS) = BS$ or, equivalently, $B[D(S)] \subseteq D(S)$ and $BSx = SBx$ for all $x \in D(S)$. Now

$$(S + iI)^{-1}H = D(S + iI) = D(S)$$

and so for every $x \in H$ we have

$$(S + iI)^{-1}x \in D(S), \quad B(S + iI)^{-1}x \in D(S)$$

$$(S + iI)B(S + iI)^{-1}x = B(S + iI)(S + iI)^{-1}x = Bx$$

because S and B commute. This last equation gives

$$B(S + iI)^{-1}x = (S + iI)^{-1}Bx$$

for every $x \in H$. Thus

$$\begin{aligned} BVx &= B(S - iI)(S + iI)^{-1}x = (S - iI)B(S + iI)^{-1}x \\ &= (S - iI)(S + iI)^{-1}Bx = VBx \end{aligned}$$

showing that B commutes with V.

Now suppose that we know that $BV = VB$. If $x \in D(S) = D[(I-V)^{-1}]$ is given, then setting $y = (I-V)^{-1}x$ gives us

$$\begin{aligned}
x &= (I-V)y \\
Bx &= B(I-V)y = (I-V)By \in D[(I-V)^{-1}] = D(S) \\
(I-V)^{-1}Bx &= By = B(I-V)^{-1}x \\
SBx &= i(I+V)(I-V)^{-1}Bx = i(I+V)B(I-V)^{-1}x \\
&= Bi(I+V)(I-V)^{-1}x = BSx
\end{aligned}$$

Corollary 1

Let S and V be as in Theorem 3 and let $\{Q(\mu)|0 \leq \mu \leq 2\pi\}$ be the spectral family of V. If $0 \leq \mu < \mu' \leq 2\pi$, then the subspace corresponding to the orthogonal projection $Q(\mu') - Q(\mu)$ reduces S.

Proof. Theorem 2 and Section 6, Theorem 3.

Corollary 2

Referring to Lemma 1 each of the subspaces H_m reduces the operator S.

For every m the operator S induces (see Lemma 2) a self-adjoint operator S_m in the space H_m.

Theorem 4

Let H_m be as in Lemma 1. Then $H_m \subseteq D(S)$ and S_m is a bounded, self-adjoint operator on H_m given by

$$S_m x = \int_{\mu_{m-1}}^{\mu_m} (-\cot \tfrac{1}{2}\mu) dQ(\mu) x$$

for all $x \in H_m$.

Proof. Let $x \in H_m$ be given and note that $Q(\mu)x = x$ for $\mu \geq \mu_m$, $Q(\mu)x = 0$ for $x \leq \mu_{m-1}$, by the definition of H_m. Using μ_{m-1} and μ_m as fixed subdivision points for our Riemann-Stieltjes sums we get, for any continuous function $\varphi(z)$ on the unit circle

$$\varphi(V)x = \int_{\mu_{m-1}}^{\mu_m} \varphi(e^{i\mu}) dQ(\mu) x \in H_m$$

In particular, if we define, for every integer m, $\zeta_m = e^{i\mu_m}$

$$\varphi_m(\zeta) = \begin{matrix} (1-\zeta)^{-1} \text{ for } \zeta = e^{i\mu}, \mu_{m-1} \le \mu \le \mu_m \\ (\zeta_m - \zeta_{m-1})^{-1}[(1-\zeta_{m-1})^{-1}(\zeta_m - \zeta) + (1-\zeta_m)^{-1}(\zeta - \zeta_{m-1})] \end{matrix}$$

for all other ζ on the unit circle, then $\varphi_m(z)$ is continuous on the unit circle and so

$$\begin{aligned} (I-V)\varphi_m(V)x &= \int_{\mu_{m-1}}^{\mu_m} (1-e^{i\mu})\varphi_m(e^{i\mu})dQ(\mu)x \\ &= \int_{\mu_{m-1}}^{\mu_m} dQ(\mu)x = Q(\mu_m)x - Q(\mu_{m-1})x = x \end{aligned}$$

We conclude that $x \in (I-V)H = D[(I-V)^{-1}] = D(S)$. This proves $H_m \subseteq D(S)$. Since S_m is self-adjoint and defined on all of H_m it is a bounded operator (Section 2, Theorem 2, and the discussion after Lemma 1 of that section). Moreover,

$$\begin{aligned} (I-V)^{-1}x &= \varphi_m(V)x \\ S_m x &= Sx = i(I+V)(I-V)^{-1}x = i(I+V)\varphi_m(V)x \\ &= \int_{\mu_{m-1}}^{\mu_m} i\frac{1+e^{i\mu}}{1-e^{i\mu}}dQ(\mu)x = \int_{\mu_{m-1}}^{\mu_m} (-\cot\frac{1}{2}\mu)dQ(\mu)x. \end{aligned}$$

Corollary 3

Let H_m and S_m be as in Theorem 4. Then

$$\|S_m x\|^2 = \int_{\mu_{m-1}}^{\mu_m} |\cot\frac{1}{2}\mu|^2 d\|Q(\mu)x\|^2 \text{ for all } x \text{ in } H_m$$

Proof. We apply Corollary 2 to Theorem 2 of Section 6 to the function $\varphi(z) = i(1+z)\varphi_m(z)$ and $x \in H_m$ to get

$$\begin{aligned} \|S_m x\|^2 = \|Sx\|^2 &= \|i(I+V)\varphi_m(V)x\|^2 \\ &= \int_{\mu_{m-1}}^{\mu_m} |\cot\frac{1}{2}\mu|^2 d\|\varphi(\mu)x\|^2 \end{aligned}$$

Corollary 4

Let μ_m and P_m be as in Lemma 1. Then SP_m is a bounded, self-adjoint operator on H and

$$SP_m = \int_{\mu_{m-1}}^{\mu_m} (-\cot\frac{1}{2}\mu)dQ(\mu)$$

Proof. Let $\varphi_m(z)$ be the continuous function on the unit circle defined in the proof of Theorem 4. Then

$$i(I+V)\varphi_m(V) = \int_{\mu_{m-1}}^{\mu_m} i(1+e^{i\mu})\varphi_m(e^{i\mu})dQ(\mu)$$

$$i(I+V)\varphi_m(V)P_m = \int_{\mu_{m-1}}^{\mu_m} i\frac{1+e^{i\mu}}{1-e^{i\mu}}dQ(\mu) = \int_{\mu_{m-1}}^{\mu_m} (-\cot\frac{1}{2}\mu)dQ(\mu) \quad (9)$$

$$i(I+V)\varphi_m(V)P_m x = \int_{\mu_{m-1}}^{\mu_m} (-\cot\frac{1}{2}\mu)dQ(\mu)x \quad \text{for all } x \text{ in } H$$

The bounded, linear operator $i(I+V)\varphi_m(V)P_m$ on H thus coincides with S_m on H_m while it sends $H_m^\perp$ to zero. Hence it must be self-adjoint and coincides with SP_m.

Corollary 5

Let P_m and S_m be as in Lemma 1 and Theorem 4 respectively and define $P(\lambda) = Q(2 \operatorname{arc cot}(-\lambda))$ for all $\lambda \in \mathbb{R}$. Then the family $\{P(\lambda)|\lambda \in \mathbb{R}\}$ of orthogonal projections satisfies:

 a. $P(\lambda) \leq P(\lambda')$ for $\lambda \leq \lambda'$;

 b. $\lim_{\lambda \to -\infty} P(\lambda)x = 0$, $\lim_{\lambda \to \infty} P(\lambda)x = x$ for all $x \in H$;

 c. $P(\lambda + 0) = P(\lambda)$ for all λ in $\mathbb{R}$;

 d. For every integer m we can write

$$SP_m = \int_{m-1}^{m} \lambda dP(\lambda), \quad S_m x = \int_{m-1}^{m} \lambda dP(\lambda)x,$$

$$\|S_m x\|^2 = \int_{m-1}^{m} |\lambda|^2 d\|P(\lambda)x\|^2$$

the last two holding for every $x \in H_m$. Furthermore, a bounded, linear operator B on H commutes with S if, and only if, $P(\lambda)B = BP(\lambda)$ for all $\lambda \in \mathbb{R}$.

Proof. The properties (a)–(c) follow from the properties of the family $\{Q(\mu)|0 \leq \mu \leq 2\pi\}$ (Section 6, Theorem 2). Property (d) follows from Corollaries 1 and 2 to Theorem 4 after the substitution $-\cot\frac{1}{2}\mu = \lambda$ has been used in the Riemann-Stieltjes sums which approximate these integrals. The last assertion is equivalent to the commutativity of B and V and, by Theorem 3, it is then equivalent to the commutativity of B and S.

We began with a fixed, unbounded, self-adjoint operator S on H and its Cayley transform V. Since V is a unitary operator it has a spectral family $\{Q(\mu)|0 \le \mu \le 2\pi\}$. For each integer m we set $\mu_m = 2 \text{ arc cot } (-m)$ and we took H_m to be the subspace of H corresponding to the orthogonal projection $P_m = Q(\mu_m) - Q(\mu_{m-1})$; see Lemma 1. We found that these subspaces H_m are pairwise orthogonal, and that their sum is H.

Next we noted that each H_m reduces the operator S (Corollary 2 to Theorem 3) and that S induces a bounded, self-adjoint operator S_m on H_m, for every m (Theorem 4). Our next result says that, in a situation like this, the S_m can be "put back together" to get S.

Theorem 5

Suppose that H is a Hilbert space, $\{H_m\}_{-\infty}^{\infty}$ is a sequence of pairwise orthogonal linear subspaces of H such that

$$H = \sum_{m=-\infty}^{\infty} H_m, \qquad (10)$$

for every m, P_m is the orthogonal projection of H onto H_m, and S_m is a bounded self-adjoint operator on H_m. Then there is a unique self-adjoint operator S in H which is reduced by every H_m and whose restriction to H_m coincides with S_m for every m. Furthermore,

$$D(S) = \{x \in H | \sum_{m=-\infty}^{\infty} \|S_m P_m x\|^2 < \infty\} \qquad (11)$$

and

$$Sx = \sum_{m=-\infty}^{\infty} S_m P_m x \text{ for all } x \in D(S). \qquad (12)$$

Proof. We first note that $D(S)$, as defined in (11), is a linear manifold in H and, since $\sum_{m=-n}^{n} H_m \subseteq D(S)$ for each n, equation (10) implies that $D(S)$ is dense in H. Let Sx be defined by (13). Then S is a linear operator with domain $D(S)$ and

$$P_m Sx = P_m(\sum_{m=-\infty}^{\infty} S_n P_n x) = S_m P_m x = S P_m x \text{ for all } x \in D(S).$$

Hence $P_m S \subseteq S P_m$ showing that H_m reduces S (Theorem 2). Let x, y be in $D(S)$ and write

$$\langle Sx, y \rangle = \sum_{m=-\infty}^{\infty} \langle S_m P_m x, P_m y \rangle = \sum_{m=-\infty}^{\infty} \langle P_m x, S_m P_m y \rangle = \langle x, Sy \rangle$$

showing that S is a symmetric operator (Section 4, Definition 1). Let us show that S is self-adjoint. Choose $y = \sum_{m=-\infty}^{\infty} P_m y \in D(S^*)$. For each $x \in H_m \subseteq D(S)$ we have

$$\begin{aligned}\langle x, S_m P_m y\rangle &= \langle S_m x, P_m y\rangle = \langle S_m x, y\rangle = \langle Sx, y\rangle \\ &= \langle x, S^* y\rangle = \langle x, P_m(S^* y)\rangle\end{aligned}$$

Thus $S_m P_m y = P_m(S^* y)$ and

$$\sum_{m=-\infty}^{\infty} \|S_m P_m y\|^2 = \sum_{m=-\infty}^{\infty} \|P_m(S^* y)\|^2 = \|S^* y\|^2$$

showing that $y \in D(S)$.

Finally, suppose that S' is some other self-adjoint operator in H which is reduced by each H_m and whose restriction to each H_m coincides with S_m. Then for any $x \in D(S)$ we have

$$x = \lim_{n\to\infty} \sum_{m=-n}^{n} P_m x$$

and

$$\lim_{n\to\infty} S'(\sum_{-n}^{n} P_m x) = \lim_{n\to\infty} \sum_{-n}^{n} S_m P_m x = Sx$$

But S' is a self-adjoint, hence closed, operator and so $x \in D(S')$ and $S'x = Sx$. Thus $S \subseteq S'$. But then $S = S^* \supseteq (S')^* = S'$ showing that $S = S'$.

We shall now apply Theorem 5 to the subspace H_m defined in Lemma 1 and the operators S_m defined in H_m (just after Corollary 2 to Theorem 3). Remember we started here with a self-adjoint operator S and we obtained the S_m from this operator. By the uniqueness part of Theorem 5 we know that when we piece together the S_m's we get S back. Hence

$$\begin{aligned}D(S) &= \{x \in H | \sum_{m=-\infty}^{\infty} \|S_m P_m x\|^2 < \infty\} \quad (13)\\ &= \{x \in H | \sum_{m=-\infty}^{\infty} \int_{m-1}^{m} |\lambda|^2 d\|P(\lambda)x\|^2 < \infty\}\end{aligned}$$

(by Corollary 3 to Theorem 4).

Theorem 6

Let S be a self-adjoint operator in H. Then there exists a family of orthogonal projections $\{P(\lambda) | \lambda \in \mathbb{R}\}$, which we shall call the spectral family of S, with the following properties:

a. $P(\lambda) \leq P(\lambda')$ for $\lambda \leq \lambda'$;
b. $\lim_{\lambda \to -\infty} P(\lambda)x = 0$, $\lim_{\lambda \to +\infty} P(\lambda)x = x$ for all x in H;
c. $P(\lambda + 0) = P(\lambda)$ for all $\lambda \in \mathbf{R}$;
d. $x \in D(S)$ if, and only if, $\int_{-\infty}^{\infty} \lambda^2 d\|P(\lambda)x\|^2 < \infty$.

Furthermore,
$$Sx = \int_{-\infty}^{\infty} \lambda dP(\lambda)x \text{ for all } x \text{ in } D(S).$$

Proof. Statements (a), (b), and (c) were proved before (Corollary 3 to Theorem 4). Referring to (13) just above we must show that the sum

$$\sum_{m=-\infty}^{\infty} \int_{m-1}^{m} |\lambda|^2 d\|P(\lambda)x\|^2$$

can be replaced by the integral written in (d). For any integers k, ℓ and for $k < 2 \leq \mu \leq k-1, \ell \leq \nu \leq \ell+1$ we have

$$\sum_{m=k}^{\ell} \int_{m-1}^{m} |\lambda|^2 d\|P(\lambda)x\|^2 \leq \int_{\mu}^{\nu} |\lambda|^2 d\|P(\lambda)x\|^2 \leq \sum_{m=k-1}^{\ell+1} \int_{m-1}^{m} |\lambda|^2 d\|P(\lambda)x\|^2$$

and so

$$\sum_{m=-\infty}^{\infty} \int_{m-1}^{m} |\lambda|^2 d\|P(\lambda)x\|^2 = \lim_{\substack{\mu \to -\infty \\ \nu \to \infty}} \int_{\mu}^{\nu} |\lambda|^2 d\|P(\lambda)x\|^2 = \lim_{\substack{\mu \to -\infty \\ \nu \to \infty}}$$
$$= \int_{-\infty}^{\infty} |\lambda|^2 d\|P(\lambda)x\|^2$$

This proves (d). Finally, from Theorem 5 we have

$$Sx = \sum_{m=-\infty}^{\infty} S_m P_m x$$

and combining this with (d) of Corollary 3 to Theorem 4 we see that

$$Sx = \sum_{m=-\infty}^{\infty} \int_{m-1}^{m} \lambda dP(\lambda)x = \lim_{\substack{k \to -\infty \\ \ell \to \infty}} \int_{k}^{\ell} \lambda dP(\lambda)x$$

THE SPECTRAL THEOREM FOR UNBOUNDED OPERATOR 283

Let $[\mu]$ be the greatest integer which is less than or equal to μ. Then

$$\int_\mu^\nu \lambda dP(\lambda)x = \int_{[\mu]}^{[\nu]} \lambda dP(\lambda)x + \int_{[\nu]}^\nu \lambda dP(\lambda)x - \int_{[\mu]}^\mu \lambda dP(\lambda)x$$

and

$$\lim_{\mu \to -\infty} \left\| \int_{[\mu]}^\mu \lambda dP(\lambda)x \right\|^2 = \lim_{\mu \to -\infty} \int_{[\mu}^\mu |\lambda|^2 d\|P(\lambda)x\|^2 \leq \lim_{\mu \to -\infty} \|S_{[\mu]+1}x\|^2 = 0$$

$$\lim_{\nu \to +\infty} \left\| \int_{[\nu]}^\nu \lambda dP(\lambda)x \right\|^2 = \lim_{\nu \to +\infty} \int_{[\nu]}^\nu |\lambda|^2 d\|P(\lambda)x\|^2 \leq \lim_{\nu \to +\infty} \|S_{[\nu]+1}x\|^2 = 0$$

hence

$$\int_{-\infty}^\infty \lambda dP(\lambda)x = \lim_{\substack{\mu \to -\infty \\ \nu \to \infty}} \int_\mu^\nu \lambda dP(\lambda)x = \lim_{\substack{\mu \to -\infty \\ \nu \to \infty}} \sum_{m=[\mu]+1}^{[\nu]} S_m x = Sx.$$

Theorem 7

Let $\{P'(\lambda) | \lambda \in \mathbf{R}\}$ be a family of orthogonal projections satisfying all of the assertions of Theorem 6. Then for each $\lambda \in \mathbf{R}$, $P'(\lambda) = P(\lambda)$ where the $P(\lambda)$ are defined as in Corollary 3 to Theorem 4.

8

NON-NORMALIZABLE EIGENVECTORS

Perhaps the best way to introduce this chapter is to consider again the differential equation

$$-\frac{d^2u}{dx^2} - \lambda_0 u = f(x), \quad 0 \le x \le \ell, \quad u(0) = 0 = u(\ell) \tag{1}$$

already discussed in Chapter 7. Recall that λ_0 is a given complex number, ℓ is a given positive real number, and $f(x)$ is a given function in the space $L^2[0,\ell]$. The first thing we did in solving this problem was to find the eigenvalues and normalized eigenfunctions of the operator $(-d^2/dx^2)$. We found these to be

$$\lambda_n = \frac{n^2\pi^2}{\ell^2}, \quad n = 1, 2, 3, \ldots \tag{2}$$

and

$$\varphi_n(x) = \left(\frac{2}{\ell}\right)^{1/2} \sin\frac{n\pi x}{\ell}, \quad n = 1, 2, 3, \ldots \tag{3}$$

respectively. Next we expanded the given function $f(x)$ in a series

$$f(x) = \sum_{n=1}^{\infty} f_n \varphi_n(x) \tag{4}$$

where

$$f_n = \langle f, \varphi_n \rangle = \left(\frac{2}{\ell}\right)^{1/2} \int_0^\ell f(x) \sin \frac{n\pi x}{\ell} dx, \quad n = 1, 2, 3, \ldots \quad (5)$$

Finally, we found the unique solution to (1), when λ_0 is *not* an eigenvalue, to be

$$u(x) = \sum_{n=1}^\infty \left[\frac{2/\ell}{\left(\frac{n^2\pi^2}{\ell^2} - \lambda_0\right)}\right] \left(\int_0^\ell f(x) \sin \frac{n\pi x}{\ell} dx\right) \sin \frac{n\pi x}{\ell} \quad (6)$$

Suppose now that we want to solve our differential equation over the interval $[0, \infty)$ instead of $[0, \ell]$. More specifically, we want to find a function $u(x) \in L^2[0, \infty)$ which satisfies

$$-\frac{d^2u}{dx^2} - \lambda_0 u = f(x), \quad 0 \le x < \infty, \quad u(0) = 0 \quad (7)$$

where λ_0 is a given complex constant, and $f(x)$ is a given function in the space $L^2[0, \infty)$. We have not yet discussed the space L^2 over an infinite interval but, below, we shall give a systematic treatment of $L^2(\mathbf{R}) = L^2(-\infty, \infty)$ which can easily be modified to the case of $[0, \infty)$.

Let us see if the solution to (1) sketched above can be generalized so as to give us a solution to (7). The first thing we must do is find the eigenvalues and eigenfunctions of the operator $(-d^2/dx^2)$ on $L^2[0, \infty)$; i.e., we must find those λ for which

$$-\frac{d^2u}{dx^2} = \lambda u, \quad u(0) = 0 \quad (8)$$

has a non-trivial solution in $L^2[0, \infty)$. Setting $\lambda = \nu^2$, for any fixed $\nu \in [0, \infty)$ equation (8) has the obvious solution

$$\varphi_\nu(x) = \sin \nu x \quad (9)$$

but, unfortunately, $\sin \nu x$ is *not* in $L^2[0, \infty)$ as one can easily check. So we have plenty of eigenfunctions, one for each $\nu \in [0, \infty)$, however they are not in the space of interest. Ignoring, for now, the theoretical status of such eigenvectors, let us show that we can in fact use them to solve problem (7). Referring to (4) and (5) above we saw that we could write the given function $f(x)$ in terms of the eigenfunctions of problem (1) as follows:

$$f(x) = \sum_{n=1}^\infty \left[\left(\frac{2}{\ell}\right)^{1/2} \int_0^\ell f(y) \sin \frac{n\pi y}{\ell} dy\right] \left(\frac{2}{\ell}\right)^{1/2} \sin \frac{n\pi x}{\ell} \quad (10)$$

Equation (10) contains a summation because problem (1) has only countably many eigenfunctions. In the case of problem (7) we have an uncountable family of eigenfunctions one for each $\nu \in [0, \infty)$, and so we expect the sum in (10) to be replaced by an integral with respect to ν. Let us give an informal sketch of how this might be done.

Set $\nu_n = \frac{n\pi}{\ell}$, $\Delta \nu = \nu_{n+1} - \nu_n = \pi/\ell$ and define

$$H(x,\nu) = \int_0^\ell f(y)\sin \nu y \sin \nu x \, dy \qquad (11)$$

Then equation (10) may be written as

$$f(x) = \frac{2}{\ell}\sum_{n=1}^\infty H(x,\nu_n) = \frac{2}{\ell}\frac{1}{\Delta\nu}\sum H(x,\nu_n)\Delta\nu$$
$$= \frac{2}{\pi}\sum H(x,\nu_n)\Delta\nu \qquad (12)$$

Now as $\ell \to \infty$ the quantity $\Delta\nu = \pi/\ell \to 0$ and so (12) and (11) yield

$$f(x) = \frac{2}{\pi}\int_0^\infty H(x,\nu)d\nu = \frac{2}{\pi}\int_0^\infty \left(\int_0^\infty f(y)\sin \nu y \, dy\right)\sin \nu x \, d\nu \qquad (13)$$

This suggests that we might define the Fourier sine transform $F_s(\nu)$ of any $f \in L^2[0,\infty)$ as follows:

$$F_s(\nu) = \left(\frac{2}{\pi}\right)^{1/2}\int_0^\infty f(x)\sin \nu x \, dx \qquad (14)$$

Equation (13) then shows us how to invert this transform

$$f(x) = \left(\frac{2}{\pi}\right)^{1/2}\int_0^\infty F_s(\nu)\sin \nu x \, d\nu \qquad (15)$$

We now have the tools we need to solve problem (7) when the constant $\lambda_0 \notin [0,\infty)$. Observe that any solution $u(x)$ must be in $L^2[0,\infty)$ and must have two derivatives. Hence both u and u' are continuous functions. One can show from equations (14) and (15) that $\lim_{x\to\infty} u(x) = 0 = \lim_{x\to\infty} u'(x)$. We will use this in a moment. Now multiply equation (7) by $(2/\pi)^{1/2}\sin \nu x$ and integrate from 0 to ∞ with respect to x. We then integrate by parts and use the facts that $u(x)$, $\sin \nu x$ vanish at $x = 0$ while u, u' tend to zero as $x \to \infty$ to get

$$\nu^2 U_s(\nu) - \lambda_0 U_s(\nu) = F_s(\nu) \qquad (16)$$

or

$$U_s(\nu) = \frac{F_s(\nu)}{\nu^2 - \lambda_0} \qquad (17)$$

We may invert this last formula using (15) to get

$$\begin{aligned} u(x) &= \left(\frac{2}{\pi}\right)^{1/2} \int_0^\infty \left(\frac{F_s(\nu)}{\nu^2 - \lambda_0}\right) \sin \nu x \, d\nu \\ &= \frac{2}{\pi} \int_0^\infty \left(\frac{\sin \nu x}{\nu^2 - \lambda_0}\right) \left(\int_0^\infty f(x) \sin \nu x \, dx\right) d\nu \end{aligned} \qquad (18)$$

which is our analogue of equation (6).

Our discussion illustrates how one may use operator-theoretic methods to solve differential equations on infinite intervals, and it shows what kinds of problems arise in using these methods. In the remainder of the chapter we shall investigate these problems more closely. For another approach to the solution of problem (7), we refer the reader to [11; pp. 283–290].

8.1 The Space $L^2(\mathbb{R})$

We consider the set of all complex-valued, measurable functions on $\mathbb{R}$ and we define two such functions to be equal if they agree almost everywhere; i.e., if the set of points at which they differ has measure zero (Section 1.8, Definition 2). Following the usual convention we shall from now on refer to the equivalence classes obtained in this way as functions. With this convention we define

$$L^2(\mathbb{R}) = \{f \mid \int_{-\infty}^\infty |f(t)|^2 dt < \infty\}$$

This is a vector space under the usual algebraic operations and

$$\langle f, g \rangle = \int_{-\infty}^\infty f(t)\overline{g(t)} dt$$

defines an inner product on this space (Section 1.4, Definition 1). One can show that

$$(L^2(\mathbb{R}), \langle , \rangle)$$

is a Hilbert space and we note that, for $f \in L^2(\mathbb{R})$,

$$\|f\|_2^2 = \int_{-\infty}^\infty |f(t)|^2 dt$$

Now suppose that $f \in L^2(\mathbb{R})$ is arbitrary but fixed. For each positive integer n let

$$f_n(t) = \begin{cases} f(t) & \text{for } |t| \leq n \\ 0 & \text{for } n < |t| \end{cases}$$

Then $\{f_n(t)\}_{n=1}^\infty$ is a sequence in $L^2(\mathbb{R})$. Also, each of the functions $|f_n(t)|^2$ is integrable, $\lim_n |f_n(t)|^2 = |f(t)|^2$ pointwise a.e. and $|f_n(t)|^2 \le |f(t)|^2$ for all n and almost every t. Thus by the Lebesque dominated convergence theorem (Section 1.8, Theorem 6) we have

$$\int_{-\infty}^\infty |f(t)|^2 dt = \lim_n \int_{-\infty}^\infty |f_n(t)|^2 dt$$

We will need this fact in just a moment.

For each n and any $f \in L^2[-n, n]$ we can extend f to a function $\tilde f$ in $L^2(\mathbb{R})$ by letting $\tilde f(t) = f(t)$ for $|t| \le n$, $\tilde f(t) = 0$ for $n < |t|$. In this way we may regard $L^2[-n, n]$ as a linear subspace of $L^2(\mathbb{R})$; it is easy to see that $L^2[-n, n]$ is closed in $L^2(\mathbb{R})$ (see problem 1 below).

Lemma 1

The linear manifold $\cup_{n=1}^\infty L^2[-n, n]$ is dense in $L^2(\mathbb{R})$.

Proof. Given $\varepsilon > 0$ and any $f \in L^2(\mathbb{R})$ we may choose N so that

$$\int_{-\infty}^\infty |f(t)|^2 dt - \int_{-N}^N |f(t)|^2 dt < \varepsilon^2$$

hence

$$\int_{-\infty}^{-N} |f(t)|^2 dt + \int_N^\infty |f(t)|^2 dt < \varepsilon^2$$

Thus we have

$$\|f - f_N\|_2^2 = \int_{-N}^N |f(t) - f_N(t)|^2 dt + \int_{-\infty}^{-N} |f(t)|^2 dt + \int_N^\infty |f(t)|^2 dt < \varepsilon^2$$

Corollary 1

The Hilbert space $L^2(\mathbb{R})$ is separable.

Proof. We must show that $L^2(\mathbb{R})$ contains a countable dense subset (Section 1.9, Definition 1). Recall that for each n we have a countable dense subset $S_n \subseteq L^2[-n, n]$ (Section 1.9, after Lemma 2). Clearly $S = \cup_{n=1}^\infty S_n$ is a countable set which is dense in $L^2(\mathbb{R})$.

It follows from our corollary that $L^2(\mathbb{R})$ must have a countable orthonormal basis (Section 1.9, Theorem 1). In the next few results we shall identify one such basis.

Lemma 2

For each $n = 0, 1, 2, \ldots$ the function $x^n \exp(-\frac{1}{2}x^2)$ is in $L^2(\mathbb{R})$ and the sequence $\{x^n \exp(-\frac{1}{2}x^2) | n = 0, 1, 2, \ldots\}$ is a linearly independent subset of this Hilbert space.

Proof. For any fixed n it is easy to see that

$$\lim_{|x| \to \infty} x^n \exp(-\frac{1}{2}x^2) = 0$$

(see problem 2 below). Thus this function is bounded by some constant which we shall call λ_n. Hence

$$\int_{-\infty}^{\infty} |x^n \exp(-\frac{1}{2}x^2)|^2 dx = \int_{-\infty}^{\infty} x^{2n} \exp(-x^2) dx \leq \lambda_n \int_{-\infty}^{\infty} \exp(-x^2) dx = \lambda_n \sqrt{2\pi}$$

(this last integral is evaluated in problem 2 below). This proves that each of our functions is in $L^2(\mathbb{R})$.

In order to prove that our sequence is linearly independent it suffices to show that for any fixed integer m

$$\sum_{n=0}^{m} \alpha_n x^n \exp(-\frac{1}{2}x^2) = 0$$

in $L^2(\mathbb{R})$ implies $\alpha_0 = \alpha_1 = \ldots = \alpha_m = 0$; Section 1.1, Definition 3. This says, since the zero here is the zero function in $L^2(\mathbb{R})$, that

$$\int_{-\infty}^{\infty} |\sum_{n=0}^{m} \alpha_n x^n \exp(-\frac{1}{2}x^2)|^2 dx = 0$$

and this means (see Section 1.8 just before Definition 2) that

$$\sum_{n=0}^{m} \alpha_n x^n \exp(-\frac{1}{2}x^2) = 0 \text{ a.e. on } \mathbb{R}.$$

Now the exponential function is never zero so we must have

$$\sum_{n=0}^{m} \alpha_n x^n = 0 \text{ a.e. on } \mathbb{R},$$

and clearly that means $\alpha_0 = \alpha_1 = \ldots = \alpha_m = 0$ since our sum is a polynomial.

Lemma 3

The linear span of the set $\{x^n \exp(-\frac{1}{2}x^2) | n = 0, 1, 2, \ldots\}$ is dense in $L^2(\mathbb{R})$; i.e., the closure of $\lim\{x^n \exp(-\frac{1}{2}x^2) | n = 0, 1, 2, \ldots\}$ is $L^2(\mathbb{R})$.

Proof. The linear span of a set is defined in Section 1.1, Definition 2, and the closure of a set is defined in Section 3.1, Definition 2. See also Exercises 3.1 problem 4(e). The proof of this lemma requires much more measure theory than we have covered and so we shall omit it. The interested reader can find a proof in [7; Theorem 5, p. 57].

Theorem 1

For each integer $n = 0, 1, 2, \ldots$ let

$$h_n(x) = \exp(\frac{1}{2}x^2) \frac{d^n}{dx^n} \exp(-x^2)$$

Then the set $\{h_n(x)/\|h_n(x)\|_2 | n = 0, 1, 2, \ldots\}$ is an orthonormal basis for $L^2(\mathbb{R})$.

Proof. We leave it as an exercise to show that

$$\frac{d}{dt}[t^m \exp(-t^2)] = [-2t^{m+1} + m t^{m-1}] \exp(-t^2)$$

for $m = 1, 2, \ldots$; see problem 3(a) below. It follows from this that every h_n is in the linear span of $\{t^n \exp(-\frac{1}{2}t^2) | n = 0, 1, 2, \ldots\}$, and one can also see from this that

$$t^m \exp(-\frac{1}{2}t^2) \in \lim\{h_n | n = 0, 1, 2, \ldots\}$$

for each fixed m. Thus $\lim\{h_n(x)/\|h_n\|_2 | n = 0, 1, 2, \ldots\}$ is dense in $L^2(\mathbb{R})$ by Lemma 3.

Let us show that $\{h_n\}$ is an orthogonal set. Using the differentiation formula stated above this will follow once we have shown that

$$\int_{-\infty}^{\infty} t^m \frac{d^n}{dt^n}[\exp(-t^2)] dt = 0$$

for $0 \le m < n$. We can prove this by repeated integration by parts:

$$\begin{aligned}
\int_{-\infty}^{\infty} t^m \frac{d^n}{dt^n}[\exp(-t^2)] dt &= t^m \frac{d^{n-1}}{dt^{n-1}} \exp(-t^2) \Big|_{-\infty}^{\infty} - m \int_{-\infty}^{\infty} t^{m-1} \frac{d^{n-1}}{dt^{n-1}}[\exp(-t^2)] dt \\
&= -m \int_{-\infty}^{\infty} t^{m-1} \frac{d^{n-1}}{dt^{n-1}}[\exp(-t^2)] dt \\
&= \ldots \\
&= (-1)^m m! \int_{-\infty}^{\infty} \frac{d^{n-m}}{dt^{n-m}}[\exp(-t^2)] dt \\
&= (-1)^m m! \frac{d^{n-m-1}}{dt^{n-m-1}}[\exp(-t^2)] \Big|_{-\infty}^{\infty} = 0.
\end{aligned}$$

Corollary 2

The Hilbert space $L^2(\mathbf{R})$ is separable (as we have already seen above) and it is isometrically isomorphic with ℓ^2.

Proof. See Section 1.9, Theorem 1, and its corollary.

We should mention that $h_n(x)$, defined above to be

$$\exp(\frac{1}{2}x^2)\frac{d^n}{dx^n}[\exp(-x^2)]$$

is called the nth Hermite function. Using the differentiation formula given in the proof of Theorem 1 it is easy to see that

$$H_n(x) = h_n(x)\exp(\frac{1}{2}x^2)$$

is a polynomial of degree n. We call this the Hermite polynomial of degree n.

Exercises 1

1. Show that for each fixed integer n the subset $L^2[-n,n]$ is closed in $L^2(\mathbf{R})$; see Exercises 2.4, problem 2(a).

2. a. Use L'Hospital's rule to show that, for fixed n,
$$\lim x^n \exp(-\frac{1}{2}x^2) = 0$$
as x tends to infinity or to minus infinity.

   b. Evaluate
$$\int_{-\infty}^{\infty} \exp(-x^2)dx$$
as follows: First note that
$$[\int_{-\infty}^{\infty}\exp(-x^2)dx]^2 = \int_{-\infty}^{\infty}\int_{-\infty}^{\infty}\exp[-(x^2+y^2)]dx\,dy$$
and then evaluate the double integral by using polar coordinates.

3. a. Show that $\frac{d}{dt}[t^m \exp(-t^2)] = [-2t^{m+1} + mt^{m-1}]\exp(-t^2)$ for $m = 1, 2, \ldots$.

   b. Use the result in (a) to show that $H_n(x)$ is a polynomial of degree n.

c. Show that $2^{-\frac{1}{2}n}H_n(-y/\sqrt{2})$ is a polynomial of degree n in y which has integer coefficients and whose highest power term is y^n. Some authors call this the Hermite polynomial of degree n.

4. Let f be a continuous function defined on $\mathbb{R}$. Let supp(f), read "support of f" be the closure of $\{x \in \mathbb{R} | f(x) \neq 0\}$. We shall say that f has compact support if supp(f) is a compact set and we shall denote by $C_0(\mathbb{R})$ the family of all continuous functions on $\mathbb{R}$ having compact support.

   a. Show that $C_0(\mathbb{R})$ is dense in $(L^2(\mathbb{R}), \|\cdot\|_2)$. Hint: Use Lemma 1 and Theorem 4 of Section 1.8.

   b. Show that $L^2(\mathbb{R})$ is the completion of the inner product space $(C_0(\mathbb{R}), \langle,\rangle)$ where
   $$\langle f, g \rangle = \int_{-\infty}^{\infty} f(t)\overline{g(t)}dt$$
   for all f, g in $C_0(\mathbb{R})$. See Section 1.9.

5. Let $a, b \in \mathbb{R}$, $a < b$. Show that $L^2[a,b] \leq L^1[a,b]$ as follows: If $f \in L^2[a,b]$, then since $1 \in L^2[a,b]$, $\int_a^b |f(t)|dt \leq (\int_a^b |f(t)|dt)^{1/2}(\int_a^b 1^2 dt)^{1/2}$.

8.2 The Fourier Transform on $L^2(\mathbb{R})$

We shall show here how one can define the multiplication operator, $Q(f) = xf(x)$, and in the next section, the operator $Tf = if'$ on linear manifolds in $L^2(\mathbb{R})$. Using the Fourier transform, which we shall also discuss here, we shall see that these two operators are "equivalent."

Let us begin with the operator Q. First consider
$$D(Q) = \{f \in L^2(\mathbb{R}) | xf(x) \in L^2(\mathbb{R})\}.$$
It is easy to see that $D(Q)$ is a linear manifold and, since it contains $L^2[-n,n]$ for each $n = 1, 2, 3, \ldots$, it is dense in $L^2(\mathbb{R})$ (Section 1, Lemma 1). For each $f \in D(Q)$ set
$$Q(f) = xf(x)$$
It is clear that Q is a linear operator in $L^2(\mathbb{R})$.

(1) The operator Q is unbounded.
Proof. For each integer n let $f_n(x) = 1$ for $n \leq x < n+1$ and let $f_n(x) = 0$ for $x \notin [n, n+1)$. Observe that $f_n \in L^2(\mathbb{R})$ and that $\|f_n\|_2 = 1$ for each n. However,
$$\|Q(f_n)\|_2^2 = \int_{-\infty}^{\infty} |tf(t)|^2 dt = \int_n^{n+1} t^2 dt = \frac{1}{3}[(n+1)^3 - n^3] \geq n^2$$

(2) The operator Q is symmetric.
 Proof. $\langle Q(f), g \rangle = \int_{-\infty}^{\infty} tf(t)\overline{g(t)}dt = \int_{-\infty}^{\infty} f(t)\overline{tg(t)}dt = \langle f, Q(g) \rangle$ for every $f, g \in D(Q)$.

(3) The operator Q is self-adjoint.
 Proof. From (2) we see that $g \in D(Q)$ implies $g \in D(Q^*)$ and $Q^*(g) = Q(g)$. Thus $Q \subseteq Q^*$ (Section 7.1, Definition 3). To prove that these operators are equal we need only show that $D(Q^*) \subseteq D(Q)$. So let $g \in D(Q^*)$ and let f be an arbitrary element of $D(Q)$. Then

$$\int_{-\infty}^{\infty} tf(t)\overline{g(t)}dt = \langle Q(f), g \rangle = \langle f, Q^*(g) \rangle = \int_{-\infty}^{\infty} f(t)\overline{Q^*(g)(t)}dt$$

and so

$$\int_{-\infty}^{\infty} f(t)\overline{[tg(t) - Q^*(g)(t)]}dt = 0$$

Let $[a, b]$ be any finite subinterval of $\mathbb{R}$ and define

$$f_1(x) = xg(x) - Q^*(g)(x), \quad \text{for } a \leq x \leq b$$

$$f_1(x) = 0 \quad \text{for } x \notin [a, b]$$

It is clear that $f_1(x) \in D(Q)$ and so our last integral gives, upon replacing f by f_1,

$$\int_a^b |tg(t) - Q^*(g)(t)|^2 dt = 0$$

Thus (Section 1.8 just before Definition 2).

$$Q^*(g)(t) = tg(t) \quad \text{a.e. in } [a, b].$$

Since this holds in any interval having finite length and since $\mathbb{R}$ is the union of countably many such intervals

$$Q^*(g)(t) = tg(t) \quad \text{a.e. in } \mathbb{R}$$

(Section 1.8, Theorem 2). It follows that $tg(t) \in L^2(\mathbb{R})$ because $Q^*(g)(t)$ is in this space. Hence $g \in D(Q)$ as was to be shown.

(4) The operator Q has no eigenvalues, its spectrum coincides with $\mathbb{R}$ and each point in this set is a generalized eigenvalue of this operator.
 Proof. This can be proved the same way we did it for the multiplication operator on $L^2[a, b]$. See Section 2.4, Theorem 4, and the discussion just after Definition 1.

(5) Let us briefly mention that the spectral family of the operator Q can be computed as follows: For each $\lambda \in \mathbb{R}$ regard $L^2(-\infty, \lambda) = \{f | \int_{-\infty}^{\lambda} |f(t)|^2 dt <$

$\infty\}$ as a linear subspace of $L^2(\mathbb{R})$. Then let $P(\lambda)$ be the orthogonal projection of $L^2(\mathbb{R})$ onto this subspace. It is easy to see that the family $\{P(\lambda)|\lambda \in \mathbb{R}\}$ has properties (a)–(d) of Theorem 6 (Section 7.7).

(6) For any $f \in D(Q)$ we have
$$\int_{-\infty}^{\infty} |f(t)|dt < \infty$$

Proof. We know, of course, that $f \in L^2(\mathbb{R})$ and hence that
$$\int_{-\infty}^{\infty} |f(t)|^2 dt < \infty$$
but that does *not* imply that $|f(t)|$ is integrable because $\mathbb{R}$ does not have finite measure (Exercises 7.3, problem 1(b); see also Excercises 8.1 problem 5). Now observe that for any $g, h \in L^2(\mathbb{R})$ we do have
$$\int_{-\infty}^{\infty} |g(t)h(t)|dt < \infty$$
because $2|g(t)h(t)| \leq |g(t)|^2 + |h(t)|^2$ a.e. in $\mathbb{R}$. Our function $f \in D(Q)$ so f and $xf(x)$ are in $L^2(\mathbb{R})$. Thus
$$\int_{-\infty}^{-1} |xf(x)|^2 dt \leq \int_{-\infty}^{\infty} |xf(x)|^2 dx < \infty$$
and since $\frac{1}{x}$ is also square integrable over $(-\infty, -1)$ we must have
$$\int_{-\infty}^{-1} |f(x)|dx = \int_{-\infty}^{-1} |\frac{1}{x}xf(x)|dx < \infty;$$
here we have $g(x) = \frac{1}{x}$ and $h(x) = xf(x)$. Similarly,
$$\int_{1}^{\infty} |f(x)|dx < \infty$$

Finally, $\int_{-\infty}^{\infty} |f(x)|dx = \int_{-\infty}^{-1} |f(x)|dx + \int_{-1}^{1} |f(x)|dx + \int_{1}^{\infty} |f(x)|dx$. The first and third terms on the right are, as we have just shown, finite and the middle term is finite because $f \in L^2(\mathbb{R})$ implies f restricted to $[-1, 1]$ is in $L^2[-1, 1]$ (Exercises 8.1, problem 5).

Definition 1

Let $L^1(\mathbb{R}) = \{f| \int_{-\infty}^{\infty} |f(t)|dt < \infty\}$. For each f in this set we take $\|f\|_1$ to be the integral over $\mathbb{R}$ of $|f(t)|$.

It is easy to see that $L^1(\mathbb{R})$ is a vector space, that $\|\cdot\|_1$ is a norm on this vector space and one can show that $(L^1(\mathbb{R}), \|\cdot\|_1)$ is a Banach space (Section 1.3, Definition 2). We leave these results to the exercises. This particular Banach space plays a fundamental role in harmonic analysis. Since we want to discuss the Fourier transform on $L^2(\mathbb{R})$ anyway, it is worth our time to briefly look at this tranform on $L^1(\mathbb{R})$. Let us start by showing that the latter space has a natural "multiplication."

Lemma 1

If $f, g \in L^1(\mathbb{R})$ then the integral

$$\int_{-\infty}^{\infty} f(x-t)g(t)\,dt$$

exists for almost all x and is an integrable function of x.

Proof. For each t we have

$$\int_{-\infty}^{\infty} |f(x-t)|\,dx = \int_{-\infty}^{\infty} |f(x)|\,dx < \infty$$

because $f \in L^1$. Thus

$$\int_{-\infty}^{\infty} dt \int_{-\infty}^{\infty} |f(x-t)g(t)|\,dx = \int_{-\infty}^{\infty} |g(t)|\,dt \int_{-\infty}^{\infty} |f(x)|\,dx < \infty$$

Thus the double integral

$$\int_{-\infty}^{\infty}\int_{-\infty}^{\infty} f(x-t)g(t)\,dx\,dt$$

is absolutely convergent and our result follows from Fubini's theorem (Section 1.8, Theorem 7).

Definition 2

For each $f, g \in L^1(\mathbb{R})$ we define

$$(f * g)(x) = \int_{-\infty}^{\infty} f(x-t)g(t)\,dt$$

and we call $f * g$ the convolution of f and g.

By Lemma 1 the convolution of any two L^1-functions exists and is also an L^1-function. The next lemma can be proved by direct calculation:

Lemma 2

for any f, g, h in $L^1(\mathbf{R})$ we have:

(a) $f * g = g * f;$ (b) $(f * g) * h = f * (g * h)$

Theorem 1

If f, g are in $L^1(\mathbf{R})$ then $f * g \in L^1(\mathbf{R})$ and $\|f * g\|_1 \leq \|f\|_1 \|g\|_1$.

Proof. We have already noted that $f * g \in L^1(\mathbf{R})$. Let

$$h(x) = \int_{-\infty}^{\infty} f(x-t)g(t)dt.$$

Then

$$\begin{aligned}
\|h\|_1 &= \int_{-\infty}^{\infty} |h(t)|dx \leq \int_{-\infty}^{\infty} dx \int_{-\infty}^{\infty} |f(x-t)g(t)|dt \\
&= \int_{-\infty}^{\infty} |g(t)|dt \int_{-\infty}^{\infty} |f(x-t)|dx = \|g\|_1 \|f\|_1
\end{aligned}$$

where the change in the order of integration is justified by the Fubini theorem (Section 1.8, Theorem 7).

Readers who are familiar with the terminology of algebra will recognize that $L^1(\mathbf{R})$ together with convolution is an example of what is called a commutative algebra. Since the multiplication (i.e., convolution) is "tied to" the norm via the inequality $\|f * g\|_1 \leq \|f\|_1 \|g\|_1$, the algebra $(L^1(\mathbf{R}), \|\cdot\|_1, *)$ is called a commutative Banach algebra. It is of great interest to find the maximal ideals in such an algebra. We shall not pursue this matter but the Fourier transform, which we shall now discuss, can be used to do this; see the exercises.

Definition 3

For each $f \in L^1(\mathbf{R})$ we define the Fourier transform of f, $\hat{f}(x)$, as follows:

$$\hat{f}(x) = \frac{1}{\sqrt{2\pi}} \int_{-\infty}^{\infty} e^{-ixt} f(t)dt.$$

It is very helpful in this connection to have in mind two real lines, $\mathbf{R}$ and $\hat{\mathbf{R}}$. One should think of $f(t)$ as a function in $L^1(\mathbf{R})$ and of $\hat{f}(x)$ as a function defined on $\hat{\mathbf{R}}$.

Theorem 2

For each $f \in L^1(\mathbf{R})$ the function $\hat{f}$ is both continuous and bounded on $\hat{\mathbf{R}}$.

Proof. For any $x \in \hat{\mathbf{R}}$ we have

$$|\hat{f}(x)| \leq \frac{1}{\sqrt{2\pi}} \int_{-\infty}^{\infty} |e^{-ixt} f(t)| dt = \frac{1}{\sqrt{2\pi}} \int_{-\infty}^{\infty} |f(t)| dt = \frac{1}{\sqrt{2\pi}} \|f\|_1$$

Thus $\hat{f}$ is defined for every $x \in \hat{\mathbf{R}}$ and it is bounded on $\hat{\mathbf{R}}$. Furthermore, for any x, h in $\hat{\mathbf{R}}$ we can write:

$$\hat{f}(x+h) - \hat{f}(x) = \frac{1}{\sqrt{2\pi}} \int_{-\infty}^{\infty} e^{-ixt}(e^{-iht} - 1) f(t) dt$$

$$|\hat{f}(x+h) - \hat{f}(x)| \leq \frac{1}{\sqrt{2\pi}} \int_{-\infty}^{\infty} |e^{-iht} - 1| |f(t)| dt$$

Now this last integrand is bounded, for all h, by $2|f(t)|$ and it (i.e., the integrand) tends to zero as h tends to zero; for each fixed t. Thus by the Lebesque dominated convergence theorem (Section 1.8, Theorem 6), generalized slightly since we do not have a sequence here, our integral must tend to zero as h does. This shows that $\hat{f}$ is continuous at each point of $\hat{\mathbf{R}}$.

One can also show [6; Theorem 4a, p. 7] that whenever $f \in L^1(\mathbf{R})$, $\lim_{|x| \to \infty} \hat{f}(x) = 0$; we say that $\hat{f}$ vanishes at infinity on $\hat{\mathbf{R}}$. Let $S(\hat{\mathbf{R}})$ be the set of all continuous, complex-valued functions defined on $\hat{\mathbf{R}}$ which vanish at infinity on this set. It is clear that $S(\hat{\mathbf{R}})$ is a vector space under the usual definitions of addition and scalar multiplication for functions, and that $\|g\|_\infty = \sup\{|g(x)| | x \in \hat{\mathbf{R}}\}$ defines a norm on this space. Furthermore, the product of any two functions in $S(\hat{\mathbf{R}})$ is also in $S(\hat{\mathbf{R}})$; i.e., if f, g are in this space then so is $h(x) = f(x)g(x)$ for all $x \in \hat{\mathbf{R}}$.

Observe that we can define a map $\mathcal{F}$ from $L^1(\mathbf{R})$ into $S(\hat{\mathbf{R}})$ by setting

$$\mathcal{F}(f) = \hat{f}$$

for every $f \in L^1(\mathbf{R})$. This map is linear and, since

$$\|\mathcal{F}(f)\|_\infty = \|\hat{f}\|_\infty \leq \frac{1}{\sqrt{2\pi}} \|f\|_1$$

(see the second line in the proof of Theorem 2), it is continuous. It also "preserves" multiplication:

Theorem 3

For any $f, g \in L^1(\mathbf{R})$ we have

$$\frac{1}{\sqrt{2\pi}} \mathcal{F}(f * g) = \mathcal{F}(f) \mathcal{F}(g).$$

Proof. Let $h(t) = (f * g)(t)$ and note that for any $x \in \hat{\mathbb{R}}$,

$$\begin{aligned}
\hat{h}(x) &= \frac{1}{\sqrt{2\pi}} \int_{-\infty}^{\infty} e^{-ixt} h(t) dt = \frac{1}{\sqrt{2\pi}} \int_{-\infty}^{\infty} e^{-ixt} dt \int_{-\infty}^{\infty} f(t-u) g(u) du \\
&= \frac{1}{\sqrt{2\pi}} \int_{-\infty}^{\infty} g(u) du \int_{-\infty}^{\infty} e^{-ixt} f(t-u) dt \\
&= \frac{1}{\sqrt{2\pi}} \int_{-\infty}^{\infty} g(u) du \int_{-\infty}^{\infty} e^{-ix(t+u)} f(t) dt \\
&= \frac{1}{\sqrt{2\pi}} \int_{-\infty}^{\infty} e^{-ixu} g(u) du \int_{-\infty}^{\infty} e^{-ixt} f(t) dt \\
&= \sqrt{2\pi} \hat{g}(x) \hat{f}(x)
\end{aligned}$$

The interchange in the order of integration is justified because the last integral is absolutely convergent.

Now what is our interest in all of this? We have already discussed the operator Q, where $Q(f) = xf(x)$, in $L^2(\mathbb{R})$ and we shall soon see that $Tf = if'$ defines a self-adjoint, unbounded operator in $L^2(\mathbb{R})$. These operators appear to be totally different. However, the next theorem indicates that, perhaps, they are not so different after all.

Theorem 4

Suppose that both $f(t)$ and $tf(t)$ are in $L^1(\mathbb{R})$. Then $\hat{f}(x)$ is differentiable and

$$i \frac{d}{dx} \hat{f}(x) = \widehat{tf(t)}(x)$$

or

$$i \frac{d}{dx} \mathcal{F}[f(t)](x) = \mathcal{F}[tf(t)](x)$$

Proof.

$$\begin{aligned}
\frac{\hat{f}(x+h) - \hat{f}(x)}{h} &= \frac{1}{\sqrt{2\pi}} \int_{-\infty}^{\infty} f(t) \left[\frac{e^{-i(x+h)t} - e^{-ixt}}{h} \right] dt \\
&= \frac{1}{\sqrt{2\pi}} \int_{-\infty}^{\infty} f(t) e^{-ixt} \left[\frac{e^{-iht} - 1}{h} \right] dt
\end{aligned}$$

Now as h tends to zero,

$$\frac{e^{-iht} - 1}{h}$$

tends to the derivative of e^{-iht}; i.e., it tends to $-ite^{-iht}$. Hence our last integrand is dominated by $|tf(t)|$ which is in $L^1(\mathbb{R})$ by hypothesis. But as h

tends to zero this integrand converges pointwise to $itf(t)e^{-ixt}$. Thus by the Lebesque dominated convergence theorem (Section 1.8, Theorem 6), again slightly generalized since we do not have a sequence here, the difference quotient converges to the Fourier transform of $-itf(t)$.

This theorem seems to be saying that if we could extend the Fourier transform in such a way that it becomes an operator on $L^2(\mathbb{R})$, call the extension $\mathcal{F}$, then we would have

$$T\mathcal{F}(f) = \mathcal{F}Q(f)$$

for every suitable f. This can, in fact, be done and the operator $\mathcal{F}$ that we get this way turns out to be a unitary operator on $L^2(\mathbb{R})$. Carrying out the details of this extension is a lengthy and delicate process that requires much more integration theory than we have done, and so some of the proofs will only be sketched. The reader who is interested in seeing the details can consult the very fine Cambridge tract by R.R. Goldberg [6].

Given $f \in L^2(\mathbb{R})$ we recall the functions f_n defined by

$$f_n(t) = \begin{cases} f(t) & \text{for } t \in [-n,n] \\ 0 & \text{for } t \notin [-n,n] \end{cases}$$

$n = 1, 2, \ldots$. It is easy to see that $f_n \in L^1(\mathbb{R}) \cap L^2(\mathbb{R})$ and it can be shown that $\hat{f}_n \in L^2(\mathbb{R})$, $\|\hat{f}_n\|_2 = \|f_n\|_2$; see [6; Theorem 12b, p. 44] and note that our definition of $\hat{f}$ differs slightly from the one given there.

Lemma 3

For each $f \in L^2(\mathbb{R})$ the sequence $\{\hat{f}\} \subseteq L^2(\mathbb{R})$ converges for the norm of this space to function $\hat{f} \in L^2(\mathbb{R})$. Also, $\|f\|_2 = \|\hat{f}\|_2$.

Proof. We have $f_m - f_n$ in $L^2[-m,m] \subseteq L^1[-m,m]$ by when $m \geq n$. Hence $f_m - f_n$ is in $L^1(\mathbb{R}) \cap L^2(\mathbb{R})$ and

$$\|\hat{f}_m - \hat{f}_n\|_2^2 = \|f_m - f_n\|_2^2 = \int_{-\infty}^{\infty} |(f_m - f_n)(t)|^2 dt$$

$$= \int_{-m}^{-n} |f(t)|^2 dt + \int_{n}^{m} |f(t)|^2 dt$$

and this tends to zero as $m, n \to \infty$. Since $(L^2(\mathbb{R}), \|\cdot\|_2)$ is a Hilbert space the sequence $\{\hat{f}_n\}$ has a limit, which we shall call $\hat{f}$, in this space. By Exercises 1.3, problem 1, $\|\hat{f}\|_2 = \lim \|\hat{f}_n\|_2 = \lim \|f_n\|_2 = \|f\|_2$.

Definition 4

For each $f \in L^2(\mathbb{R})$ the function $\hat{f}$, which is the limit of the sequence $\{\hat{f}_n\} \subseteq L^2(\mathbb{R})$, will be called the Fourier transform of f. We may define a linear operator $\mathcal{F}$ on $L^2(\mathbb{R})$ by setting $\mathcal{F}(f) = \hat{f}$ for every f in this space. $\mathcal{F}$ will be called the Fourier-Plancherel operator on $L^2(\mathbb{R})$.

Notice that by Lemma 3 the Fourier-Plancherel operator preserves norm. Hence this operator is continuous and one-to-one. It is actually a unitary operator as we shall now show:

Lemma 4

For any $f, g \in L^2(\mathbb{R})$ we have

$$\langle f, g \rangle = \langle \hat{f}, \hat{g} \rangle = \langle \mathcal{F}(f), \mathcal{F}(g) \rangle$$

Proof. By Lemma 3 we have $\|\hat{f} + \hat{g}\|_2^2 = \|f + g\|_2^2$ and so

$$\int_{-\infty}^{\infty} (\hat{f} + \hat{g})\overline{(\hat{f} + \hat{g})}dx = \int_{-\infty}^{\infty} (f + g)\overline{(f + g)}dx$$

Expanding both sides we get

$$\int |\hat{f}|^2 dx + \int |\hat{g}|^2 dx + \int \hat{f}\overline{\hat{g}}dx + \int \overline{\hat{f}}\hat{g}dx = \int |f|^2 dx + \int |g|^2 dx + \int f\overline{g}dx + \int \overline{f}g dx$$

But again using Lemma 3 we have some cancellation:

$$(*) \qquad \int \hat{f}\overline{\hat{g}}dx + \int \overline{\hat{f}}\hat{g}dx = \int f\overline{g}dx + \int \overline{f}g dx$$

Since g is an arbitrary element of $L^2(\mathbb{R})$ we may replace g by ig, and hence $\hat{g}$ by $i\hat{g}$, in this last equation to get

$$\int \hat{f}\overline{(i\hat{g})}dx + \int \overline{\hat{f}}(i\hat{g})dx = \int f\overline{(ig)}dx + \int \overline{f}(ig)dx$$
$$- i\int \hat{f}\overline{\hat{g}}dx + i\int \overline{\hat{f}}\hat{g}dx = -i\int f\overline{g}dx + i\int \overline{f}g\,dx$$

We may divide this last equation through by i and add the result to $(*)$ and thereby obtain the desired relation.

Since $\mathcal{F}$ is a unitary operator it has an inverse $\mathcal{F}^{-1}$. For these two operators we have:

Theorem 5

For every $f \in L^2(\mathbb{R})$ we have:

$$\mathcal{F}[f(x)] = \frac{1}{\sqrt{2\pi}} \frac{d}{dx} \int_{-\infty}^{\infty} \frac{e^{-ixt} - 1}{-it} f(t) dt \text{ a.e. in } x,$$

$$\mathcal{F}^{-1}[f(x)] = \frac{1}{\sqrt{2\pi}} \frac{d}{dx} \int_{-\infty}^{\infty} \frac{e^{+ixt} - 1}{+it} f(t) dt \text{ a.e. in } x.$$

Proof. We recall that the linear span of the sequence of Hermite functions $\{h_n(x) | n = 0, 1, 2, \ldots\}$, call it M, is a dense linear manifold in $L^2(\mathbb{R})$ (Section 1, Lemma 3). Thus for any fixed $f \in L^2(\mathbb{R})$ we can find a sequence $\{f_n\} \subseteq M$ which converges to f. Note that we must also have $\{\mathcal{F}(f_n)\}$ converging to $\mathcal{F}(f)$. For each fixed $\rho \geq 0$ we define

$$c_\rho(x) = \begin{cases} 1 & \text{for } 0 \leq x \leq \rho \\ 0 & \text{for } x \notin [0, \rho] \end{cases}$$

Then

$$\int_0^\rho \mathcal{F}[f(x)] dx = \langle \mathcal{F}(f), c_\rho \rangle = \lim_{n \to \infty} \langle \mathcal{F}(f_n), c_\rho \rangle$$

$$= \lim_{n \to \infty} \int_0^\rho \frac{1}{\sqrt{2\pi}} \int_{-\infty}^{\infty} e^{-ixt} f_n(t) dt \, dx$$

By Fubini's Theorem (Section 1.8, Theorem 7) we may interchange the order of integration to get

$$\int_0^\rho \mathcal{F}[f(x)] dx = \lim_{n \to \infty} \frac{1}{\sqrt{2\pi}} \int_{-\infty}^{\infty} \left[\int_0^\rho e^{-ixt} dx \right] f_n(t) dt$$

$$= \lim_{n \to \infty} \frac{1}{\sqrt{2\pi}} \int_{-\infty}^{\infty} \frac{e^{-i\rho t} - 1}{-it} f_n(t) dt$$

Let us leave this for a moment and consider $h_\rho(t) = \frac{e^{-i\rho t} - 1}{-it}$. This function is continuous because

$$\lim_{t \to 0} h_\rho(t) = \rho$$

and since it tends to zero as $|t| \to \infty$ it is even in $L^2(\mathbb{R})$. Hence our last equation can be written

$$\int_0^\rho \mathcal{F}[f(x)] dx = \lim \frac{1}{\sqrt{2\pi}} \langle h_\rho, f_n \rangle = \frac{1}{\sqrt{2\pi}} \langle h_\rho, f \rangle$$

$$= \frac{1}{\sqrt{2\pi}} \int_{-\infty}^{\infty} \frac{e^{-i\rho t} - 1}{-it} f(t) dt$$

Now we know that $\mathcal{F}[f(x)] \in L^2(\mathbb{R})$, hence it is in $L^2[0,\rho]$ for every fixed $\rho \geq 0$. It follows that $\mathcal{F}(f)$ is integrable on this interval (Exercises 7.3, problem 1) and so, by Section 7.3 just after Definition 2,

$$\int_0^\rho \mathcal{F}[f(x)]dx$$

is absolutely continuous as a function of ρ. Thus

$$\mathcal{F}[f(\rho)] = \frac{d}{d\rho}\int_0^\rho \mathcal{F}[f(x)]dx = \frac{1}{\sqrt{2\pi}}\frac{d}{d\rho}\int_{-\infty}^\infty \frac{e^{-i\rho t}}{-it}f(t)dt$$

for almost all ρ (Section 7.3, Theorem 1). To complete the proof we should consider the case when ρ is negative. But this is very similar to what we just did. The formula for $\mathcal{F}^{-1}$ is proved in an analogous way.

Theorem 6

Suppose that $f \in L^2(\mathbb{R}) \cap L^1(\mathbb{R})$. Then

$$\mathcal{F}[f(x)] = \frac{1}{\sqrt{2\pi}}\int_{-\infty}^\infty e^{-ixt}f(t)dt \quad \text{a.e. in } x,$$

$$\mathcal{F}^{-1}[f(x)] = \frac{1}{\sqrt{2\pi}}\int_{-\infty}^\infty e^{+ixt}f(t)dt \quad \text{a.e. in } x$$

Proof. In the proof of the last theorem we saw that

$$\mathcal{F}[f(\rho)] = \frac{1}{\sqrt{2\pi}}\frac{d}{d\rho}\int_{-\infty}^\infty \frac{e^{-i\rho t}-1}{-it}f(t)dt \quad \text{a.e. in } \rho.$$

Now since we are assuming that $f \in L^1(\mathbb{R})$ it can be shown that the integral and the derivative can be interchanged giving us

$$\mathcal{F}[f(\rho)] = \frac{1}{\sqrt{2\pi}}\int_{-\infty}^\infty e^{-i\rho t}f(t)dt$$

To get the formula for $\mathcal{F}^{-1}$ we need only replace i by $-i$.

Corollary 1

For any $f \in D(Q)$ (i.e., any $f \in L^2(\mathbb{R})$ which is in the domain of the multiplication operator) we have

$$\mathcal{F}(f) = \frac{1}{\sqrt{2\pi}}\int_{-\infty}^\infty e^{-ixt}f(t)dt \quad \text{a.e. in } x.$$

Proof. This follows from our theorem and (6) above.

Exercises 2

1. Let $\alpha(t)$ be a complex-valued, continuous functions on $\mathbb{R}$ and define $D(A_\alpha) = \{f \in L^2(\mathbb{R}) | \alpha(t)f(t) \in L^2(\mathbb{R})\}$. For each f in this last set take $A_\alpha(f)$ to be $\alpha(t)f(t)$.

    a. Show that $D(A_\alpha)$ is dense in $L^2(\mathbb{R})$, that $D(A_\alpha^*) = D(A_\alpha)$ and that $A_\alpha^{**} = A_\alpha$.

    b. Show that A_α is self-adjoint iff $\alpha(t) \in \mathbb{R}$ for all t, and that A_α is unitary iff $|\alpha(t)| = 1$ for all t.

    c. Show that A_α is bounded on $L^2(\mathbb{R})$ if $\alpha(t)$ is bounded on $\mathbb{R}$. Hint: If $\alpha(t_n) \to \infty$, $t_n \to \infty$ then we can construct a step function $f(t) \in L^2(\mathbb{R})$ such that $\int_{-\infty}^{\infty} |\alpha(t)f(t)|^2 dt = \infty$.

2. Show that $L^1(\mathbb{R})$ is a vector space, that $\|\cdot\|_1$ is a norm on this space and that $(L^1(\mathbb{R}), \|\cdot\|_1)$ is a Banach space. Hint: The last assertion may be troublesome to those who have had little integration theory.

3. Prove Lemma 2.

4. Choose and fix $x_0 \in \hat{\mathbb{R}}$. Let $I = \{f \in L^1(\mathbb{R}) | \hat{f}(x_0) = 0\}$.

    a. Show that I is an ideal in $L^1(\mathbb{R})$; i.e., it is a linear manifold and for any $f \in L^1(\mathbb{R})$, $g \in I$ we have $f * g \in I$.

    b. Show that I is a closed subset of $L^1(\mathbb{R})$.

5. a. Show that $f(t) = (1 + |t|)^{-1}$ is in $L^2(\mathbb{R})$ but not in $L^1(\mathbb{R})$.

    b. Show that $\mathcal{F}$ does *not* map $L^1(\mathbb{R})$ into itself as follows: Take $g(t) = e^{-t}$ for $t \geq 0$, $= 0$ for $t < 0$ and show that $g \in L^1(\mathbb{R})$. Next show that $\hat{g}(x) = (1 - ix)^{-1}$ and that this is not in $L^1(\mathbb{R})$.

6. Recall that $L^2[-n, n]$ can be regarded as a linear subspace of $L^2(\mathbb{R})$ (see Section 1). Let P_n be the orthogonal projection of $L^2(\mathbb{R})$ onto this subspace; here $n = 1, 2, \ldots$. Using this sequence show that

$$\mathcal{F}(f) = \lim_{n \to \infty} \frac{1}{\sqrt{2\pi}} \int_{-n}^{n} e^{-ixt} f(t) dt, \quad \mathcal{F}^{-1}(f) = \lim_{n \to \infty} \frac{1}{\sqrt{2\pi}} \int_{-n}^{n} e^{ixt} f(t) dt$$

for every $f \in L^2(\mathbb{R})$.

7. Consider the unitary operator $\mathcal{F}$ on $L^2(\mathbb{R})$. Show that $\sigma(\mathcal{F})$ consists of the four points $\pm 1, \pm i$ and that each of these is an eigenvalue of the operator.

8.3 More About the Fourier Transform

Let us begin by discussing differentiation in $L^2(\mathbb{R})$. We consider the linear manifold $D(T)$ taken to be:

$\{f \in L^2(\mathbb{R}) | f$ is absolutely continuous on any finite sub-interval of $\mathbb{R}$, and $f' \in L^2(\mathbb{R})\}$

and we refer the reader to Section 7.3 for the definition of absolute continuity and the relevent theorem (Definition 2 and Theorem 1). Since $D(T)$ contains $x^n \exp(-\frac{1}{2}x^2)$ for $n = 0, 1, 2, \ldots$ we see that this linear manifold is dense in $L^2(\mathbb{R})$ (Section 8.1, Lemma 3). For each $f \in D(T)$ we set

$$T(f) = if'$$

where $f'(t) = \frac{d}{dt}f(t)$ and $i = \sqrt{-1}$. It is clear that T is a linear operator in $L^2(\mathbb{R})$.

(1) The operator T is unbounded.

Proof. Our operator contains the unbounded differential operators defined in Section 7.3.

(2) For any $f \in D(T)$ we have

$$\lim_{|t| \to \infty} f(t) = 0$$

Proof. Given f restrict it to the interval $[0, \infty)$. We have

$$\langle f, f' \rangle + \langle f', f \rangle = \lim_{x \to \infty} \int_0^x [f(t)\overline{f'(t)} + f'(t)\overline{f(t)}]dt$$
$$= f(t)\overline{f(t)} \Big|_0^x = |f(x)|^2 - |f(0)|^2.$$

Thus the limit of $|f(x)|^2$, as x tends to infinity, exists and is equal to, say, $\alpha^2 \geq 0$. Now given $\varepsilon > 0$ we can choose x_0 so that

$$\left||f(x)|^2 - \alpha^2\right| < \varepsilon \text{ for all } x \geq x_0.$$

Then

$$\int_0^\infty |f(t)|^2 dt \geq \int_{x_0}^\infty |f(t)|^2 dt \geq \int_{x_0}^\infty (\alpha^2 - \varepsilon)dt = \infty$$

unless $\alpha^2 = 0$. But since $f \in L^2[0, \infty)$, our integral must be finite showing that $\alpha = 0$.

(3) The operator T is self-adjoint.

Proof. For any $f, g \in D(T)$ we have, by (2),

$$\langle if', g\rangle - \langle f, ig'\rangle = i\int_{-\infty}^{\infty} f'(t)\overline{g(t)}dt + i\int_{-\infty}^{\infty} f(t)\overline{g'(t)}dt$$

$$= if(t)\overline{g(t)}\,|_{-\infty}^{\infty} = i\left[\lim_{t\to\infty} f(t)\overline{g(t)} - \lim_{t\to-\infty} f(t)\overline{g(t)}\right] = 0.$$

Thus T is a symmetric operator. In order to show that it is self-adjoint we need only show that $D(T^*) \subseteq D(T)$; see Section 7.4, Definition 1, and Section 7.1, Definition 3. But this is entirely similar to what we did in Section 7.4 just after Definition 1.

(4) The operator $T(= if')$ and $Q(= tf(t))$ are related as follows: (a) $D(TQ - QT)$ is dense in $L^2(\mathbb{R})$; (b) On this dense set $TQ - QT = iI$.

Proof. To prove (a) it is enough to note that each of the functions $x^n \exp(-\frac{1}{2}x^2)$, $n = 0, 1, 2, \ldots$ is in $D(TQ - QT)$ by Exercises 1, problem 3(a), and the definition of these operators; see also Section 8.1, Lemma 3. Next let f be an element of this dense set. We have:

$$(TQ - QT)f = T[tf(t)] - tif'(t) = i[tf'(t) + f(t)] - tif'(t)$$
$$= if(t) = (iI)f.$$

The next theorem gives precise meaning to the statement made earlier that T and Q are "equivalent." Recall that $\mathcal{F}$ denotes the Fourier-Plancherel operator on $L^2(\mathbb{R})$.

Theorem 1

The operators T, Q, $\mathcal{F}$ on $L^2(\mathbb{R})$ are related as follows: $T\mathcal{F} = \mathcal{F}Q$, equivalently $T = \mathcal{F}Q\mathcal{F}^{-1}$ or $Q = \mathcal{F}^{-1}T\mathcal{F}$.

Proof. Note first that $D(\mathcal{F}Q) = \{f \in D(Q) | Q(f) \in D(\mathcal{F})\}$. Since $D(\mathcal{F}) = L^2(\mathbb{R})$ we have $D(\mathcal{F}Q) = D(Q)$. Now $D(\mathcal{F}Q\mathcal{F}^{-1}) = \{f \in D(\mathcal{F}^{-1}) = L^2(\mathbb{R}) | \mathcal{F}^{-1}(f) \in D(\mathcal{F}Q) = D(Q)\}$. But since $\mathcal{F}$ is one-to-one $\mathcal{F}^{-1}(f) \in D(Q)$ if and only if $f \in \mathcal{F}[D(Q)]$. So $D(\mathcal{F}Q\mathcal{F}^{-1}) = \mathcal{F}[D(Q)]$. Thus we must show two things: (i) $D(T) = \mathcal{F}[D(Q)]$; (ii) $Tf = (\mathcal{F}Q\mathcal{F}^{-1})f$ for all $f \in D(T)$.

Let $f \in D(T)$ and recall that (Section 2, Theorem 5)

$$\mathcal{F}^{-1}[(Tf)](x) = \frac{1}{\sqrt{2\pi}} \frac{d}{dx} \int_{-\infty}^{\infty} \frac{e^{ixy} - 1}{iy} i f'(y) dy$$

Since $(e^{ixy}-1)/y$ is absolutely continuous on every finite interval we can apply integration by parts to this last integral (Section 7.3 just after Theorem 1) to get

$$= \frac{1}{\sqrt{2\pi}} \frac{d}{dx}\left[\frac{e^{ixy}-1}{y}f(y)\,|_{-\infty}^{\infty} - \int_{-\infty}^{\infty} \frac{ixye^{ixy} - (e^{ixy}-1)}{y^2} f(y) dy\right]$$

The first term vanishes by (2) and second term can be written as:

$$= \frac{1}{\sqrt{2\pi}} \frac{d}{dx} x \int_{-\infty}^{\infty} \frac{e^{ixy}-1}{iy} f(y) dy + \frac{1}{\sqrt{2\pi}} \frac{d}{dx} \int_{-\infty}^{\infty} \frac{e^{ixy}-1-ixy}{y^2} f(y) dy$$

$$= x \frac{d}{dx} \frac{1}{\sqrt{2\pi}} \int_{-\infty}^{\infty} \frac{e^{ixy}-1}{iy} f(y) dy + \frac{1}{\sqrt{2\pi}} \int_{-\infty}^{\infty} \frac{e^{ixy}-1}{iy} f(y) dy$$

$$+ \frac{1}{\sqrt{2\pi}} \int_{-\infty}^{\infty} \frac{iy e^{ixy} - iy}{y^2} f(y) dy$$

where, in the last summand, we have interchanged integration and differentiation (justifying this requires some technical knowledge of the Lebesque integral).

$$= x(\mathcal{F}^{-1} f)(x) \text{ a.e.}$$

by (Section 2, Theorem 5). Now $\mathcal{F}(Tf) \in L^2(\mathbb{R})$ and this is what we started our long equation with, so $x\mathcal{F}^{-1} f$ is also in $L^2(\mathbb{R})$ and hence $\mathcal{F}^{-1} f \in D(Q)$. Also

$$\begin{aligned} \mathcal{F}^{-1} T f &= Q \mathcal{F}^{-1} f \quad \text{for all } f \in D(T), \\ Tf &= \mathcal{F} Q \mathcal{F}^{-1} f \quad \text{for all } f \in D(T), \\ (*) \quad T &\subseteq \mathcal{F} Q \mathcal{F}^{-1} \quad \text{(Section 7.1, Definition 3).} \end{aligned}$$

In order to prove the reverse inclusion to that given in (*) we must first observe that $\mathcal{F} Q \mathcal{F}^{-1}$ is symmetric. To see this, let $g, h \in D(\mathcal{F} Q \mathcal{F}^{-1}) = \mathcal{F}[D(Q)]$. We have $\mathcal{F}^{-1} g \in D(Q)$ and $\mathcal{F}^{-1} h \in D(Q)$. Since $\mathcal{F}$ is unitary and Q is self-adjoint (Section 2, (3)) we find that

$$\begin{aligned} \langle \mathcal{F} Q \mathcal{F}^{-1} g, h \rangle &= \langle Q \mathcal{F}^{-1} g, \mathcal{F}^{-1} h \rangle = \langle \mathcal{F}^{-1} g, Q \mathcal{F}^{-1} h \rangle \\ &= \langle g, \mathcal{F} Q \mathcal{F}^{-1} h \rangle. \end{aligned}$$

It now follows from (*) that, since T is self-adjoint (by (3)),

$$T = T^* \supseteq (\mathcal{F} Q \mathcal{F}^{-1})^* \supseteq \mathcal{F} Q \mathcal{F}^{-1} \supseteq T$$

Thus $D(T) = D(\mathcal{F} Q \mathcal{F}^{-1})$ and $T = \mathcal{F} Q \mathcal{F}^{-1}$.

Definition 1

Let H_1, H_2 be two Hilbert spaces over the same field and let A_1, A_2 be two linear operators acting in each of these spaces. We shall say that A_1 and A_2 are eqivalent if there is an isometric isomorphism U from H_1 onto H_2 such that (Section 1.6, Definition 2) $A_1 = U^{-1} A_2 U$. More explicitly, $U[D(A_1)] = D(A_2)$ and $A_1 x = U^{-1}[A_2(Ux)]$ for every $x \in D(A_1)$.

Let us take a moment here to discuss the meaning of Theorem 1. Once again it is useful to consider two "copies" of the real line which we shall denote by $\mathbb{R}$ and $\hat{\mathbb{R}}$ (see Section 2 just after Definition 3), and two "different" Hilbert spaces $L^2(\mathbb{R})$ and $L^2(\hat{\mathbb{R}})$. We have

$$\mathcal{F} : L^2(\mathbb{R}) \to L^2(\hat{\mathbb{R}})$$

and

$$\mathcal{F}^{-1} : L^2(\hat{\mathbb{R}}) \to L^2(\mathbb{R}).$$

Let us now regard T ($= if'(x)$) as an operator in $L^2(\hat{\mathbb{R}})$ and Q ($= tf(t)$) as an operator in $L^2(\mathbb{R})$. Then, by Theorem 1, $f \in L^2(\hat{\mathbb{R}})$ is in the domain of T if, and only if, $\mathcal{F}^{-1}(f)$ is in the domain of Q and $\mathcal{F}^{-1}[T(f)] = Q[\mathcal{F}^{-1}(f)]$. Thus $Q\mathcal{F}^{-1} = \mathcal{F}^{-1}T$.

This is a special case of a very general version of the spectral which can be stated as follows: Let H be a Hilbert space and let A be a self-adjoint operator in H. Think of H and A as "abstract" objects. Then there is a set M and a measure μ on M (in our examples M will be $\mathbb{R}$ and μ will be a variant of Lebesque measure) for which:

i. There is an isometric, isomorphism $U : H \to L^2(M, \mu)$;

ii. There is a real-valued, μ-measurable function α on M such that

$$U[A(x)] = \alpha(x)[U(x)]$$

for each $x \in D(A)$. Thus

$$(\alpha I)U = UA$$

where I is the identity operator.

In Theorem 1 our "abstract" Hilbert space is $L^2(\hat{\mathbb{R}})$ and our self-adjoint operator in this space is T. Our "concrete" Hilbert space is $L^2(\mathbb{R})$, so $M = R$ and μ = Lebesque measure. Our isometric isomorphism is $\mathcal{F}^{-1} : L^2(\hat{\mathbb{R}}) \to L^2(\mathbb{R})$ and our real-valued, Lebesque measurable function α on $M = \mathbb{R}$ is $\alpha(t) = t$ for all t. The conclusion of Theorem 1 is

$$(tI)\mathcal{F}^{-1} = \mathcal{F}^{-1}T$$

where Q has been written tI here.

It turns out that the isometric isomorphism U is not unique but we can still define $\varphi(A)$, for any real-valued, measurable function φ, to be the operator which is equivalent to multiplication by $\varphi(\alpha(t))$.

Now we would like to investigate this representation of the operator T in more detail. We want, more specifically, to study the relationship between T and $\mathcal{F}^{-1}$. It would appear that there is a rather close relationship between these operators because

$$\mathcal{F}^{-1}(f) = \frac{1}{\sqrt{2\pi}} \int_{-\infty}^{\infty} f(t)e^{ixt} dt = \langle f(t), \frac{e^{-ixt}}{\sqrt{2\pi}} \rangle$$

CHAPTER 8 NON-NORMALIZABLE EIGENVECTORS

for all f in $L^1(\hat{\mathbb{R}}) \cap L^2(\hat{\mathbb{R}})$ certainly, and

$$T[e^{-ixt}] = i\frac{d}{dx}(e^{-ixt}) = i(-it)e^{-ixt} = te^{-ixt}$$

We would like to say that e^{-ixt} is an "eigenvector" for T corresponding to the eigenvalue t. Now, of course, there is a difficulty here in that the function e^{-ixt} is not in $L^2(\hat{\mathbb{R}})$. In the following we shall describe one way to interpret these observations within the framework we have developed in our study of operators.

Let us begin with some specific spaces. We shall write $L^2(\hat{\mathbb{R}})$ as $L^2(dx)$ for reasons which will become clear in a moment. Choose a function ρ on $\hat{\mathbb{R}}$,

$$\rho(x) = (1+x^2)^2$$

and note that ρ is continuous on $\hat{\mathbb{R}}$, it is positive and

$$1/\rho(x) \in L^1(\hat{\mathbb{R}})$$

Next we let $L^2(\rho dx) = \{$ measurable f on $\hat{\mathbb{R}}|\int_{-\infty}^{\infty}|f(x)|^2\rho(x)dx = \int_{-\infty}^{\infty}|f(x)|^2(1+x^2)^2 dx < \infty\}$. In this way we obtain a vector space which is a Hilbert space for the inner product

$$\langle f, g \rangle_\rho = \int_{-\infty}^{\infty} f(x)\overline{g(x)}\rho(x)dx$$

(5) Each $f \in L^2(\rho dx)$ is in $L^2(dx)$ and so we have a natural inclusion map $\mathcal{I} : L^2(\rho dx) \to L^2(dx)$, $\mathcal{I}(f) = f$. The map $\mathcal{I}$ is continuous, and it maps $L^2(\rho dx)$ onto a dense linear manifold in $L^2(dx)$.

Proof. For any $f \in L^2(\rho dx)$ we have

$$|f(x)|^2 \leq (1+x^2)^2|f(x)|^2 = \rho(x)|f(x)|^2$$

for all x. Thus

$$\int_{-\infty}^{\infty}|f(x)|^2 dx \leq \int_{-\infty}^{\infty}|f(x)|^2\rho(x)dx < \infty$$

by hypothesis. Thus $f \in L^2(dx)$. It also follows that

$$\|f\|_\rho \geq \|f\|_2$$

where

$$\|f\|_\rho^2 = \langle f, f \rangle_\rho,$$

and so $\mathcal{I}$ is continuous as claimed.

Finally, choose and fix n and take any $f \in L^2[-n, n]$. Then

$$\int_{-\infty}^{\infty} |f(x)|^2 \rho dx = \int_{-\infty}^{\infty} |f(x)|^2 (1+x^2)^2 dx$$
$$= \int_{-n}^{n} |f(x)|^2 (1+x^2)^2 dx \leq (n^2+1)^2 \int_{-n}^{n} |f(x)|^2 dx$$

and so $f \in L^2(\rho dx)$. This says that the image of $\mathcal{I}$ contains

$$\cup_{n=1}^{\infty} L^2[-n, n] \subseteq L^2(dx)$$

and the union is known to be dense in this space (Section 1, Lemma 1).

(6) The function $f(x) = (1+x^2)^{-1}$ is in $L^2(dx)$, but not in $L^2(\rho(x)dx)$.
Proof. It is clear that this function is in $L^2(dx)$ because

$$\int_{-\infty}^{\infty} |f(x)|^2 dx = \int_{-\infty}^{\infty} \frac{dx}{(1+x^2)^2} < \infty$$

However,

$$\int_{-\infty}^{\infty} |f(x)|^2 \rho(x) dx = \int_{-\infty}^{\infty} \frac{1}{(1+x^2)^2} (1+x^2)^2 dx = \infty$$

(7) For any $f \in L^2(\rho dx)$ the function $f(x)\sqrt{\rho(x)}$ is in $L^2(dx)$.
Proof. $f(x)\sqrt{\rho(x)} = f(x)(1+x^2)$ and so

$$\int_{-\infty}^{\infty} |f(x)\sqrt{\rho(x)}|^2 dx = \int_{-\infty}^{\infty} |f(x)|^2 \rho(x) dx < \infty$$

Let us now consider the set

$$L^2(\frac{dx}{\rho}) = \{\text{measurable } f \text{ on } \mathbb{\hat{R}}| \int_{-\infty}^{\infty} |f(x)|^2 \frac{dx}{\rho(x)} = \int_{-\infty}^{\infty} \frac{|f(x)|^2}{(1+x^2)^2} dx < \infty\}$$

Again we have a vector space which is a Hilbert space for

$$\langle f, g \rangle_{1/\rho} = \int_{-\infty}^{\infty} f(x)\overline{g(x)} \frac{dx}{\rho(x)}$$

(8) Each $f \in L^2(dx)$ is in $L^2(\frac{dx}{\rho})$ and so we have a natural inclusion map $\mathcal{J}: L^2(dx) \to L^2(\frac{dx}{\rho})$, $\mathcal{J}(f) = f$. The map $\mathcal{J}$ is continuous, and it maps $L^2(dx)$ onto a dense linear manifold in $L^2(\frac{dx}{\rho})$.
Proof. For any $f \in L^2(dx)$ we have

$$\frac{|f(x)|^2}{(1+x^2)^2} \leq |f(x)|^2$$

and so
$$\int_{-\infty}^{\infty} |f(x)|^2 \frac{dx}{\rho(x)} = \int_{-\infty}^{\infty} \frac{|f(x)|^2}{(1+x^2)^2} dx \le \int_{-\infty}^{\infty} |f(x)|^2 dx < \infty$$
by hypothesis. Thus $f \in L^2(\frac{dx}{\rho})$ and we also have
$$\|f\|_{1/\rho}^2 \le \|f\|_2$$
where
$$\|f\|_{1/\rho}^2 = \langle f, f \rangle_{1/\rho},$$
and so $\mathcal{J}$ is a continuous map.

As in the proof of (6), the image of $\mathcal{J}$ contains
$$\cup_{n=1}^{\infty} L^2[-n, n] \subseteq L^2(\frac{dx}{\rho})$$

Here we have to show that the union is dense in our space. We can do this as follows: For each $f \in L^2(\frac{dx}{\rho})$ the function
$$f_n(x) = \begin{cases} f(x) & \text{for } |x| \le n \\ 0 & \text{for } n < |x| \end{cases}$$
is not only in $L^2([-n, n], \frac{dx}{\rho})$ but actually in $L^2[-n, n]$ because ρ is bounded away from zero. Clearly
$$\int_{-\infty}^{\infty} |f(x)|^2 \frac{dx}{\rho} = \lim_n \int_{-\infty}^{\infty} |f_n(x)|^2 \frac{dx}{\rho},$$
see Section 1.

(9) The function $f(x) = 1$ for all $x \in \hat{\mathbb{R}}$ is in $L^2(\frac{dx}{\rho})$ but is not in $L^2(dx)$.

(10) Every bounded, measurable function on $\hat{\mathbb{R}}$ is in $L^2(\frac{dx}{\rho})$. In particular, $e^{-ixt} \in L^2(\frac{dx}{\rho})$ for each fixed t.

Proof. If $f(x)$ is any bounded, measurable function on $\hat{\mathbb{R}}$ with $|f(x)| \le M$ for all x (or even just for almost all x),
$$\int_{-\infty}^{\infty} |f(x)|^2 \frac{dx}{\rho} \le M \int_{-\infty}^{\infty} \frac{dx}{(1+x^2)^2} < \infty.$$

(11) For any $f \in L^2(\frac{dx}{\rho})$ the function $f/\sqrt{\rho} = \frac{f(x)}{1+x^2}$ is in $L^2(dx)$.

Let us pause a moment here to see what we have. We fixed $\rho(x) = (1+x^2)^2$ and defined two Hilbert spaces $L^2(\rho dx)$ and $L^2(\frac{dx}{\rho})$. We found that there are continuous, inclusion mappings, $\mathcal{I} : L^2(\rho dx) \to L^2(dx)$ and $\mathcal{J} : L^2(dx) \to L^2(\frac{dx}{\rho})$, which are not onto but do have dense images. Three Hilbert spaces and mapping related in this way will be called a scale of Hilbert spaces. Notice another thing. The self-adjoint operator T ($= if'$) in $L^2(dx)$ has a domain $D(T)$ which contains $x^n \exp(-\frac{1}{2}x^2)$ for $n = 0, 1, 2, \ldots$ and these functions are dense in $L^2(dx)$ (Section 8.1, Lemma 3). But clearly these functions are in $L^2(\rho dx)$ because

$$\int |x^n exp(-\frac{1}{2}x^2)|^2 \rho(x) dx = \int |x^n exp(-\frac{1}{2}x^2)|^2 (1+x^2)^2 dx < \infty$$

Thus, regarding $L^2(\rho dx)$ as a linear manifold in $L^2(dx)$, as we may do, $L^2(\rho dx) \cap D(T)$ is dense in $L^2(\rho dx)$.

(12) For any $f \in L^2(\rho dx)$ and any $g \in L^2(\frac{dx}{\rho})$ we have

$$|\langle f, g \rangle| = |\int_{-\infty}^{\infty} f(x)\overline{g(x)} dx| \leq \|f\|_\rho \|g\|_{1/\rho}$$

Thus each $g \in L^2(\frac{dx}{\rho})$ defines a continuous, linear functional on $L^2(\rho dx)$.

Proof. By (7) $f\sqrt{\rho} \in L^2(dx)$ and by (11) $g/\sqrt{\rho} \in L^2(dx)$. Thus

$$|\langle f, g \rangle| = |\int_{-\infty}^{\infty} f(x)\overline{g(x)} dx| = |\int_{-\infty}^{\infty} f(x)\sqrt{\rho} \frac{\overline{g(x)}}{\sqrt{\rho}} dx| \leq \|f\sqrt{\rho}\|_2 \|g/\sqrt{\rho}\|_2$$

by the C.S.B. inequality (Section 1.4, Theorem 1). However,

$$\|f\sqrt{\rho}\|_2 = \|f\|_\rho$$

and

$$\|g/\sqrt{\rho}\|_2 = \|g\|_{1/\rho}.$$

Let us now extend our self-adjoint operator T to an operator $\tilde{T}$ in $L^2(\frac{dx}{\rho})$; note, $\tilde{T}$ is not, in general, self-adjoint. We may do this as follows:

We have seen that $D(T) \cap L^2(\rho dx)$ is dense in this space; i.e., in $L^2(\rho dx)$. Given $g \in L^2(\frac{dx}{\rho})$ it may happen that there is an $h \in L^2(\frac{dx}{\rho})$ such that

$$\langle Tf, g \rangle = \langle f, h \rangle$$

for all $f \in D(T) \cap L^2(\rho dx)$. Since $f \to \langle f, h \rangle$ is a continuous, linear functional on $L^2(\rho dx)$, by (12), and since our equation is to hold on a dense set, we see that this equation uniquely determines h (if it determines any $h \in L^2(\frac{dx}{\rho})$ is determined exactly one). So we may set

$$\tilde{T} g = h.$$

Observe that if g and h both happen to be in $L^2(dx)$ then

$$\tilde{T}g = h$$

merely says

$$T^*g = Tg = h$$

and so $\tilde{T}$ is an extension of T.

We have seen that $e^{-ixt} \in L^2(\frac{dx}{\rho})$. Let us compute $\tilde{T}(e^{-ixt})$. We must find $h \in L^2(\frac{dx}{\rho})$ such that

$$\langle Tf, e^{-ixt}\rangle = \langle f, h\rangle$$

for all $f \in D(T) \cap L^2(\rho dx)$. This becomes, integrating by parts,

$$\begin{aligned}
\langle Tf, e^{-ixt}\rangle &= \int_{-\infty}^{\infty} if'(x)e^{ixt}dt &= if(x)e^{ixt}|_{-\infty}^{\infty} - i\int_{-\infty}^{\infty} f(x)ite^{ixt}dx\\
&= \int_{-\infty}^{\infty} f(x)\overline{te^{-ixt}}dx &= \langle f, te^{-ixt}\rangle\\
&= &\langle f, i\frac{d}{dx}(e^{-ixt})\rangle
\end{aligned}$$

where we have used the fact that

$$\lim f(x) = 0$$

as $|x| \to \infty$ because $f \in D(T)$; see (2) above. Thus each of our functions e^{-ixt} is an eigenvector of the operator $\tilde{T}$ corresponding to the eigenvalue t. We call such functions non-normalizable eigenvectors of T. The results discussed above can be formulated abstractly and then applied to other operators. We cannot do so because we lack the necessary measure theory. The interested reader is referred to[1; Chapter 5] and [5].

Exercises

1. Let H be a Hilbert space and let A, B be linear operators in H which are equivalent. Show that:

   a. A complex number λ is an eigenvalue of A iff it is an eigenvalue of B;

   b. If λ is not an eigenvalue of A then $(A - \lambda I)^{-1}$ is equivalent to $(B - \lambda I)^{-1}$;

   c. $\sigma(A) = \sigma(B)$ and λ is a generalized eigenvalue of A iff it is a generalized eigenvalue of B.

THE AXIOMS FOR A VECTOR SPACE

We shall, as usual, use the letter $\mathcal{K}$ to denote either the set of real numbers or the set of complex numbers. A non-empty set V is said to be a vector space over $\mathcal{K}$ if it has all of the following properties.

1. There is a rule that assigns to each pair of elements u, v of V a unique element of V called their sum and denoted by $u + v$. This rule, or single-valued binary operation, is called vector addition.

2. Vector addition satisfies:

   a. The Commutative Law; i.e., $u + v = v + u$ for all u, v in V.

   b. The Associative Law; i.e., $u + (v + w) = (u + v) + w$ for all u, v, w in V;

   c. There is an element $0 \in V$ which is an "additive identity"; i.e., $u + 0 = 0 + u = u$ for all u in V;

   d. For each $u \in V$ there is an element $(-u) \in V$ such that $u + (-u) = 0$.

3. There is a rule that assigns to each pair α, v, where $\alpha \in \mathcal{K}$ and $v \in V$, a unique element of V called the product of α and v and denoted by αv. This rule is called scalar multiplication.

4. Scalar multiplication satisfies:

   a. $\alpha(\beta u) = (\alpha\beta)u$ for all α, β in $\mathcal{K}$ and all u in V;
   b. $\alpha(u+v) = \alpha u + \alpha v$ for all α in $\mathcal{K}$ and all u, v in V;
   c. $(\alpha + \beta)u = \alpha u + \beta u$ for all α, β in $\mathcal{K}$ and all u in V;
   d. $1u = u$ for all u in V.

The elements of a vector space V over $\mathcal{K}$ are called vectors.

B

SOLUTIONS TO STARRED PROBLEMS

Here we present detailed solutions to the starred problems given in the introductory chapters of the text (Chapters 1–4). We have also included the solutions to certain starred problems given in Chapters 5 and 6.

Exercises 1.1

3(a) $v_1 \neq 0$, $v_2 \in V \setminus lin\{v_1\}$, $v_3 \in V \setminus lin\{v_1, v_2\}, \ldots, V \setminus lin\{v_1, \ldots, v_p\} = \emptyset$.
Clearly $\{v_1\}$ is linearly independent. So also is $\{v_1, v_2\}$ since $\alpha v_1 + \beta v_2 = 0$ implies $v_2 = \frac{-\alpha}{\beta} v_1 \in lin\{v_1\}$ contradicting the way v_2 was chosen.
If $\{v_1, \ldots, v_p\}$ is not independent then there is a first integer $r \leq p$ such that $\{v_1, \ldots, v_r\}$ is linearly dependent; i.e., there are scalars, not all zero, such that $\sum_{j=1}^{r} \alpha_j v_j = 0$. Now $\alpha_r = 0$ would imply that $\sum_{j=1}^{r-1} \alpha_j v_j = 0$ and some $\alpha_j \neq 0$, $j < r$, and this contradicts the way r was chosen. So $\alpha_r \neq 0$. Thus $v_r = \sum_{j=1}^{r-1} \left(\frac{-\alpha_j}{\alpha_r} \right) v_j$ showing that $v_r \in lin\{v_1, \ldots, v_{r-1}\}$. However, this contradicts the fact that v_r was chosen so that $v_r \in V \mid$

$lin\{v_1,\ldots,v_{r-1}\}$. This same proof shows that the sequence constructed in (b) is linearly independent.

To finish part (a) we must show that $\{v_1,\ldots,v_p\}$ spans V. But we have here $V \setminus lin\{v_1,\ldots,v_p\} = \emptyset$ which means $V = lin\{v_1,\ldots,v_p\}$ and this says the latter set spans V (Definition 2).

8. Let V, W be two vector spaces over the same field and let $T : V \to W$ be an isomorphism from V onto W (Definition 5). Suppose that V has dimension $n < \infty$ and let $v_1,\ldots,v_n$ be a Hamel basis for this space (Definition 4). Consider the set $\{Tv_j\}_{j=1}^n \subseteq W$. Now T is one-to-one and no v_j is 0 so no Tv_j is zero. Also, $\sum_{j=1}^n \alpha_j(Tv_j) = 0 \Leftrightarrow \sum_{j=1}^n T(\alpha_j v_j) = 0 \Leftrightarrow T(\sum_{j=1}^n \alpha_j v_j) = 0 \Leftrightarrow \sum \alpha_j v_j = 0$ by the linearity of T and the fact that this map is one-to-one. But $\{v_j\}_{j=1}^n$ is a Hamel basis for V hence it is an independent set and so every $\alpha_j = 0$. This shows that $\{Tv_j\}$ is a linearly independent subset of W. Next we show that $\{Tv_j\}$ spans W. If $w \in W$ then $w = T(v)$ for some $v \in V$ because T is onto. Now $v = \sum_{j=1}^n \beta_j v_j$ for some scalars β_j because $\{v_j\}$ is a Hamel basis for V. Thus $w = T(v) = T(\sum \beta_j v_j) = \sum \beta_j T(v_j)$. We have shown that an isomorphism maps a Hamel basis for V, one-to-one, to a Hamel basis for W. Thus $\dim W$ is the number of vectors in $\{Tv_j\}_{j=1}^n$ and this is $n = \dim V$.

We have shown that if T is a linear map which takes an n-dimensional vector V space onto a vector space W and if this map is one-to-one, then $\dim W = n$. Suppose now that $T : V \to W$ is an isomorphism but that we now only know that W has (finite) dimension n. This problem can be reduced to the previous one as follows: Define $S : W \to V$ by setting $S(w) = v$ where, given $w \in W$ we find the unique $v \in V$ which T takes to w and we set $S(w)$ equal to this. Since T is one-to-one and onto the same is true of S; for those familiar with the notation, $S = T^{-1}$. Also, a direct calculation shows that S is linear: $S(\alpha w_1 + \beta w_2) = v$ where $T(v) = \alpha w_1 + \beta w_2$. But there are vectors v_1, v_2 such that $T(v_1) = w_1$, $T(v_2) = w_2$ so, by the linearity of T, $T(\alpha v_1 + \beta v_2) = \alpha w_1 + \beta w_2$. But T is one-to-one so $v = \alpha v_1 + \beta v_2$ which says $S(\alpha w_1 + \beta w_2) = \alpha v_1 + \beta v_2 = \alpha S(w_1) + \beta S(w_2)$. Thus S is an isomorphism from W to V.

Finally, if one of these vector spaces is infinite dimensional, say V to be specific, then we must have an infinite, linearly independent set $v_1,\ldots$ in this space. But our discussion above then shows that $Tv_1,\ldots$ is an infinite linearly independent subset of W showing that W is infinite dimensional.

9(a) Suppose that $v = \sum_{j=1}^n \alpha_j v_j = \sum_{j=1}^n \beta_j v_j$ where $\{v_j\}_{j=1}^n$ is a Hamel basis for V and $n < \infty$. Then $0 = \sum_{j=1}^n (\alpha_j - \beta_j) v_j$ where this is the zero vector in V. Since $\{v_j\}$ is linearly independent it follows that $\alpha_j - \beta_j = 0$ for every j; i.e., $\alpha_j = \beta_j$ for each j.

(b) V is n-dimensional with Hamel basis $v_1,\ldots,v_n$ and W is n-dimensional. We set $Tv_j = w_j$ (here $\{w_j\}$ is a Hamel basis for W) and extend by linearity as stated explicitly in the problem. Let us show that T is an isomorphism in steps.

(i) T is one-to-one. Suppose that $u = \sum \alpha_j v_j$, $v = \sum \beta_j v_j$ and $T(u) = T(v)$. Then $\sum \alpha_j w_j = \sum \beta_j w_j$ by the definition of T. But by (a), since $\{w_j\}$ is a Hamel basis for W, this implies $\alpha_j = \beta_j$ for all j. But then $u = v$ and we are done.

(ii) T is onto. Given $w \in W$, $w = \sum \alpha_j w_j$ because $\{w_j\}$ spans W. But then $\sum \alpha_j v_j = v \in V$ and, from the way T was defined, $Tv = w$.

(iii) T is linear. Take $u = \sum \alpha_j v_j$, $v = \sum \beta_j v_j$ and scalars α, β. Then $T(\alpha u + \beta v) = T[\sum (\alpha \alpha_j) v_j + \sum \beta \beta_j v_j] = T[\sum (\alpha \alpha_j + \beta \beta_j) v_j] = \sum (\alpha \alpha_j + \beta \beta_j) w_j = \alpha (\sum \alpha_j w_j) + \beta (\sum \beta_j w_j) = \alpha T(u) + \beta T(v)$.

Exercises 1.2

1(c) For the Euclidean norm it is just the unit disk; i.e., the unit circle centered at $(0,0)$ together with all of its interior points. For the norm $\|\cdot\|_n$ it is the "diamond" formed by the line segments joining the points $(1,0)$, $(0,1)$, $(-1,0)$, $(0,-1)$ together with all of its interior points. For the norm $\|\cdot\|_s$ it is the square formed by the line segments: $x = 1$ from $(1,-1)$ to $(1,1)$; $y = 1$ from $(1,1)$ to $(-1,1)$; $x = -1$ from $(-1,1)$ to $(-1,-1)$; $y = -1$ from $(-1,-1)$ to $(1,-1)$ together with all of its interior points.

2 First note that $\sum_{j=1}^n \alpha_j \vec{e}_j = (\alpha_1, \alpha_2, \ldots, \alpha_n) = \vec{0} = (0, 0, \ldots, 0)$ implies $\alpha_1 = 0$, $\alpha_2 = 0$, ..., $\alpha_n = 0$. Thus $\{\vec{e}_j\}_{j=1}^n$ is linearly independent. Next given $\vec{v} = (v_1, v_2, \ldots, v_n) \in \mathbf{R}^n$ we have $\vec{v} = \sum_{j=1}^n v_j \vec{e}_j$ so $\{\vec{e}_j\}_{j=1}^n$ spans our space.

4 Let V be an n-dimensional vector space over $\mathbf{R}$ and let $v_1, \ldots, v_n$ be a Hamel basis for V. Define a map $T : V \to \mathbf{R}^n$ as follows: $Tv_j = \vec{e}_j$, $j = 1, 2, \ldots, n$ and extend T by linearity. In Exercises 1.1, problem 9, we shows that this definition gives us an isomorphism from V onto $\mathbf{R}^n$. In an entirely analogous manner we see that any n-dimensional vector space over $\mathbf{C}$ must be isomorphic to $\mathbf{C}^n$.

5 We have an isomorphism $\varphi : E \to F$ and a norm $\|\cdot\|_F$ on F. For each $x \in E$ we define $\|x\|_E = \|\varphi(x)\|_F$. We show $\|\cdot\|_E$ is a norm.

(a)

$$\begin{aligned}\|x+y\|_E &= \|\varphi(x+y)\|_F = \|\varphi(x) + \varphi(y)\|_F \leq \|\varphi(x)\|_F + \|\varphi(y)\|_F \\ &= \|x\|_E + \|y\|_E\end{aligned}$$

and here we have used the linearity of φ, to say $\varphi(x+y) = \varphi(x) + \varphi(y)$, and the triangle inequality for the norm $\|\cdot\|_F$.

(b) $\|\alpha x\|_E = \|\varphi(\alpha x)\|_F = \|\alpha\varphi(x)\|_F = |\alpha|\|\varphi(x)\|_F = |\alpha|\|x\|_E$ again by the linearity of φ and property (b) of the norm $\|\cdot\|_F$.

(c) $\|x\|_E = 0 \Rightarrow \|\varphi(x)\|_F = 0 \Rightarrow \varphi(x) = 0$ because $\|\cdot\|_F$ is a norm. But φ is an isomorphism so $\varphi(x) = 0$ implies $x = 0$.

Exercises 1.3

1 For any x_1, x_2 in E we have $x_1 = x_2 + (x_1 - x_2)$ and so $\|x_1\| = \|x_2 + (x_1 - x_2)\| \leq \|x_2\| + \|x_1 - x_2\|$. Similarly, $\|x_2\| \leq \|x_1\| + \|x_2 - x_1\|$ and so $\|x_1\| - \|x_2\| \leq \|x_1 - x_2\|$, and $\|x_2\| - \|x_1\| \leq \|x_2 - x_1\|$. But $\|(x_1 - x_2)\| = \|(-1)(x_1 - x_2)\| = \|x_2 - x_1\|$ by property (b) of a norm. Thus $\|x_1\| - \|x_2\|$ and $\|x_2\| - \|x_1\|$ are both $\leq \|x_1 - x_2\|$; i.e., $|\|x_1\| - \|x_2\|| \leq \|x_1 - x_2\|$. In this problem we were given a sequence $\{x_k\}$ which converges to x_0 for the norm. Thus $|\|x_0\| - \|x_k\|| \leq \|x_0 - x_k\|$ and this tends to zero proving our result.

2 In Chapter 2 just after Definition 3 we do parts (a) – (c). We shall do parts (d) – (f) here. We have $\|\cdot\|_1 \equiv \|\cdot\|_2$ and so there are non-zero scalars λ, μ such that $\lambda\|x\|_1 \leq \|x\|_2 \leq \mu\|x\|_1$ for all $x \in E$ by (c). Suppose that $\{x_k\} \subseteq E$ and that $\{x_k\}$ converges to x_0 for $\|\cdot\|_1$. Then $\|x_k - x_0\|_2 \leq \mu\|x_k - x_0\|_1 \to 0$ showing that $\{x_k\}$ converges to x_0 for $\|\cdot\|_2$. Now suppose that $\{x_k\}$ converges to x_0 for $\|\cdot\|_2$. Then $\|x_k - x_0\|_1 \leq \frac{1}{\lambda}\|x_k - x_0\|_2 \to 0$.

(e) In problem 1(c) of Exercises 2 we draw the unit ball of $\mathbf{R}^2$ for each of these norms. It is clear from these pictures that we can multiply any one of these balls by a small enough scalar to get it to fit into any other one of these balls and that proves the norms are equivalent.

(f) We have μ_1, λ_1 and μ_2, λ_2 non-zero scalars such that $\lambda_1\|x\|_1 \leq \|x\| \leq \mu_1\|x\|_1, \lambda_2\|x\|_2 \leq \|x\| \leq \mu_2\|x\|_2$ for all x. Then $\frac{\lambda_1}{\lambda_2}\|x\|_1 \leq \|x\|_2 \leq \frac{\mu_1}{\mu_2}\|x\|_1$.

3 $\{z_n\}$ is a subsequence of $\{y_k\}$ so each $z_n = y_{k(n)}$ where, by the definition of the term subsequence, we know that $n \leq k(n) < k(n+1)$ for all n. Let $\varepsilon > 0$ be given. Choose N_1 so that $\|y_n - y_m\| < \varepsilon/2$ for all $n, m \geq N_1$ and choose N_2 so that $\|z_0 - z_n\| < \varepsilon/2$ for all $n \geq N_2$. Let $N = \max\{N_1, N_2\}$. Now $\|z_0 - y_n\| \leq \|z_0 - z_{k(n)}\| + \|z_{k(n)} - y_n\|$ and if $n \geq N$ then $k(n) \geq n \geq N$ so both of these norms are $< \varepsilon/2$.

Exercises 1.4

1(a) $\langle u, 0 \rangle = \langle u, 20 \rangle = \overline{\langle 2 \cdot 0, u \rangle} = 2\overline{\langle 0, u \rangle} = 2\langle u, 0 \rangle$ if $\langle u, 0 \rangle \neq 0$ we could then divide obtaining $1 = 2$. Similarly, $\langle 0, u \rangle = 0$.

(b) $\langle u, \alpha v + \beta w \rangle = \overline{\langle \alpha v + \beta w, u \rangle} = \overline{\alpha \langle v, u \rangle + \beta \langle w, u \rangle} = \overline{\alpha} \overline{\langle v, u \rangle} + \overline{\beta} \overline{\langle w, u \rangle} = \overline{\alpha} \langle u, v \rangle + \overline{\beta} \langle u, w \rangle$

(c)
$$\begin{aligned} \|u+v\|^2 &= \langle u+v, u+v \rangle = \langle u,u \rangle + \langle u,v \rangle + \langle v,u \rangle + \langle v,v \rangle \\ &= \|u\|^2 + \langle u,v \rangle + \langle v,u \rangle + \|v\|^2 \\ \|u-v\|^2 &= \langle u-v, u-v \rangle = \|u\|^2 - \langle v,u \rangle - \langle u,v \rangle + \|v\|^2 \end{aligned}$$

Adding $\|u+v\|^2 + \|u-v\|^2 = 2[\|u\|^2 + \|v\|^2]$.

(d) We have just computed $\|u+v\|^2$ and $\|u-v\|^2$ subtracting we get $\|u+v\|^2 - \|u-v\|^2 = 2\langle u,v \rangle + 2\langle v,u \rangle$. Now

$$\begin{aligned} \|u+iv\|^2 &= \langle u+iv, u+iv \rangle = \langle u, u+iv \rangle + i\langle v, u+iv \rangle \\ &= \langle u,u \rangle - i\langle u,v \rangle + i[\langle v,u \rangle - i\langle v,v \rangle] \end{aligned}$$

(remember $\overline{i} = -i$) $= \|u\|^2 - i\langle u,v \rangle + i\langle v,u \rangle + \|v\|^2$

$$\begin{aligned} i\|u+iv\|^2 &= i\|u\|^2 + \langle u,v \rangle - \langle v,u \rangle + i\|v\|^2 \\ \|u-iv\|^2 &= \langle u-iv, u-iv \rangle = \langle u, u-iv \rangle - i\langle v, u-iv \rangle \\ &= \langle u,u \rangle + i\langle u,v \rangle - i[\langle v,u \rangle + i\langle v,v \rangle] \\ &= \|u\|^2 + i\langle u,v \rangle - i\langle v,u \rangle + \|v\|^2 \\ i\|u-iv\|^2 &= i\|u\|^2 - \langle u,v \rangle + \langle v,u \rangle + i\|v\|^2 \end{aligned}$$

So $i\|u+iv\|^2 - i\|u-iv\|^2 = 2\langle u,v \rangle - 2\langle v,u \rangle$. Hence

$$\begin{aligned} \|u+v\|^2 - \|u-v\|^2 &= 2\langle u,v \rangle + 2\langle v,u \rangle \\ i\|u+iv\|^2 - i\|u-iv\|^2 &= 2\langle u,v \rangle - 2\langle v,u \rangle \\ \text{sum} &= 4\langle u,v \rangle \end{aligned}$$

Thus

$$\frac{\|u+v\|^2 - \|u-v\|^2 + i\|u+iv\|^2 - i\|u-iv\|^2}{4} = \frac{4\langle u,v \rangle}{4} = \langle u,v \rangle$$

(e)
$$\begin{aligned} \|u+v\|^2 &= \langle u+v, u+v \rangle = \langle u,u \rangle + \langle u,v \rangle + \langle v,u \rangle + \langle v,v \rangle \\ &= \|u\|^2 + \text{zero} + \overline{\langle u,v \rangle} + \|v\|^2 = \|u\|^2 + \|v\|^2 \end{aligned}$$

since $\overline{\text{zero}} = \text{zero}$.

(f)
$$|\langle u, v_n \rangle - \langle u, v \rangle| = |\langle u, v_n - v \rangle| \le \|u\| \|v_n - v\| \to 0$$

We have used the C.S.B. inequality.

(g)
$$\begin{aligned}|\langle u_n, v_n\rangle - \langle u, v\rangle| &= |\langle u_n, v_n\rangle - \langle u_n, v\rangle + \langle u_n, v\rangle - \langle u, v\rangle| \\ &= |\langle u_n, v_n - v\rangle + \langle u_n - u, v\rangle| \le |\langle u_n, v_n - v\rangle| \\ &\quad + |\langle u_n - u, v\rangle| \\ &\le \|u_n\|\|v_n - v\| + \|u_n - u\|\|v\|\end{aligned}$$

Now $\|u_n - u\| \to 0$ so $\|u_n - u\|\|u\| \to 0$. Also $\|v_n - v\| \to 0$ but we note that $\|u_n\|$ is bounded as a function of n; because $\|u_n\| = \|u_n - u + u\| \le \|u_n - u\| + \|u\|$ and $\|u_n - u\| \to 0$. Thus $\|u_n\|\|v_n - v\| \to 0$ and we are done.

2(a)(i)

$$\begin{aligned}\langle f, g\rangle &= \int_a^b f(t)\overline{g(t)}dt \text{ and } \overline{\langle g, f\rangle} = \overline{\int_a^b g(t)\overline{f(t)}dt} \\ &= \int_a^b \overline{g(t)\overline{f(t)}}dt = \int_a^b f(t)\overline{g(t)}dt = \langle f, g\rangle\end{aligned}$$

(ii) $\langle \alpha f + \beta h, g\rangle = \int_a^b [\alpha f(t) + \beta h(t)]\overline{g(t)}dt = \alpha\langle f, g\rangle + \beta\langle h, g\rangle$.

(iii) $\langle f, f\rangle = \int_a^b f(t)\overline{f(t)}dt = \int_a^b |f(t)|^2 dt$. Now f is continuous so if $|f(t_0)| \ne 0$ for some t_0 then $|f(t)| > 0$ for all t is some small interval $(t_0 - \delta, t_0 + \delta)$. But then
$$\langle f, f\rangle = \int_a^b |f(t)|^2 dt \ge \int_{t_0 - \delta}^{t_0 + \delta} |f(t)|^2 dt > 0.$$

Exercises 1.5

1(a) $\langle \vec{e}_n, \vec{e}_m\rangle = \sum_{j=1}^{\infty} \delta_{nj}\delta_{mj} =$ zero unless $n = m$ because $\delta_{nj} = 1$ only when $j = n$ and $\delta_{mj} = 1$ only when $j = m$.

(b)
$$\begin{aligned}\|\vec{e}_n - \vec{e}_m\|^2 &= \langle \vec{e}_n - \vec{e}_m, \vec{e}_n - \vec{e}_m\rangle = \langle \vec{e}_n, \vec{e}_n - \vec{e}_m\rangle \\ &\quad - \langle \vec{e}_m, \vec{e}_n - \vec{e}_m\rangle = \langle \vec{e}_n, \vec{e}_n\rangle - \langle \vec{e}_n, \vec{e}_m\rangle - \langle \vec{e}_m, \vec{e}_n\rangle \\ &\quad + \langle \vec{e}_m, \vec{e}_m\rangle = 1 + 1 = 2\end{aligned}$$

Thus $\|\vec{e}_n - \vec{e}_m\| = 0$ if $n = m$ and $= \sqrt{2}$ if $n \ne m$.

2(a)
$$\begin{aligned}A_r(\alpha\vec{z} + \beta\vec{w}) &= A_r(\{\alpha z_1 + \beta w_1, \alpha z_2 + \beta w_2, \ldots\}) \\ &= \{0, \alpha z_1 + \beta w_1, \alpha z_2 + \beta w_2, \ldots\} = \alpha\{0, z_1, z_2, \ldots\}\end{aligned}$$

$$
\begin{aligned}
&\qquad\qquad + \ \beta\{0,w_1,w_2,\ldots\} = \alpha A_r(\vec{z}) + \beta A_r(\vec{w})\\
A_\ell(\alpha\vec{z}+\beta\vec{w}) &= A_\ell(\{\alpha z_1+\beta w_1, \alpha z_2+\beta w_2,\ldots\})\\
&= \{\alpha z_2+\beta w_2,\ldots\} = \alpha\{z_2,\ldots\} + \beta\{w_2,\ldots\}\\
&= \alpha A_\ell(\vec{z}) + \beta A_\ell(\vec{w})
\end{aligned}
$$

(b)
$$
\begin{aligned}
A_\ell \circ A_r(\vec{z}) &= A_\ell[A_r(\vec{z})] = A_\ell[A_r(\{z_1,z_2,z_3,\ldots\})]\\
&= A_\ell[\{0,z_1,z_2,z_3,\ldots\}] = \{z_1,z_2,\ldots\} = \vec{z} \text{ for any } \vec{z}.\\
A_r \circ A_\ell(\vec{z}) &= A_r[A_\ell(\vec{z})] = A_r[A_\ell(\{z_1,z_2,z_3,\ldots\})]\\
&= A_r[\{z_2,z_3,\ldots\}] = \{0,z_2,z_3,\ldots\} \neq \{z_1,z_2,\ldots\} = \vec{z}.
\end{aligned}
$$

In particular, $A_r \circ A_\ell(\vec{e}_1) = \vec{0} \neq \vec{e}_1$.

Exercises 1.6

1(a) Let $\mathcal{S}$ be an orthonormal subset of V. Choose any finite set of vectors $v_1,\ldots,v_n$ from $\mathcal{S}$ and consider the sum $\sum_{j=1}^n \alpha_j v_j = 0$. We can show that this equation implies that every $\alpha_k = 0$ as follows: $\langle\sum_{j=1}^n \alpha_j v_j, v_k\rangle = \sum_{j=1}^n \alpha_j\langle v_j,v_k\rangle = \alpha_k$, but $\langle 0, v_k\rangle = 0$ so $\alpha_k = 0$.

(b) Let $\mathcal{B}$ be any orthonormal set. Suppose $\mathcal{B}$ is maximal. Then for any $u \in V$ if $\langle u,v\rangle = 0$ for all $v \in \mathcal{B}$ and if $u \neq 0$, then $\{\frac{u}{\|u\|}\} \cup \mathcal{B}$ is an orthonormal set which properly contains $\mathcal{B}$. This contradiction, of the maximality of $\mathcal{B}$, shows u must be the zero vector.

Next suppose that $\langle u,v\rangle = 0$ for all $v \in \mathcal{B}$ implies $u = 0$. If $\mathcal{B}' \supsetneq \mathcal{B}$ and $\mathcal{B}'$ is orthonormal then choose $u \in \mathcal{B}' \setminus \mathcal{B}$. We must then have $\langle u,v\rangle = 0$ for all $v \in \mathcal{B}$ because $\mathcal{B}'$ is orthogonal. However, our assumption then gives $u = 0$ which is impossible because $u \in \mathcal{B}'$ and $\mathcal{B}'$ is an orthonormal set; i.e., $\|u\| = 1$.

2(b) Let $v_1,\ldots,v_m$ be a maximal orthonormal subset of $\mathbf{R}^n$. Since any such set is linearly independent by 1(a), $m \leq n$. Suppose $m < n$. Then $\text{lin}\{v_1,\ldots,v_m\} \subsetneq \mathbf{R}^n$ so we can choose a vector $w \in \mathbf{R}^n$ but not in the linear hull. As in the proof of lemma 1 above we set $u = w - \sum_{j=1}^m \langle w,v_j\rangle v_j$ and we note that $u \neq 0$ while $\langle u,v_j\rangle = 0$ for $j = 1,2,\ldots,m$. It follows that $\{v_1,\ldots,v_m, u/\|u\|\}$ is an orthonormal set which properly contains $\{v_1,\ldots,v_m\}$ contradicting the fact that the latter is maximal.

3 Suppose $m \neq n$ and compute

$$\langle \exp(int), \exp(imt)\rangle = \int_{-\pi}^{\pi} e^{int}\overline{e^{imt}}dt = \int_{-\pi}^{\pi} e^{i(n-m)t}dt = \frac{e^{i(n-m)}}{i(n-m)}\Big|_{-\pi}^{\pi}$$

$$= \frac{1}{i(n-m)}\{\cos(n-m)t + i\sin(n-m)t\}\,|_{-\pi}^{\pi}$$

$$= \frac{1}{i(n-m)}[\{\cos(n-m)\pi t + i\sin(n-m)\pi\}$$
$$\quad - \{\cos(n-m)\pi - i\sin(n-m)\pi\}]$$

$$= \text{zero}$$

so our set is orthogonal. However,

$$\langle \exp(int), \exp(int)\rangle = \int_{-\pi}^{\pi} dt = 2\pi$$

so this set is not orthonormal.

Since $\|\exp(int)\| = \sqrt{2\pi}$ we see that the set $\left\{\frac{\exp(int)}{\sqrt{2\pi}}\,|\,n=0,\pm 1, \pm 2, \ldots\right\}$ is an orthonormal set.

4 We have an orthonormal basis $v_1, v_2, \ldots$ in H and scalars $a_j, j = 1, 2, \ldots$.

(i) Suppose that for some $u \in H$ we have $\langle u, v_j\rangle = a_j$ for every j. Then $\sum |a_j|^2 < \infty$ by Theorem 2.

(ii) Suppose that $\sum |a_j|^2$ is known to be convergent. For each n let $u_n = \sum_{j=1}^{n} a_j v_j$. Clearly these vectors are in H since our sums are finite. Next for $n < m$

$$\|u_m - u_n\|^2 = \langle \sum_{j=n+1}^{m} a_j v_j, \sum_{j=n+1}^{n} a_j v_j\rangle = \sum_{j=n+1}^{m} |a_j|^2$$

and the latter sum tends to zero as $n, m \to \infty$. Thus $\{v_n\}$ is a Cauchy sequence in H and hence, since H is a Hilbert space, converges to some vector $u \in H$. Thus

$$u = \lim_{m\to\infty} \sum_{j=1}^{m} a_j v_j = \sum_{j=1}^{\infty} a_j v_j$$

Finally, we must show that $\langle u, v_k\rangle = a_k$ for all k. But

$$\langle u, v_k\rangle = \langle \sum_{j=1}^{\infty} a_j v_j, v_k\rangle = \lim_{m\to\infty} \langle \sum_{j=1}^{m} a_j v_j, v_k\rangle$$

$$= lim_{m\to\infty} \sum_{j=1}^{m} a_j \langle v_j, v_k\rangle = a_k$$

Exercises 1.7

2(a) For each n choose a polynomial $p_n(t)$ such that $|f(t) - p_n(t)| < \frac{1}{n}$ for $a \le t \le b$. Then $\|f - p_n\|_\infty < \frac{1}{n}$ and so $\{p_n(t)\}$ converges to f for this norm; note that we can always choose such a p_n by Corollary 3 with $\varepsilon = \frac{1}{n}$.

(b) We have $\|f - p_n\|_\infty < \frac{1}{2n}$ for each n and p_n has real coefficients. Fix n and suppose $p_n(t) = \sum_{j=1}^m a_j t^j$. Let $q_n(t) = \sum_{j=1}^m q_j t^j$ and note $|p_n(t) - q_n(t)| \le \sum_{j=1}^m |a_j - q_j||t|^j$ for any t. Now $|t| \le \max\{|a|, |b|\} \equiv \ell$ and we may choose rational numbers q_j such that $|a_j - q_j| < \frac{1}{2^n \ell^m (m+1)}$. Then $\frac{|t|^j}{\ell^m} \le 1$ on $[a,b]$ for each j and there are $m+1$ terms so our sum is $\le \frac{m+1}{2^n(m+1)} = \frac{1}{2n}$. Finally $\|f - q_n\|_\infty \le \|f - p_n\|_\infty + \|p_n - q_n\|_\infty < \frac{1}{n}$.

(c) Again $\|f - p_n\|_\infty < \frac{1}{2n}$ for each n. Since p_n has complex coefficients we may write $p_n(t) = p'_n(t) + ip"_n(t)$ where p'_n, $p"_n$ have real coefficients. Using (b) we find q'_n, $q"_n$, polynomials with rational coefficients such that $\|(p'_n + ip"_n) - (q'_n + iq"_n)\|_\infty \le \|p'_n - q'_n\|_\infty + \|p"_n - q"_n\|_\infty < \frac{1}{n}$.

Exercises 1.8

4 We have $K(s,t) \in L^2(S)$ and $0 = \langle K(s,t), \psi_{m,n}(s,t)\rangle$ for all m,n. Now $\psi_{m,n}(s,t) = \varphi_m(s)\varphi_n(t)$ where $\{\varphi_n(t)\}$ is an orthonormal basis for $L^2[a,b]$. We have

$$0 = \int\int_S [K(s,t)\varphi_m(s)\varphi_n(t)] ds\, dt = \int_a^b \left(\int_a^b K(s,t)\varphi_m(s) ds\right)\varphi_n(t) dt$$

Now by Fubini's theorem $K(s,t) \in L^2[a,b]$ for almost all fixed values of t. Thus $K(s,t)\varphi_m(s) \in L^2[a,b]$ for almost all fixed t. $\int K(s,t)\varphi_m(s) ds \in L^2[a,b]$ as a function of t. So our equation says that, as a function of t, this integral is zero a.e. because $\{\varphi_n(t)\}$ is an orthonormal basis. Thus $K(s,t) = 0$ a.e. in t and $\int_a^b K(s,t)\varphi_m(s) ds = 0$ which says $\langle K(s,t), \varphi_m(s)\rangle = 0$ all m. We conclude that $K(s,t) = 0$ a.e. in s.

Exercises 1.9

2(a) Given $\varepsilon > 0$ and any $x \in E$ we find $t \in T$ such that $\|x - t\| < \frac{\varepsilon}{2}$. Next we find $s \in S$ such that $\|t - s\| < \frac{\varepsilon}{2}$. Then $\|x - s\| \le \|x - t\| + \|t - s\| < \varepsilon$. Since x and $\varepsilon > 0$ were arbitrary we are done.

(b) If $(E, \|\cdot\|)$ is separable then it has a countable dense set S. Then $S \subseteq linS \subseteq E$ and S is dense in E so clearly $linS$ is dense in E. Hence S is a countable total set.

Now suppose that E contains a countable, total set G. We want to show that E is separable hence we must find a countable, dense subset of E. We know that lin G is dense in E but lin G is not countable. Suppose that our field of scalars is $\mathbb{R}$ and let $\lin_Q G$ be the set of all (finite) linear combinations of elements of G with rational coefficients. This is a countable set and it is dense in lin G. Thus, by (a), $\lin_Q G$ is a countable set which is dense in E showing that E is separable.

4(a) This was done in the course of our solution to problem 1.8.

(b) We have $\langle u, v \rangle_1 = \langle Tu, Tv \rangle_2$ for all $u, v \in V_1$. Given w, w^* in V_2 we find u, u^* in V_1 such that $Tu = w$, $Tu^* = w^*$. Then $\langle w, w^* \rangle_2 = \langle Tu, Tu^* \rangle_2 = \langle u, u^* \rangle_1 = \langle T^{-1}w, T^{-1}w^* \rangle_1$ and we are done.

(c) Let $(V_1, \langle,\rangle_1)$ and $(V_2, \langle,\rangle_2)$ both be isometrically isomorphic to $(W, \langle,\rangle^*)$. Then we have $T_1 : V_1 \to W$ and $T_2 : V_2 \to W$ both isometric isomorphisms. As we saw in (b) the map $T_2^{-1} : W \to V_2$ is an isometric isomorphism. Thus $T_2^{-1} \cdot T_1 : V_1 \to V_2$ and we will be finished once we show that the composition of two isometric isomorphisms is an isometric isomorphism. So let $V_1 \xrightarrow{S_1} V_2 \xrightarrow{S_2} V_3$ be isometric isomorphisms. $S_2 \circ S_1$ is linear since S_1 and S_2 are linear and it is one-to-one and onto because S_1 and S_2 have these properties. For $u, v \in V_1$ $\langle u, v \rangle_1 = \langle S_1 u, S_2 v \rangle_2 = \langle S_2(S_1 u), S_2(S_1 v) \rangle_3$ and we are done.

Exercises 2.1

2 Let $\{x_n\}$ be a Cauchy sequence in $(E, \|\cdot\|)$. Choose $\varepsilon = 1$ and find N so that $\|x_n - x_m\| \leq 1$ for all $m, n \geq N$. Then for any $n \geq N$ we have $\|x_n\| \leq \|x_n - x_N\| + \|x_N\| \leq 1 + \|x_N\|$ so this last number is a bound on all terms beyond the Nth. But there are only a finite number of terms x_n with $n < N$. So let $K = \max\{\|x_n\| \mid 1 \leq n \leq N\}$ and then $\|x_n\| \leq K + \|x_N\|$ for *all* n, since we may take $K > 1$

3(a) We have $\|Tx\| \leq \|T\|\|x\|$ and $\|Sx\| \leq \|S\|\|x\|$ for all $x \in E$. Thus $\|S \circ T(x)\| = \|S[T(x)]\| \leq \|S\|\|T(x)\| \leq (\|S\|\|T\|)\|x\|$ for all $x \in E$. Hence $S \circ T$ is bounded and $\|S \circ T\| \leq \|S\|\|T\|$.

(b) We just proved this inequality in (a). The matrices $\begin{pmatrix} 0 & 0 \\ 0 & 1 \end{pmatrix}$ and $\begin{pmatrix} 0 & 1 \\ 0 & 0 \end{pmatrix}$ each define a bounded, non-zero linear operator on $\mathbb{R}^3$ so they have norms N_1 and N_2 say. But their composition is the zero operator.

(c) This follows from (b).

4 If T is an isometry on an inner product space $(V, \langle,\rangle)$ then $\|Tv\|^2 = \langle Tv, Tv \rangle = \langle v, v \rangle = \|v\|^2$ for all $v \in V$. Thus $\|Tv\| = \|v\|$ for all $v \in V$ showing T has norm one. In case T maps $(V_1, \langle,\rangle_1)$ onto $((V_2, \langle,\rangle_2)$ and is an isometry, then $\|Tv\|_2^2 = \langle Tv, Tv\rangle_2 = \langle v, v\rangle_1 = \|v\|_1^2$ again showing that T has norm one.

5. We had $A_r(\vec{z}) = A_r(\{z_1, z_2, \ldots\}) = \{0, z_1, z_2, \ldots\}$ and $A_\ell(\vec{z}) = A_\ell(\{z_1, z_2, z_3, \ldots\}) = \{z_2, z_3, \ldots\}$. Also we saw in Exercises 1.5, problem 2 (done earlier), that $A_\ell \circ A_r = I$ but not conversely. Now $\|A_r(\vec{z})\|^2 = \langle A_r\vec{z}, A_r\vec{z}\rangle = \langle \{0, z_1, \ldots\}, \{0, z_1, \ldots\}\rangle = \sum |z_j|^2 = \|\vec{z}\|^2$ showing $\|A_r\| = 1$.
Now $I = A_\ell \circ A_r$ so $1 = \|I\| = \|A_\ell \circ A_r\| \le \|A_\ell\|\|A_r\| = \|A_\ell\|$. But clearly $\|A_\ell\| \le 1$ from the definition. Thus $\|A_\ell\| = 1$.

Exercises 2.2

2. We have $(E, \|\cdot\|)$, $(F, \|\cdot\|_F)$ and a topological isomorphism $T: E \to F$.

(a) For $x \in E$ we set $|\|x\|| = \|Tx\|_F$. This was shown to be a norm on E in Exercises 1.2, problem 5. Let us show that it is equivalent to $\|\cdot\|_E$. We have $\||x\|| = \|Tx\|_F \le \|T\|\|x\|_E$ and, also, $\|x\|_E = \|T^{-1}(Tx)\|_E \le \|T^{-1}\|\|Tx\|_F = \|T^{-1}\|\||x\||$. Combining these

$$\left(\frac{1}{\|T\|}\right)\||x\|| \le \|x\|_E \le (\|T^{-1}\|)\||x\||$$

showing these norms are equivalent.

(b) This is just (a) again for $\||y\|| = \|T^{-1}y\|_E$.

(c) E just a vector space, $T: E \to F$ an isomorphism but F has a norm $\|\cdot\|_F$. Then $\||x\|| = \|Tx\|_F$ gives us a norm on E. Note that this equation, which we have forced to be true, says that T is an isometry.

4. $T: E \to F$. The set $C \subseteq F$ is closed iff $F \backslash C$ is open. Now $T^{-1}(F \backslash C) = \{x \in E | Tx \in F \backslash C\} = E \backslash T^{-1}(C)$. By Theorem 1 we have: (i) T continuous implies $T^{-1}(F \backslash C)$ is open whenever C is closed and so $E \backslash T^{-1}(C)$ is open whenever C is closed and so $T^{-1}(C)$ is closed whenever C is closed; (ii) $T^{-1}(C)$ closed whenever C is closed implies $E \backslash T^{-1}(C)$ is open whenever C is closed and this implies $T^{-1}(F \backslash C)$ is open whenever C is closed. If $\mathcal{O}$ is any open subset of F then $F \backslash \mathcal{O} \equiv C$ is closed and $T^{-1}(\mathcal{O}) = T^{-1}(F \backslash C)$ is open showing that T is continuous by Theorem 1.

Exercises 2.3

2. $(S - \lambda I)^{-1} - (S - \mu I)^{-1} = (S - \lambda I)^{-1}[(S - \mu I) - (S - \lambda I)](S - \mu I)^{-1} = (\lambda - \mu)(S - \lambda I)^{-1}(S - \mu I)^{-1}$ each step can be verified by direct calcula-

tion. Thus $(S - \lambda I)^{-1}(S - \mu I)^{-1} = \frac{1}{\lambda - \mu}[(S - \lambda I)^{-1} - (S - \mu I)^{-1}]$ and, interchanging λ and μ

$$(S - \mu I)^{-1}(S - \lambda I)^{-1} = \frac{1}{\mu - \lambda}[(S - \mu I)^{-1} - (S - \lambda I)^{-1}]$$
$$= \frac{1}{\lambda - \mu}[(S - \lambda I)^{-1} - (S - \mu I)^{-1}]$$

Hence we have $(S - \mu I)^{-1}(S - \lambda I)^{-1} = (S - \lambda I)^{-1}(S - \mu I)^{-1}$ as claimed.

3(a) Suppose that $Av = \lambda v$ for some non-zero vector v. Then $\langle v, v \rangle = \langle Av, Av \rangle = \langle \lambda v, \lambda v \rangle = |\lambda|^2 \langle v, v \rangle$ giving $|\lambda| = 1$.

(b) Suppose that $Av = \lambda v$, $Aw = \mu w$ where $\lambda \neq \mu$. Then $\langle v, w \rangle = \langle Av, Aw \rangle = \langle \lambda v, \mu w \rangle = \lambda \overline{\mu} \langle v, w \rangle$ giving $(1 - \lambda \overline{\mu})\langle v, w \rangle = 0$. Now $\lambda \overline{\lambda} = 1$ and $\mu \overline{\mu} = 1$ but $\lambda \neq \mu$ so $\lambda \overline{\mu} \neq 1$. Thus $\langle v, w \rangle = 0$.

B.1 Exercises 2.4

2(a) Let $\{x_n\} \subseteq C$ converge to a point $y \in E$. We need only show that $y \in C$ (Theorem 1). Now given $\varepsilon > 0$ we can choose N so that $\|x_n - y\| < \varepsilon/2$ whenever $n \geq N$. Thus for $m, n \geq N$ $\|x_n - x_m\| \leq \|x_n - y\| + \|y - x_m\| < \varepsilon$ showing that $\{x_n\}$ is a Cauchy sequence. By our assumption on C we see $y \in C$.

(b) Let G be a finite dimensional, linear manifold in E. Then $(G, \|\cdot\|)$ where we mean that we have given G the norm from E is a finite dimensional normed space and hence it is complete (Theorem 2.2, Corollary 2). Thus every Cauchy sequence of points of G converges to a point of G. By (a) the set G is closed in $(E, \|\cdot\|)$.

(c) Let $T : E \to F$ be a bounded, linear map from $(E, \|\cdot\|)$ to $(F, \|\cdot\|_F)$. Let $G = \{x \in E | Tx = 0 \in F\}$. If $\{x_n\} \subseteq G$ and converges to $y \in E$ then $\lim Tx_n = Ty$ by Section 2.1, Lemma 1(e). But $Tx_n = 0$ for all n and so Ty is the zero vector in F. By definition of G this says $y \in G$. Thus G is closed by Theorem 1.

(d) Let $\{x_n\} \subseteq C$ be a Cauchy sequence. Then $\{x_n\}$ must converge to some point $y \in B$ because B is a Banach space. But C is a closed set so $y \in C$ by Theorem 1.

3 We have: (i) For any $\varepsilon > 0$ there is a unit vector x_ε such that $\|Tx_\varepsilon - \lambda x_\varepsilon\| < \varepsilon$; (ii) For any $\varepsilon > 0$ there is a non-zero vector x_ε such that $\|Tx_\varepsilon - \lambda x_\varepsilon\| < \varepsilon \|x_\varepsilon\|$; (iii) There is a sequence $\{x_n\}$ of unit vectors such that $\lim(Tx_n - \lambda x_n) = 0$. Clearly (i) $\Rightarrow$ (ii) because a unit vector x_ε is a non-zero vector and we have $\|Tx_\varepsilon - \lambda x_\varepsilon\| < \varepsilon \|x_\varepsilon\|$ yields $\|T\left(\frac{x_\varepsilon}{\|x_\varepsilon\|}\right) -$

$\lambda \frac{x_\varepsilon}{\|x_\varepsilon\|} \| < \varepsilon$ and $\frac{x_\varepsilon}{\|x_\varepsilon\|}$ is a unit vector. Now from (i) we can get (iii) by taking $\varepsilon = \frac{1}{n}$, $n = 1, 2, \ldots$. In this way we get a sequence $\{x_n\}$ of unit vectors (we have changed notation slightly) such that $\|Tx_n - \lambda x_n\| < \frac{1}{n}$ hence $\lim \|Tx_n - \lambda x_n\| = 0$. Finally, assume (iii). Then we have a sequence of unit vectors $x_1, x_2, \ldots$ such that $\lim(Tx_n - \lambda x_n) = 0$. Thus given $\varepsilon > 0$ there is a smallest integer n such that $\|Tx_n - \lambda x_n\| < \varepsilon$, so we set $x_\varepsilon = x_n$; of course many different ε's will correspond to the same x_n but no one said that the x_ε's have to be different. Thus (iii) $\Rightarrow$ (i).

4 Let λ be a generalized eigenvalue of the bounded, linear operator A. Then we have a sequence $\{x_n\}$ of unit vectors such that $(A - \lambda I)x_n \to 0$ as $n \to \infty$. Suppose that $\lambda \notin \sigma(A)$ so that $(A - \lambda I)$ has a bounded (hence continuous, of course) linear inverse $(A - \lambda I)^{-1}$. Then setting $y = (A - \lambda I)x_n$, $\lim y_n = 0$ and so $\lim(A - \lambda I)^{-1} y_n = 0$ by Section 2.1, Lemma 1(e). However, $(A - \lambda I)^{-1} y_n = (A - \lambda I)^{-1}[(A - \lambda I)x_n] = x_n$ and so we are saying that $\lim x_n = 0$. But this is impossible because then $\lim \|x_n\| = 0$ (Exercises 1.3, problem 1) and $\|x_n\| = 1$ for all n. Thus $(A - \lambda I)$ cannot have a bounded, linear inverse which says $\lambda \in \sigma(A)$.

5 We have $\sigma(A_r) = \{\lambda \in \mathbb{C} \mid |\lambda| \le 1\}$. We claim that the set of generalized eigenvalues of the operator coincides with $\{\lambda \in \mathbb{C} \mid |\lambda| = 1\}$.
Suppose first that $\lambda \in \mathbb{C}$ and that we have $|\lambda| = 1$. Let us set $x_n = \frac{1}{n}(1, \frac{1}{\lambda}, \frac{1}{\lambda^2}, \ldots, \frac{1}{\lambda^{n-1}}, 0, 0, \ldots)$ for $n = 1, 2, \ldots$. Observe that $\|x_n\| = 1$ for each n. Also

$$A_r x_n - \lambda x_n = \frac{1}{n}(0, 1, \frac{1}{\lambda}, \frac{1}{\lambda^2}, \ldots, \frac{1}{\lambda^{n-1}}, 0, \ldots) - \frac{1}{n}(\lambda, 1, \frac{1}{\lambda}, \ldots, \frac{1}{\lambda^{n-2}}, 0, \ldots)$$

$$= \frac{1}{n}(-\lambda, 0, \ldots, 0, \frac{1}{\lambda^{n-1}}, 0, \ldots)$$

Hence

$$\|A_r x_n - \lambda x_n\|^2 = \frac{1}{n^2}\left(|\lambda|^2 + \frac{1}{|\lambda^{n-1}|^2}\right) = \frac{2}{n^2} \to 0$$

as $n \to \infty$. Thus every point on the unit circle is a generalized eigenvalue of our operator.
Next suppose that $\lambda \in \sigma(A_r)$, so $|\lambda| \le 1$, and that λ is known to be a generalized eigenvalue of this operator. We shall show that $|\lambda| = 1$ thus proving that every point within the unit circle is *not* a generalized eigenvalue of A_r.
By assumption we have a sequence $\{x_n\}$ of unit vectors such that $\lim \|A_r x_n - \lambda x_n\| = 0$. Since $|\langle A_r x_n, x_n \rangle| \le \|A_r x_n\| \|x\|$ by the C.S.B. inequality, and this is ≤ 1 for all n, we see that $\{\langle A_r x_n, x_n \rangle\}_{n=1}^\infty$ s a bounded sequence of complex numbers. Any such sequence has a convergent subsequence and so, by passing to a subsequence if necessary, we have both $\lim \langle A_r x_n, x_n \rangle = \alpha$, say, and $\lim \|A_r x_n - \lambda x_n\| = 0$. So $0 = \lim \|A_r x_n - \lambda x_n\|^2 = \lim \langle A_r x_n - \lambda x_n, A_r x_n - \lambda x_n \rangle = \lim \{\langle A_r x_n, A_r x_n \rangle -$

$\lambda \langle x_n, A_r x_n \rangle - \overline{\lambda} \langle A_r x_n, x_n \rangle - |\lambda|^2 \langle x_n, x_n \rangle \}$. Now $\langle x_n, x_n \rangle = 1$ by the definition of A_r so we have

$$0 = \lim\{1 + |\lambda|^2 - \lambda \overline{\langle A_r x_n, x_n \rangle} - \overline{\lambda} \langle A_r x_n, x_n \rangle\}$$

or

$$\lambda \overline{\alpha} + \overline{\lambda} \alpha = 1 + |\lambda|^2$$

Let us write $\lambda = \rho e^{i\mathcal{O}}$, $|\lambda| = \rho$ and note that the C.S.B. inequality tells us that $|\alpha| \leq 1$, $|\overline{\alpha}| \leq 1$. Then $\rho(e^{i\mathcal{O}}\overline{\alpha} + e^{-i\mathcal{O}}\alpha) = 1 + \rho^2$ or

$$|\frac{1+\rho^2}{\rho}| \leq |e^{i\mathcal{O}}\overline{\alpha} + e^{-i\mathcal{O}}\alpha| \leq |\overline{\alpha}| + |\alpha| \leq 2$$

$1 + \rho^2 \leq 2\rho$, $\rho^2 - 2\rho + 1 \leq 0$, $(\rho - 1)^2 \leq 0$ giving us $\rho = 1$ since a square cannot be negative.

Exercises 3.1

2 E is a normed space. T is a bounded, linear operator on E and S is a compact operator on E.

(a) Let $\{x_n\}$ be a bounded sequence in E. Then $\{Tx_n\}$ is also bounded because $\|Tx_n\| \leq \|T\|\|x_n\|$. Thus $\{S \circ T(x_n)\} = \{S[T(x_n)]\}$ has a convergent subsequence because $\{T(x_n)\}$ is bounded and S is compact.
Next consider $\{T \circ S(x_n)\}$. Since $\{S(x_n)\}$ has a convergent subsequence we may assume that $\lim S(x_n) = y$; we can always use the convergent subsequence and change our notation. But then $\lim T[S(x_n)] = T(y)$ by Lemma 1(e) of Section 2.1. Thus $T \circ S$ is a compact operator.

(b) $(I-S)^n = I^n - nI^{n-1}S + \frac{n(n-1)}{2!}I^{n-1}S^2 - \ldots \pm S^n$. The operator $I^n = I$ and is not compact (unless, in fact iff, the space is finite dimensional) but all other summands are compact operators by (a). Thus $(I-S)^n = I - S_n$ where S_n is the sum of all the other terms.

3(b) Let C be a compact set and let $\{x_n\} \subseteq C$ be any sequence. Assume that $\{x_n\}$ is a Cauchy sequence. Since C is compact some subsequence of $\{x_n\}$ converges to a point $y \in C$. But $\{x_n\}$ is a Cauchy sequence and so by Exercises 1.3, problem 3, $\{x_n\}$ must converge to y.

4 S a non-empty subset of a normed space $(E, \|\cdot\|)$

(a) By definition $y \in E$ is an adherent point of S iff every neighborhood of y contains infinitely many points of S. So for each fixed n the set $U_n = \{x \in E \mid \|x - y\| < \frac{1}{n}\}$ is a neighborhood of y and so contains infinitely many points of S. Choose $s_1 \in U_1$, $s_2 \in U_2$, $s_2 \neq s_1$, $s_3 \in U_3$, $s_3 \neq s_1$ or s_2, etc. Then $\{s_n\} \subseteq S$, $\|s_n - y\| < \frac{1}{n}$ so $\{s_n\}$ converges to y and, by construction, these points are distinct.

(b) If $y \in (S')'$ then for any $\varepsilon > 0$ the set $\{x \in E | \|x - y\| < \varepsilon\}$ contains infinitely many points of S'. Choose s_0 in here. Then $s_0 \in S'$ so $\{x \in E | \|x - s_0\| < \varepsilon\}$ contains infinitely many points of S. For any such point, say s_1, $\|s_1 - y\| \leq \|s_1 - s_0\| + \|s_0 - y\| < 2\varepsilon$ showing that the neighborhood $\{x \in E | \|x - y\| < 2\varepsilon\}$ of y contains infinitely many points of S. Since $\varepsilon > 0$ was arbitrary we have shown that $y \in S'$. Thus $(S')' \subseteq S'$. Finally, the closure of S' is $S' \cup (S')'$ and as we have just seen this is S'. So S' is a closed set by (c) below.

(c) Perhaps the easiest way to do this is to show that any sequence of points of $\overline{S}$ which converges has its limit in $\overline{S}$. So let $\{t_n\} \subseteq \overline{S}$ and let it converge to t_0. If $\{t_n\} \subseteq S$ then $t_0 \in S' \subset S' \cup S = \overline{S}$, so we may assume $\{t_n\} \subseteq S' \setminus S$. For each n we can choose $s_n \in S$ such that $\|t_n - s_n\| < \frac{1}{n}$. Then $\|t_0 - s_n\| \leq \|t_0 - t_n\| + \|t_n - s_n\| \to 0$ showing that $\{s_n\} \subseteq S$ converges to t_0. Thus $t_0 \in S' \subseteq \overline{S}$ as was to be shown.

Next suppose that C is a closed set and that $S \subseteq C$. Then any $t \in S'$ is in C because $t = \lim s_n$ where $\{s_n\} \subseteq S$ by (a) and $\{s_n\} \subseteq S \subseteq C$, C closed, implies $\lim s_n \in C$ by Section 2.4, Theorem 1. So if $S \subseteq C$ then $S' \subseteq C \Rightarrow S \cup S' \subseteq C \Rightarrow \overline{S} \subseteq C$.

Finally, $\overline{S}$ is a closed set as we just saw. So $\cap \{C \subseteq E | C \text{ closed}, S \subseteq C\} \supseteq \overline{S}$, because every such $C \supseteq \overline{S}$, but these are equal because $\overline{S}$ is in our family of sets.

(d) As discussed in the hint to this problem we take $x, y \in \overline{H}$ and $\lambda \in \mathbb{C}$ (or $\mathbb{R}$) and show $x + y \in \overline{H}$, $\lambda x \in \overline{H}$. Since $x, y \in \overline{H}$ there are sequences $\{x_n\}$, $\{y_n\}$ in H such that $x = \lim x_n$, $y = \lim y_n$ by (a); note if $x, y \in H$ there is nothing to do so we may as well suppose they are in $H' \setminus H$. But then $\{x_n + y_n\} \subseteq H$ since this is a linear manifold and clearly $\lim(x_n + y_n) = x + y$ showing that $x + y \in \overline{H}$. Similarly, $\lambda x \in \overline{H}$ because $\{\lambda x_n\} \subseteq H$ and $\lambda x = \lim \lambda x_n$.

(e) By definition for any $x \in E$ and any $\varepsilon > 0$ there is an $s \in S$ such that $\|x - s\| < \varepsilon$. This says that, if $x \notin S$, every neighborhood of x contains a point of S different from (it can't help but be different since $x \notin S$ and the point is in S) x. But that just means that every neighborhood of x contains infinitely many points of S; $s \in \{y | \|x - y\| < \varepsilon\} \cap S$, $\|s - x\| = \varepsilon'$, find $s' \in \{y | \|x - y\| < \varepsilon'\} \cap S$ and note s' is in the first neighborhood and $s' \neq s$. Thus given $x \in E$ either $x \in S$ or $x \in S'$; i.e., $E = S \cup S' = \overline{S}$.

5 Let $T : E \to E$ be a bounded, linear operator on E whose range is a finite dimensional sub-manifold of E. We say T has finite rank. Let $E_1 = \text{range } T$, E_1 inherits a norm from E and $(E_1, \|\cdot\|)$ is a finite dimensional normed space. By Section 2.2, Theorem 2, and its corollaries: $(E_1, \|\cdot\|)$ is a Banach space (i.e., it is complete) and it is topologically isomorphic to $\mathbb{R}^n$ (or $\mathbb{C}^n$) where $n = \dim E_1$. But now if $\mathcal{B}_1$ denotes the unit ball in E then $T(\mathcal{B}_1)$ is a bounded subset of E_1 and hence it is relatively compact by Theorem

1. We conclude that T is a compact operator by Corollary 1 to Theorem 2.

6(a) $S : E \to F$, $\{x_n\} \subseteq E$ bounded $\Rightarrow \{Sx_n\}$ has convergent subsequence. Such an S is called a compact mapping. Claim: These are equivalent: (i) S is a compact mapping; (ii) S maps the unit ball of E onto a relatively compact subset of F; (iii) S maps any bounded subset of E onto a relatively compact subset of F.
Proof: (i) $\Rightarrow$ (ii) Let $\mathcal{B}_1$ be the unit ball of E. Choose $\{y_n\}$ any sequence in $S(\mathcal{B}_1)$ and find $\{x_n\} \subseteq \mathcal{B}_1$ such that $Sx_n = y_n$ for all n. Then $\{x_n\}$ is bounded so $\{Sx_n\}$, i.e. $\{y_n\}$, has a convergent subsequence. Thus $S(\mathcal{B}_1)$ is relatively compact in F by Theorem 2.
(ii) $\Rightarrow$ (iii). Given any bounded subset $\mathcal{S}$ of E, $\mathcal{S} \subseteq \ell \mathcal{B}_1$ for some constant scalar ℓ. Thus $S(\mathcal{S}) \subseteq \ell S(\mathcal{B}_1)$ showing that $S(\mathcal{S})$ is relatively compact.
(iii) $\Rightarrow$ (i) Given any bounded sequence $\{x_n\} \subseteq E$, this is a bounded set hence, by (iii), $S(\{x_n\}) = \{Sx_n\}$ is relatively compact. Thus $\{S(x_n)\}$ has a convergent subsequence.

(b) Let S, T be compact, linear mappings from E into F. Given any bounded sequence $\{x_n\}$ in E the sequence $\{Tx_n\} \subseteq F$ has a convergent subsequence. Thus $\{x_n\}$ has a subsequence, $\{y_n\}$ say for convenience, such that $\{T(y_n)\}$ is convergent. Now $\{y_n\}$ is bounded so $\{S(y_n)\}$ has a convergent subsequence. Thus there is a subsequence $\{z_n\}$ of $\{y_n\}$ such that $\{S(z_n)\}$ converges. However, $\{z_n\}$ is a subsequence of $\{y_n\}$ which is a subsequence of $\{x_n\}$ so $\{z_n\}$ is a subsequence of $\{x_n\}$. Also, $\{T(z_n)\}$ is a subsequence of the convergent sequence $\{T(y_n)\}$, hence it, too, converges. Thus both $\{S(z_n)\}$ and $\{T(z_n)\}$ converge, and so $\{(S+T)(z_n)\}$ converges. Thus $\{(S+T)(x_n)\}$ has a convergent subsequence showing that $S+T$ is a compact map. Obviously, λT is a compact map if T is such a map.

Exercises 3.2

1(c) For each n find a $\frac{1}{n}$-net $\mathcal{S}_n$ in S. Then each $\mathcal{S}_n$ is a finite set and so $S_0 = \cup_{n=1}^{\infty} \mathcal{S}_n$ is a countable set. Clearly $\overline{S_0}$ contains S because: Given $s \in S$ we must have, for each n, some $t_n \in \mathcal{S}_n$ such that $\|s - t_n\| < \frac{1}{n}$. Thus $\{t_n\} \subseteq S_0$ and $\lim t_n = s$. Since $S_0 \subseteq S$, $\overline{S_0} \subseteq \overline{S}$ and we already have $\overline{S} \subseteq \overline{S_0}$ so $\overline{S} = \overline{S_0}$.

(d) Let $\mathcal{B}_1$ denote the unit ball in E. Then $E = \cup_{k=1}^{\infty} k \mathcal{B}_1$ and so $R(T) = \cup_{k=1}^{\infty} k T(\mathcal{B}_1)$. Now T is a compact mapping and so $T(\mathcal{B}_1)$ is precompact (it is, of course, actually relatively compact). By (a) so also is $kT(\mathcal{B}_1)$ for every k. By (c) every one of these sets contains a countable set, say S_k, such that $kT(\mathcal{B}_1) \subseteq \overline{S_k}$. Now $\cup_{k=1}^{\infty} S_k \equiv S_0$ is countable and it is clearly dense in $R(T)$.

2 Let $\mathcal{B}_1$ be the ball in ℓ^2. The set $\{\vec{e}\}_{n=1}^\infty \subseteq \mathcal{B}_1$ and $\|\vec{e}_k - \vec{e}_m\| = \sqrt{2}$ if $k \neq m$. Choose $\varepsilon < \frac{\sqrt{2}}{2}$. Suppose that we had an ε-net $\vec{z}_1, \ldots, \vec{z}_n$ in $\mathcal{B}_1$; remember n must be finite. Then for $k \neq m$ we must have $\vec{z}_i$ and $\vec{z}_j$ in our net such that $\|\vec{e}_k - \vec{z}_i\|_2 < \varepsilon$, $\|\vec{e}_m - \vec{z}_j\|_2 < \varepsilon$. But there are infinitely many $\vec{e}_n$'s and only finitely many $\vec{z}_j$'s so for some $k \neq m$ we must have the same $\vec{z}_j$ such that $\|\vec{e}_k - \vec{z}_j\| < \varepsilon$ and $\|\vec{e}_m - \vec{z}_j\| < \varepsilon$. But then $\|\vec{e}_k - \vec{e}_m\| \leq \|\vec{e}_k - \vec{z}_j\| + \|\vec{z}_j - \vec{e}_m\| < 2\varepsilon < \sqrt{2}$ which contradicts the fact that $\|\vec{e}_k - \vec{e}_m\| = \sqrt{2}$.

3 $A(\{x_n\}) = \{\frac{x_n}{n}\}$; $B(\{x_n\}) = \{\frac{x_{n-1}}{n}\}$, $x_0 \equiv 0$.

(a) Since $\{x_n\} \in \ell^2$, $\sum |x_n|^2 < \infty$ and so $\sum \frac{|x_n|^2}{n^2} < \infty$. Similarly, $\sum \frac{|x_{n-1}|^2}{n^2} < \infty$ hence both A and B map ℓ^2 to ℓ^2. They are obviously linear maps. Also $\|A(\{x_n\})\|^2 = \|\{\frac{x_n}{n}\}\|^2 = \sum_{n=1}^\infty \frac{|x_n|^2}{n^2} \leq \sum |x_n|^2 = \|\{x_n\}\|^2$ hence $\|A\| \leq 1$ and, similarly, $\|B\| \leq 1$.

(b) For each m let $A_m(\{x_n\}) = (\frac{x_1}{1}, \frac{x_2}{2}, \ldots, \frac{x_m}{m}, 0, 0, 0 \ldots)$. Then A_m has finite rank. Also

$$\|A - A_m\|^2 = \sup_{\|\vec{x}\| \leq 1} \|(A - A_m)(\vec{x})\|^2 = \sum_{n \geq m} \frac{|x|^2}{n^2} \to 0 \text{ as } m \to \infty$$

showing that A is a compact operator.

4(a) Define $\overline{f}$ as follows: For $s \in S$, $\overline{f}(s) = f(s)$, for $s \in \overline{S}$ but not S, $s = \lim s_j$, where $\{s_j\} \subseteq S$, $\overline{f}(s) = \lim f(s_j)$. We must show: (i) $\{s_j\} \subseteq S$, $\{s_j\}$ convergent implies $\{f(s_j)\}$ is convergent; (ii) If $\{s_j\} \subseteq S$, $\{t_j\} \subseteq S$ and $\lim s_j = \lim t_j$ then we must show $\lim f(s_j) = \lim f(t_j)$; (iii) $\overline{f}$ is uniformly continuous on $\overline{S}$; (iv) $\overline{f}$ is unique.
To prove (i) we recall that f is uniformly continuous on S. Given $\varepsilon > 0$ there is an integer N such that $\|s_j - s_k\| < \varepsilon$ whenever $j, k \geq N$ because any convergent sequence is a Cauchy sequence. Thus given $\varepsilon > 0$ we first choose δ so that $s, t \in S$ and $\|s - t\| < \delta \Rightarrow \|f(s) = f(t)\| < \varepsilon$ (here is where we use the uniform continuity). Next we choose N so that $\|s_j - s_k\| < \delta$ for any $j, k \geq N$. Then $\|f(s_j) - f(s_k)\| < \varepsilon$ whenever $j, k \geq N$ showing $\{f(s_j)\}$ is a Cauchy sequence. So we have shown: (*) If f is uniformly continuous on S and $\{s_j\} \subseteq S$ is a Cauchy sequence, then $\{f(s_j)\}$ is a Cauchy sequence.
(i) follows from this since $(B, \|\cdot\|)$ is complete.
Part (ii) is a little bit subtle. If $\lim s_j = s = \lim t_j$ we cannot say $\lim f(s_j) = f(s) = \lim f(t_j)$ because $f(s)$ is not defined for $s \in \overline{S} \setminus S$. What we can do is this: $s_1, t_1, s_2, t_2, \ldots, s_n, t_n, \ldots$ is a Cauchy sequence because: Given $\varepsilon > 0$ there is an N such that $\|s_m - s\| < \varepsilon$, $\|t_n - s\| < \varepsilon$ for $m, n \geq N$; so $\|s_m - t_n\| \leq \|s_m - s\| + \|s - t_n\| < 2\varepsilon$ for $m, n \geq N$. By (*) above, $f(s_1), f(t_1), \ldots, f(s_n), f(t_n), \ldots$ is a Cauchy sequence. Thus given $\varepsilon > 0$ there is an N such that, in particular, $\|f(s_n) - f(t_n)\| < \varepsilon$ when $n \geq N$. Hence $\lim f(s_n) = \lim f(t_n)$.

By (i) and (ii) $\overline{f}$ is a well-defined function on $\overline{S}$. Suppose $\varepsilon > 0$ is given. We first choose $\delta > 0$ so that $s, t \in S$ and $\|s - t\| < \delta$ implies $\|f(s) - f(t)\| = \|\overline{f}(s) - \overline{f}(t)\| < \frac{\varepsilon}{3}$. Suppose $u, v \in \overline{S}$ are given with $\|u - v\| < \frac{\varepsilon}{3}$. The only hard case is that in which $u, v \in \overline{S} \setminus S$. Choose $s, t \in S$ then, so that $\|u - s\| < \frac{\delta}{3}$ and $\|\overline{f}(u) - f(s)\| < \frac{\varepsilon}{3}$, $\|v - t\| < \frac{\delta}{3}$ and $\|\overline{f}(s) - f(t)\| < \frac{\varepsilon}{3}$. Then

$$\|\overline{f}(u) - \overline{f}(v)\| \le \|\overline{f}(u) - f(s)\| + \|f(s) - f(t)\| + \|f(t) - \overline{f}(v)\|$$

and the first and last terms on the right-hand side are both $< \frac{\varepsilon}{3}$. Also $\|s - t\| \le \|s - u\| + \|u - v\| + \|v - t\| < \delta$ hence by the choice of δ, $\|f(s) - f(t)\| < \frac{\varepsilon}{3}$ and we are done; i.e., we have shown that $\overline{f}$ is uniformly continuous on $\overline{S}$.

Finally, let us show that $\overline{f}$ is unique. Suppose that there is another function $\overline{\overline{f}}$ on $\overline{S}$ which is uniformly continuous on $\overline{S}$ and satisfies $\overline{\overline{f}}|S = f$. Then $\overline{\overline{f}}(s) = \overline{f}(s) = f(s)$ for all $s \in S$. Take any $s \in \overline{S} \setminus S$ and any sequence $\{s_j\} \subseteq S$ which converges to s. Then $\{\overline{f}(s_j)\}$ and $\{\overline{f}(s_j)\}$ are both Cauchy sequences by (*) above. But $\lim \overline{f}(s_j) = \overline{f}(s)$ and $\lim \overline{\overline{f}}(s_j) = \overline{\overline{f}}(s)$ by the continuity of these functions. However, $\overline{f}(s_j) = \overline{\overline{f}}(s_j)$ so these limits must be the same.

(b) The function $f : (0, 1) \to \mathbb{R}$ defined by $f(x) = \frac{1}{x}$ is certainly continuous. It is not uniformly continuous because, if it were, it would satisfy (*) proved in (a) and clearly $\{\frac{1}{n}\}$ is a Cauchy sequence in $(0, 1)$ while $\{f(\frac{1}{n}) = n\}$ is not a Cauchy sequence. It is also clear that f cannot be extended to a continous function on $(0,1)$ since any such function would be infinite at $x = 0$; i.e., $\overline{f} : [0, 1] \to \mathbb{R}$ continuous, then $\overline{f}(0) = \lim f(x_n)$ and $x_n \to 0$ but $f(x_n) = \frac{1}{x_n}$ and this tends to ∞ if $x_n \to 0$.

Exercises 3.3

1 $A(\{x_n\}) = \{\frac{x_n}{x}\}$ is a compact operator on ℓ^2.
(a) Recall $\vec{e}_n = (0, 0, \ldots, 1, 0, 0, \ldots)$ where the 1 is in the nth place. Clearly $A\vec{e}_n = \frac{1}{n}\vec{e}_n$ showing that for every positive integer n, $\frac{1}{n}$ is an eigenvalue of A. Thus $\{\frac{1}{n}|n = 1, 2, \ldots\} \subseteq \sigma(A)$ and since $\sigma(A)$ is closed and $\lim \frac{1}{n} = 0$, zero is in $\sigma(A)$ also. Clearly zero is not an eigenvalue.
Suppose λ is a non-zero eigenvalue of A. Then there must be a non-zero vector $\vec{z} \in \ell^2$ such that $A\vec{z} = \lambda\vec{z}$. But then $\vec{z} = \{z_n\}$ and for some k, $z_k \ne 0$. Hence $A\vec{z} = A(\{z_n\}) = \{\frac{z_n}{n}\} = \{\lambda z_n\}$ gives $\frac{z_k}{k} = \lambda z_k$ or $\lambda = \frac{1}{k}$. Thus $\sigma(A) = \{0\} 0 \{\frac{1}{n}|n = 1, 2, \ldots\}$.
(b) Choose and fix $n_0 \ne 0$, an integer, and consider the null space of $(A - \frac{1}{n_0}I)$. This is $\{\vec{z} \in \ell^2 | A\vec{z} = \frac{1}{n_0}\vec{z}\}$. Thus $A\vec{z} = A(\{z_n\}) = \{\frac{z_n}{n}\} = \frac{1}{n_0}\vec{z} = \{\frac{z_n}{n_0}\}$

gives us $\frac{z_n}{n} = \frac{z_n}{n_0} \Rightarrow z_n = 0$ for $n \neq n_0$. So our null space consists of $\{\vec{z} = \{z_n\} \in \ell^2 | z_n = 0 \text{ for all } n \neq n_0\} = \{\lambda \vec{e}_{n_0} | \lambda \in \mathbb{C}\}$, this is a one-dimensional subspace of ℓ^2.
$R(s - \frac{1}{n_0}I) = \{\vec{w} \in \ell^2 | \vec{w} = (A - \frac{1}{n_0}I)\vec{z} \text{ for some } \vec{z}\}$. $(A - \frac{1}{n_0}I)\vec{z} = A(\vec{z}) - \frac{1}{n_0}\vec{z} = \{\frac{z_n}{n}\} - \{\frac{z_n}{n_0}\} = \{(\frac{1}{n} - \frac{1}{n_0})z_n\}$. Clearly this vector has a zero in the n_0th place. Given any $\vec{w} = \{w_n\}$ with $w_{n_0} = 0$, set $w_n = (\frac{1}{n_0} - \frac{1}{n_0})z_n = \frac{n_0 - n}{nn_0}z_n$ or $z_n = (\frac{nn_0}{n_0 - n})w_n$; remember $\vec{w}$ is known so every w_n is known and we are finding z_n, $n \neq n_0$. The sequence $\frac{n_0}{n_0 - 1}w_1$, $\frac{2n_0}{n_0 - 2}w_2, \frac{3n_0}{n_0 - 3}w_3, \ldots, 0, \ldots$ the zero is in the n_0th position will be mapped by $A - \frac{1}{n_0}I$ onto the given $\vec{w}$, however we must be able to prove that this sequence is in ℓ^2. Now $\vec{w} \in \ell^2$, it was given to be here, so $\sum |w_n|^2 < \infty$. Consider $\sum |\frac{nn_0}{n_0 - n}|^2 |w_n|^2$ since $\lim_{n \to \infty} \frac{nn_0}{n_0 - n} = \lim_{n \to \infty} \frac{n_0}{\frac{n_0}{n} - 1} = -n_0$ we see that $|\frac{nn_0}{n_0 - n}|^2$ is bounded and so our sequence is in ℓ^2. Thus $R(A - \frac{1}{n_0}I) = \{\vec{w} = \{w_n\} \in \ell^2 | w_{n_0} = 0\}$.
Finally, given $\vec{z} \in R(A - \frac{1}{n_0}I) \cap N(A - \frac{1}{n_0}I)$ we have $z_{n_0} = 0$ since $\vec{z}$ is in the range and all other terms of $\vec{z}$ zero since it is in the null space. For any $\vec{z} \in \ell^2$, $\vec{z} = \{z_n\} = (0, \ldots, 0, z_{n_0}, 0, 0, \ldots) + (z_1, z_2, \ldots, z_{n_0}, 0, z_{n_0+1}, \ldots)$ and the first of these is in the null space of $A - \frac{A}{n_0}I$ while the second is in the range of this operator.

(c) $B(\{x_n\}) = \{\frac{x_{n-1}}{n}\}$, $x_0 \equiv 0$ is a compact operator on ℓ^2. Since ℓ^2 is infinite dimensional $0 \in \sigma(B)$. Suppose now that $\lambda \in \sigma(B)$, $\lambda \neq 0$. Then λ must be an eigenvalue of B and so $B(\{x_n\}) = \lambda\{x_n\}$ for some non-zero vector $\{x_n\} \in \ell^2$. Then $\{\frac{x_{n-1}}{n}\} = \lambda\{x_n\}$, $n = 1, 2, \ldots$. This gives us the equations:

$$n = 1, \quad 0 = \lambda x_1,$$
$$n = 2, \quad \frac{x_1}{2} = \lambda x_2$$
$$n = 3, \quad \frac{x_2}{3} = \lambda x_3$$

Clearly, since $\lambda \neq 0$, $x_1 = 0$ from the first equation. Then the second equation becomes $0 = \lambda x_2$ giving us $x_2 = 0$ and so on. It follows that $\sigma(B) = \{0\}$. Here is a compact operator on a Hilbert space which has no eigenvalues; 0 is not an eigenvalue since $B(\{x_n\}) = 0\{x_n\}$ gives $B(\{x_n\}) = 0$ and this is impossible for $\{x_n\} \neq \vec{0}$.

5(c) $\lambda \neq 0$ is not an eigenvalue of T and so $\lambda \notin \sigma(T)$ because T is compact. Thus $T - \lambda I$ is a bounded, linear operator which has a bounded, linear inverse. Thus $N_0 = \{0\}$ and $R_0 = B$ and $p = q = 0$ in this case.

(a) $N_n = \{x | (T - \lambda I)^n x = 0\}$ so for $x \in N_n$, $(T - \lambda I)^{n+1} x = (T - \lambda I)[(T - \lambda I)^n x] = (T - \lambda I)(0) = 0$. So $N_n \subseteq N_{n+1}$.

(b) $R_{n+1} = \{y \in B\}|(T - \lambda I)^{n+1}x = y$ for some $x \in B\}$. Thus $y = (T - \lambda I)^{n+1}x = (T - \lambda I)^n[(T - \lambda I)x] = (T - \lambda I)^n z$ where $z = (T - \lambda I)x$. Thus $y \in R_n$ showing $R_{n+1} \subseteq R_n$.

Exercises 4.1

1 $(I - P)^2 = (I - P)(I - P) = I(I - P) - P(I - P) = I^2 - IP - PI + P^2 = I - P - P + P^2 = I - P - P + P = I - P$ showing $(I - P)^2 = (I - P)$. Now $y \in \text{Ker}(I - P)$ iff $(I - P)y = 0$ iff $y = Py$ and this implies $y \in R(P)$. So $\text{Ker}(I - P) \subseteq R(P)$. If $z \in R(P)$ then $z = Px$ for some x and so $Pz = P^2 x = Px = z$ shows $z \in \text{Ker}(I - P)$. Next let $x \in \text{Ker}P$. Then $Px = 0$ so $x = Ix - Px = (I - P)x$ shows $x \in R(I - P)$. If $x \in R(I - P)$ then $x = (I - P)y$ for some $y \Rightarrow x = y - Py \Rightarrow Px = Py - P^2 y = Py - Py = 0 \Rightarrow x \in \text{Ker}P$.

2(a) $\|(x_1, x_2)\|_p = \max(\|x_1\|_1, \|x_2\|_2)$. Clearly $\|\cdot\|_p \geq 0$ for all x because this is true of $\|\cdot\|_1$ and $\|\cdot\|_2$ since these are given to be norms. Also, $\|(x_1, x_2)\|_p = 0$ iff both $\|x_1\|_1 = 0$ and $\|x_2\|_2 = 0$ iff $x_1 = 0$ and $x_2 = 0$ iff $(x_1, x_2) = (0, 0)$ which is the zero vector in $E_1 \times E_2$.
$\|\lambda(x_1, x_2)\|_p = \|(\lambda x_1, \lambda x_2)\|_p = \max(\|\lambda x_1\|_1, \|\lambda x_2\|_2) = \max(|\lambda|\|x_1\|_1, |\lambda|\|x_2\|_2) = |\lambda| \max(\|x_1\|_1, \|x_2\|_2) = |\lambda|\|(x_1, x_2)\|_p$.
Finally, $\|(x_1, x_2) + (y_1, y_2)\|_p = \|(x_1 + y_1, x_2 + y_2)\|_p = \max(\|x_1 + y_1\|_1, \|x_2 + y_2\|_2) \leq \max(\|x_1\|_1 + \|y_1\|_1, \|x_2\|_2 + \|y_2\|_2) \leq \max(\|x_1\|_1, \|x_2\|_2) + \max(\|y_1\|_1, \|y_2\|_2) = \|(x_1, x_2)\|_p + \|(y_1, y_2)\|_p$.

(b) Let $\{(x_n, y_n)\}_{n=1}^\infty$ be a sequence in $E_1 \times E_2$. Claim: $\{(x_n, y_n)\}_{n=1}^\infty$ is a Cauchy sequence for $\|\cdot\|_p$ iff both $\{x_n\}$ and $\{y_n\}$ are Cauchy sequences for $\|\cdot\|_1$ and $\|\cdot\|_2$ respectively.
If $\{x_n\}$ and $\{y_n\}$ are Cauchy sequences and $\varepsilon > 0$ is given then $\|(x_n, y_n) - (x_m, y_m)\|_p = \max(\|x_n - x_m\|_y, \|y_n - y_m\|_2)$ and both $\|x_n - x_m\|_1$ and $\|y_n - y_m\|_2$ are less than ε for n, m large enough. Hence $\{(x_n, y_n)\}$ is Cauchy for $\|\cdot\|_p$.
Next suppose $\{(x_n, y_n)\}$ is Cauchy for $\|\cdot\|_p$. Given $\varepsilon > 0$ choose N so that $\|(x_n, y_n) - (x_m, y_m)\|_p < \varepsilon$ for $m, n \geq N$. Then $\|x_n - x_m\|_1 \leq \max(\|x_n - x_m\|_1, \|y_n - y_m\|_2) = \|(x_n - x_m, y_n - y_m)\|_p < \varepsilon$ for $m, n \geq N$. Thus $\{x_n\}$ is Cauchy for $\|\cdot\|_1$ and, similarly, $\{y_n\}$ is Cauchy for $\|\cdot\|_2$.
If $(E_1, \|\cdot\|_1)$ and $(E_2, \|\cdot\|_2)$ are Banach spaces then so also is $(E_1 \times E_2, \|\cdot\|_p)$ because $\{(x_n, y_n)\}$ Cauchy for $\|\cdot\|_p$ implies $\{x_n\}, \{y_n\}$ are Cauchy for $\|\cdot\|_1$ and $\|\cdot\|_2$ respectively, hence $x_n \to x_0 \in E_1$, $y_n \to y_0 \in E_2$ and clearly $\{(x_n, y_n)\} \to (x_0, y_0)$ for $\|\cdot\|_p$.

(c) $\pi_1 : E_1 \times E_2 \to E_1$, $\pi_1[(x, y)] = x$. The map is linear: $\pi_1[\alpha(x_1, y_1) + \beta(x_2, y_2)] = \pi_1[(\alpha x_1 + \beta x_2, \alpha y_1 + \beta y_2)] = \alpha x_1 + \beta x_2 = \alpha \pi_1[(x_1, y_1)] + \beta \pi_1[(x_2, y_2)]$. It is onto since given $x \in E_1$, $(x, 0) \in E_1 \times E_2$ and $\pi_1[(x, 0)] =$

x. Finally, $\|\pi_1[(x,y)]\|_1 = \|x\|_1 \leq \max(\|x\|_1, \|y\|_2) = \|(x,y)\|_p$ showing π_1 is continuous and $\|\pi_1\| \leq 1$. The proofs of π_2 are entirely analogous to those just given.

Exercises 4.2

1(a) M is a linear manifold. $M^\perp = \{y \in H | \langle x, y \rangle = 0 \text{ for all } x \in M\}$. If $y_1, y_2 \in M^\perp$ and α, β are scalars then $\langle \alpha y_1 + \beta y_2, x \rangle = \alpha \langle y_1, x \rangle + \beta \langle y_2, x \rangle = \alpha \cdot 0 + \beta \cdot 0 = 0$ for all $x \in M$, so $M^\perp$ is a linear manifold.
Next let $\{y_n\} \subseteq M^\perp$ and suppose that $\lim y_n = z$ for the norm. Then $\langle z, x \rangle = \lim \langle y_n, x \rangle = 0$ for all $x \in M$ by Exercises 1.4, problem 1(f), or simply by the C.S.B. inequality. Hence $z \in M^\perp$ showing that $M^\perp$ is a subspace.

(b) Fix $y \in H$ and define $f : M \to \mathbb{C}$ by $f(x) = \langle x, y \rangle$. Given $\varepsilon > 0$ we have $|f(x) - f(x_1)| < \varepsilon$ whenever $\|x_1 - x_2\| < \frac{\varepsilon}{\|y\|}$ because: $|f(x_1) - f(x_2)| = |\langle x_1, y \rangle - \langle x_2, y \rangle| = |\langle x_1 - x_2, y \rangle| \leq \|x_1 - x_2\|\|y\|$. Thus f is uniformly continuous on M. If $f \equiv 0$ (i.e., if $y \in M^\perp$) then f can be extended to all of $\overline{M}$ by setting $\overline{f}(x) = 0$ for all $x \in \overline{M}$. Since $\overline{f}$ when defined in this way is certainly a uniformly continuous extension of f it is the only one. Thus $y \in M^\perp$ implies $y \in (\overline{M})^\perp$ showing $M^\perp \subseteq (\overline{M})^\perp$. However, $M \subseteq \overline{M}$ and so $M^\perp \supseteq (\overline{M})^\perp$. Thus $M^\perp = (\overline{M})^\perp$.

(c) $\langle (A \pm B)x, y \rangle = \langle Ax, y \rangle \pm \langle Bx, y \rangle = \langle x, A^*y \rangle \pm \langle x, B^*y \rangle = \langle x, (A^* \pm B^*)y \rangle$ and $(A \pm B)^* = A^* \pm B^*$.

2 P, Q are orthogonal projections.

(a) $R(P)^\perp = R(I - P)$.
Let $P^\perp$ be the orthogonal projection of H onto $R(P)^\perp$. We need only show that $P^\perp = I - P$. By Corollary 1 to Theorem 1, $y - Py \in M^\perp$ for all $y \in H \setminus M$ (if $y \in M$ then $y - Py = 0 \in M^\perp$ also); here $M \equiv R(P)$. Thus $y = Py + z$ where $z \in M^\perp$. But M and $M^\perp$ are complements and so this representation of y is unique. Thus $P^\perp y = z$ showing $P^\perp y = y - Py = (I - P)y$.

(b) T.A.E. (i) $R(P) \subseteq R(Q)$; (ii) $P = PQ$; (iii) $P = QP$.
Assume (i). Then $Py \in R(P) \subseteq R(Q)$, so $QPy = Py$ for all $y \in H$. This is (iii). Assume (iii) so $P = QP$ then $P^* = (QP)^* = P^*Q^*$ but these are orthogonal projections so they are self-adjoint. Thus $P = PQ$ which is (ii); in fact, (ii) $\Rightarrow$ (iii) also. Finally, assume (ii) hence (iii). For $y \in H$, $Py \in R(P)$ but $Q(Py) = (QP)(y) = Py$ by (iii) shows that $Py \in R(Q)$. Thus $R(P) \subseteq R(Q)$ as was to be shown.

(c) PQ is a projection iff $PQ = QP$ (observe them that PQ is a projection iff QP is a projection). First assume that $PQ = QP$. Then $(PQ)^* =$

$Q^*P^* = QP = PQ$ so PQ is self-adjoint. Also $(PQ)^2 = (PQ)(PQ) = P(QP)Q = P(PQ)Q = P^2Q^2 = PQ$.
Now suppose that PQ is an orthogonal projection. Then $QP = Q^*P^* = (PQ)^* = PQ$ shows $PQ = QP$.

(d) Suppose that $PQ = QP$ so that PQ is an orthogonal projection by (c). Then $R(PQ) = R(P) \cap R(Q)$. First Observe (mini-theorem 1): If P is an orthogonal projections onto a subspace M of H then $M = \{x \in H | Px = x\} = \{x \in H | \|Px\| = \|x\|\} = \{Px | x \in H\}$. Proof. We know that $Py \in M$ for all y and $Px = x$ for all $x \in M$. Thus $M \subseteq \{x | Px = x\}$ and $M = \{Px | x \in H\}$. For $y \notin M$, $y = Py + P^\perp y$ ($P^\perp$ is the projection onto $M^\perp$) and $P^\perp y \neq 0$. Thus $\|y\|^2 = \|Py\|^2 + \|P^\perp y\|^2 > \|Py\|^2$ and this implies that y is not in any of the three sets listed as being equal to M; so they really are equal to M and do not just contain M.

Returning to our problem, we see that we must prove that $R(P) \cap R(Q) = PQ(H)$. Now $PQ = QP$ so $PQ(H) \subseteq P(H) \subseteq R(P)$ and $QP(H) \subseteq Q(H) \subseteq R(Q)$ and so $PQ(H) \subseteq R(P) \cap R(Q)$. If $x \in R(P) \cap R(Q)$ then $x = Qx = PQx$ and so $R(P) \cap R(Q) \subseteq PQ(H)$.

Now we turn to the last part of this problem. Suppose that $PQ = QP$. We shall show that $P + Q - PQ$ is then an orthogonal projection whose range is $R(P) + R(Q)$.
$(P + Q - PQ)^2 = (P + Q - PQ)(P + Q - PQ) = P^2 + QP - PQP + PQ + Q^2 - PQ^2 - P^2Q - QPQ + (PQ)^2$. Now $P^2 = P$, $Q^2 = Q$ and $PQP = QPP = QP^2 = QP$. Thus $= P+Q-PQ-PQ+PQ = P+Q-PQ$ because $QPQ = PQQ = PQ^2 = PQ$ and $(PQ)^2 = PQ$ since we know PQ is a particular (because we are assuming $PQ = QP$).
$(P + Q - PQ)^* = P^* + Q^* - (PQ)^* = P + Q - PQ$ because P, Q and PQ are orthogonal projections and hence self-adjoint. We have shown that $P + Q - PQ$ is an orthogonal projection. Let us now show that its range is $R(P) + R(Q)$. Recall that $R(PQ) = R(P) \cap R(Q)$ by (d). Thus $R(PQ) \subseteq R(P)$ and $R(PQ) \subseteq R(Q)$ also.

Mini-theorem 2. P, Q orthogonal projections then T.A.E. (i) $R(P) \perp R(Q)$; i.e., $\langle x, y \rangle = 0$ all $x \in R(P)$, all $y \in R(Q)$; (ii) $P[R(Q)] = \{0\}$; (iii) $Q[R(P)] = \{0\}$; (iv) $PQ = 0$; (v) $QP = 0$ (in (iv) and (v) 0 denotes the zero operator); (vi) $P + Q$ is an orthogonal projection. In this case $R(P + Q) = R(P) + R(Q)$.
(i) $\Rightarrow$ (ii) is trivial because (i) $\Rightarrow R(Q) \subseteq R(P)^\perp$. (ii) $\Rightarrow$ (iv) is trivial also.
(iv) $\Rightarrow$ (vi) $QP = (PQ)^* = 0$ by (iv) so $(P+Q)^2 = P^2 + PQ + QP + Q^2 = P+Q$ ((iv) gives $PQ = 0$ and we just observed that, by taking the adjoint, $QP = 0$. Also $(P + Q)^* = P^* + Q^* = P + Q$. Thus (iv) holds.
(iv) $\Rightarrow$ (i). For $x \in R(P)$, $\|x\|^2 \geq \|(P+Q)x\|^2 = \langle (P+Q)x, (P+Q)x \rangle = \langle (P+Q)x, x \rangle = \langle Px, x \rangle + \langle Qx, x \rangle = \|x\|^2 + \|Qx\|^2$ and so $Qx = 0$ (we have used the fact that $P + Q$, as an orthogonal projection, must have norm ≤ 1). For all $y \in R(Q)$ $\langle x, y \rangle = \langle x, Qy \rangle = \langle Qx, y \rangle = 0$ giving us (i). The other statements follow by symmetry.

The only thing left to prove is that $R(P+Q) = R(P) + R(Q)$ when $P+Q$ is an orthogonal projection.

The range $R(P+Q)$ is $\{x \in H | (P+Q)x = x\}$ by mini-theorem 1 part (d) of this problem. Clearly $x = (P+Q)x = Px + Qx \in R(P) + R(Q)$. So $R(P+Q) \subseteq R(P) + R(Q)$. Conversely, if $x \in R(P)$ and $y \in R(Q)$ then $Px = x$, $Qy = y$ and $\langle x, y \rangle = 0$ (assuming $P+Q$ is an orthogonal projection, which we are doing here, is equivalent to (i) which says $R(P) \perp R(Q)$). Thus $x + y = P(x+y) + Q(x+y)$, more explicitly, $P(x+y) = Px + Py = x + 0$ and $Q(x+y) = Qx + Qy = 0 + y$. But then $x + y = (P+Q)(x+y)$ shows $x + y \in R(P+Q)$. Hence $R(P) + R(Q) \subseteq R(P+Q)$.

(e) This follows from mini-theorem 2 by induction. See also Exercises 5.2, problem 4.

3 Suppose that A is a self-adjoint operator and that λ is an eigenvalue of A. Then for some non-zero vector x we have $Ax = \lambda x$. Thus $\langle Ax, x \rangle = \langle \lambda x, x \rangle$ but $\langle Ax, x \rangle = \langle x, A^*x \rangle = \langle x, Ax \rangle = \langle x, \lambda x \rangle$. Hence $\langle \lambda x, x \rangle = \langle x, \lambda x \rangle$ or $\lambda \langle x, x \rangle = \overline{\lambda} \langle x, x \rangle$. Now $\langle x, x \rangle \neq 0$ and so $\lambda = \overline{\lambda}$ showing λ is real.

4 On $L^2[a,b]$ we have $P(f) = tf(t)$.

$$\langle P(f), g \rangle = \int_a^b tf(t)\overline{g(t)}dt = \int_a^b f(t)\overline{tg(t)}dt = \langle f, P(g) \rangle$$

showing that P is self-adjoint. We may bring the t under the bar because t is a real-variable hence $\overline{t} = t$.

5 $A(\vec{z}) = A(\{z_n\}) = \{\frac{z_n}{n}\}$ so $\langle A\vec{z}, \vec{w} \rangle = \langle A(\{z_n\}), \{w_n\} \rangle = \sum_{n=1}^\infty \frac{z_n}{n} \overline{w_n} = \sum z_n \overline{\frac{w_n}{n}} = \langle \{z_n\}, \{\frac{w_n}{n}\} \rangle = \langle \vec{z}, A\vec{w} \rangle$ so A is a self-adjoint operator.

6 $B\vec{z} = B(\{z_n\}) = \{\frac{z_{n-1}}{n}\}_{n=1}^\infty$, $z_0 \equiv 0$. Thus $\langle B\vec{z}, \vec{w} \rangle = \langle B(\{z_n\}), \{w_n\} \rangle = \sum_{n=1}^\infty \frac{z_{n-1}}{n} \overline{w_n} = \frac{z_0}{1}\overline{w_1} + \frac{z_1}{2}\overline{w_2} + \frac{z_2}{3}\overline{w_3} + \cdots = \langle \{z_n\}_{n=1}^\infty, \{\frac{w_{n+1}}{n+1}\}_{n=1}^\infty \rangle$. Thus $B^*(\{w_n\}) = \{\frac{w_{n+1}}{n+1}\}_{n=1}^\infty = \{\frac{w_2}{2}, \frac{w_3}{3}, \ldots\}$.

Consider now BB^* and B^*B. Recall $\vec{e}_n = \{1, 0, \ldots\} \in \ell^2$. We have $BB^*(\vec{e}_n) = B(\{0, 0, \ldots\}) = \{0, 0, \ldots\}$ since B is linear, but $B^*B(\vec{e}_n) = B^*(\{0, \frac{1}{2}, 0, 0, \ldots\}) = \{\frac{1}{4}, 0, 0, \ldots\} \neq \vec{0}$. Thus we see that $BB^* \neq B^*B$.

We have just seen that $B^*\vec{e}_1 = \vec{0} = 0 \cdot \vec{e}_1$, $\vec{e}_1 \neq \vec{0}$. Thus zero is an eigenvalue of the operator B^*. It is the only one because $B^*\vec{z} = \lambda \vec{z}$ gives us $\{\frac{z_{n-1}}{2}, \frac{z_3}{3}, \frac{z_4}{4}, \ldots\} = \{\lambda z_1, \lambda z_2, \ldots\}$ and so $\frac{z_2}{2} = \lambda z_1$, $\frac{z_3}{3} = \lambda z_2$, $\frac{z_4}{4} = \lambda z_3$, ...

Hence $z_2 = 2\lambda z_1$, $z_3 = 3\lambda z_2 = 3 \cdot 2\lambda^2 z_1$, $z_4 = 4 \cdot 3 \cdot 2\lambda^3 z_1, \ldots$. Now $\lambda \neq 0$ any if $\vec{z} \neq \vec{0}$ then $z_1 \neq 0$ as our equations here show. But then $z_n = n!\lambda^{n-1} z_1$ and $\sum |z_n|^2 = |z_1|^2 \sum_{n=1}^\infty (n!)^2 |\lambda^{n-1}|^2$ which cannot converge for any fixed $\lambda \neq 0$; because $\lim_{n \to \infty} n! |\lambda|^{2n-2} \neq 0$ or, another way to see it, $\sum_{n=1}^\infty (n!)^2 |\lambda|^{2n-2}$ diverges by the ratio test.

7(b)(i) (i) $K^*(s,t) = \overline{K(t,s)}$ for $a \leq s, t \leq b$. Then $\|K^*\|_2^2 = \iint |K^*(s,t)|^2 ds dt = \iint |\overline{K(t,s)}|^2 ds\, dt = \iint |K(t,s)|^2 ds\, dt = \|K\|_2^2$.

(ii) $\langle T_K f, g \rangle = \int [\int K(s,t) f(t) dt] \overline{g(s)} ds = \int f(t) [\int \overline{K^*(t,s) g(s)} \, ds] dt = \langle f, T_{K^*} g \rangle$.

Exercises 4.3

1(a) $\langle Ax, y \rangle = \langle x, A^* y \rangle = \overline{\langle A^* y, x \rangle} = \overline{\langle y, (A^*)^* x \rangle} = \langle (A^*)^* x, y \rangle$ for all x, y. Since y is arbitrary this says $Ax = (A^*)^* x$ for all x, and this means $A = (A^*)^*$.

(b) $\langle \lambda A x, y \rangle = \lambda \langle Ax, y \rangle = \langle Ax, \overline{\lambda} y \rangle = \langle x, A^* (\overline{\lambda} y) \rangle = \langle x, (\overline{\lambda} A^*) y \rangle$ for all x, y. But $\langle (\lambda A) x, y \rangle = \langle x, (\lambda A)^* y \rangle$ for all x, y and so $\langle x, (\overline{\lambda} A^*) y \rangle = \langle x, (\lambda A)^* y \rangle$ all x, y. Since x is arbitrary $(\overline{\lambda} A^*) y = (\lambda A)^* y$ for all y, and this means $\overline{\lambda} A^* = (\lambda A)^*$.
$\langle ABx, y \rangle = \langle x, (AB)^* y \rangle$. But $\langle ABx, y \rangle = \langle A(Bx), y \rangle = \langle Bx, A^* y \rangle = \langle x, B^* A^* y \rangle$. Thus $\langle x, (AB)^* y \rangle = \langle x, B^* A^* y \rangle$.

(c) $AA^{-1} = A^{-1} A = I$ and so by (b) $(AA^{-1})^* = (A^{-1} A)^* = I^*$ or $(A^{-1})^* A^* = A^* (A^{-1})^* = I$ showing $(A^{-1})^*$ and A^* are inverses of each other. Thus $(A^{-1})^* = (A^*)^{-1}$.

(d) By Exercises 4.2, problem 1(c), $(A - \lambda I)^* = A^* - (\lambda I)^* = A^* - \overline{\lambda} I^* = A^* - \overline{\lambda} I$.

(e) The complex number λ is a regular value for the operator A iff $(A - \lambda I)$ has a bounded, linear inverse. By (c) this will be true iff $(A - \lambda I)^* = A^* - \overline{\lambda} I$ has a bounded, linear inverse. Thus $\lambda \in \sigma(A)$ (i.e., $(A - \lambda I)$ does *not* have a bounded, linear inverse iff $(A - \lambda I)^* = A^* - \overline{\lambda} I$ does *not* have a bounded, linear inverse) iff $\overline{\lambda} \in \sigma(A^*)$. Hence $\sigma(A^*) = \{\overline{\lambda} | \lambda \in \sigma(A)\}$.

2(a)
$$\hat{\varphi}(x+y) = \varphi(x+y, x+y) = \varphi(x,x) + \varphi(y,x) + \varphi(x,y) + \varphi(y,y)$$
$$\hat{\varphi}(x-y) = \varphi(x-y, x-y) = \varphi(x,x) - \varphi(y,x) - \varphi(x,y) + \varphi(y,y)$$

Subtracting $\hat{\varphi}(x+y) - \hat{\varphi}(x-y) = 2\varphi(x,y) + 2\varphi(y,x)$.

(b) Set $y = iy$ in this last expression to get:
$$\hat{\varphi}(x+iy) - \hat{\varphi}(x-iy) = 2\varphi(x, iy) + 2\varphi(iy, x)$$
$$= -2i\,\varphi(x,y) + 2i\,\varphi(y,x)$$

because $\overline{i} = -i$.

(c)
$$\hat{\varphi}(x+y) - \hat{\varphi}(x-y) = 2\varphi(x,y) + 2\varphi(y,x)$$
$$i\hat{\varphi}(x+iy) - i\hat{\varphi}(x-iy) = 2\varphi(x,y) - 2\varphi(y,x)$$

Adding: $\hat{\varphi}(x+y) - \hat{\varphi}(x-y) + i\hat{\varphi}(x+iy) - i\hat{\varphi}(x-iy) = 4\varphi(x,y)$.

(d) First suppose that φ is symmetric so $\varphi(x,y) = \overline{\varphi(y,x)}$. Then $\hat{\varphi}(x) = \varphi(x,x) = \overline{\varphi(x,x)} = \overline{\hat{\varphi}(x)}$ giving $\hat{\varphi}(x)$ real for all x; in the middle step we interchanged x with x and hence had to take the conjugate.

Now suppose that $\hat{\varphi}(x)$ is real for all x. Now we have $\hat{\varphi}(x) = \varphi(x,x) = \varphi(-x,-x) = \hat{\varphi}(-x) = \overline{\varphi(x,x)} = \varphi(ix,ix) = \hat{\varphi}(ix)$ directly from the definition of bilinear form. Hence

$$\varphi(y,x) = \frac{1}{4}[\hat{\varphi}(y+x) - \hat{\varphi}(y-x) + i\hat{\varphi}(y+ix) - i\hat{\varphi}(y-ix)]$$

$$= \frac{1}{4}[\hat{\varphi}(x+y) - \hat{\varphi}(x-y) + i\hat{\varphi}(x-iy) - i\hat{\varphi}(x+iy)]$$

$$= \overline{\varphi(x,y)}$$

here we have first used the polar identity (c) and the identities derived here.

(e)(i) A self-adjoint implies (ii) $\varphi(x,y) = \langle Ax, y \rangle$ is symmetric, meaning $\varphi(x,y) = \overline{\varphi(y,x)}$.

Clearly, $\varphi(x,y)$ is a bilinear form on H and $\varphi(x,y) = \langle Ax, y \rangle = \langle x, A^*y \rangle = \langle x, Ay \rangle = \overline{\langle Ay, x \rangle} = \overline{\varphi(y,x)}$ because of (i). Now (ii) is equivalent to (iii) $\hat{\varphi}(x)$ is real for all x by (d). Finally, assume (ii). We must show (i); i.e., $A = A^*$. We have $\varphi(x,y) = \langle Ax, y \rangle = \overline{\varphi(y,x)} = \overline{\langle Ay, x \rangle}$. But $\langle Ax, y \rangle = \langle x, A^*y \rangle$ and $\overline{\langle Ay, x \rangle} = \langle x, Ay \rangle$. Hence $\langle x, (A^* - A)y \rangle = 0$ for all x, y. Letting x vary we see that $(A^* - A)y = 0$ for each fixed y. Thus, since y is arbitrary, $A^* = A$.

3 $\varphi(x,y) = \psi(x,y)$ iff $\hat{\varphi}(x) = \hat{\psi}(x)$. Clearly $\varphi(x,y) = \psi(x,y)$ for all x, y implies $\hat{\varphi}(x) = \varphi(x,x) = \psi(x,x) = \hat{\varphi}(x)$ for all x. Conversely, if the two quadratic forms are equal for all x then the associated bilinear forms must be equal by the polar identity (2(c)).

4 φ is bounded iff $\hat{\varphi}$ is bounded. Suppose that φ is bounded. Then

$$\sup_{\|x\|=1} |\hat{\varphi}(x)| = \sup_{\|x\|=1} |\varphi(x,x)| \leq \sup_{\|x\|=1=\|y\|} |\varphi(x,y)| = \|\varphi\|$$

thus $\|\hat{\varphi}\| \leq \|\varphi\|$. Next suppose that $\hat{\varphi}$ is bounded. Then from the polar identity (2(c)) we have (see the way $\|\hat{\varphi}\|$ acts as discussed in the statement of problem 4)

$$|\varphi(x,y)| \leq \frac{1}{4}\|\hat{\varphi}\|[\|x+y\|^2 + \|x-y\|^2 + \|x+iy\|^2 + \|x-iy\|^2]$$

$$= \frac{1}{4}\|\hat{\varphi}\|2[\|x\|^2 + \|y\|^2 + \|x\|^2 + \|y\|^2]$$

(Exercises 1.4, problem 1())

$$= \|\hat{\varphi}\|[\|x\|^2 + \|y\|^2]$$

Thus $\sup_{\|x\|=1=\|y\|} |\varphi(x,y)| \leq 2\|\hat{\varphi}\|$.

(b) We know that $\hat{\varphi}$ is real (2(d)). By (a) we need only show $\|\varphi\| \leq \|\hat{\varphi}\|$; i.e., $|\varphi(x,y)| \leq \|\hat{\varphi}\|$ for any two unit vectors x, y. Suppose $\varphi(x,y) = \rho e^{i\alpha}$, $\rho \geq 0$ and let $x' = e^{-i\alpha}x$. Then $|\varphi(x,y)| = \rho = \varphi(e^{-i\alpha}x,y) = \frac{1}{4}[\hat{\varphi}(x'+y) - \hat{\varphi}(x'-y)]$ the imaginary terms in the polar identity must vanish because $\varphi(x,y)$ is given to be real. Thus $\|\varphi(x,y)\| \leq \frac{1}{4}\|\hat{\varphi}\|(\|x'+y\|^2 + \|x'-y\|^2) = \frac{1}{2}\|\hat{\varphi}\|(\|x'\|^2 + \|y\|^2) = \|\hat{\varphi}\|$.

(c) A is given to be self-adjoint hence by 2(e) $\varphi(x,y) = \langle Ax, y \rangle = \langle x, Ay \rangle$ because A is self-adjoint. By Lemma 1 then $\|\varphi\| = \|A\|$. Thus $\|A\| = \|\varphi\| = \|\hat{\varphi}\| = \sup_{\|x\|=1} |\langle Ax, x \rangle|$.

5(a) Given A, a bounded, linear operator we suppose that we can write $A = B + iC$ where B and C are self-adjoint. Then $A^* = B - iC$ and these equations can be solved for $B = \frac{A+A^*}{2}$, and $C = \frac{A-A^*}{2i}$. Thus if A can be written as $B + iC$ with B, C self-adjoint then there is only one way to do it.

Conversely, given A we may simply write down two new operators $B \equiv \frac{A+A^*}{2}$, $C \equiv \frac{A-A^*}{2i}$. A direct calculation shows that B and C, as just defined, are self-adjoint and if we compute $B + iC$ we get A.

(b) Again A is given but now, because of (a), we know that we can always find two unique self-adjoint operators B, C such that $A = B + iC$. Let us compute BC and CB.

$$BC = \left(\frac{A+A^*}{2}\right)\left(\frac{A-A^*}{2i}\right) = \frac{1}{4i}[A^2 - AA^* + A^*A - (A^*)^2]$$

$$CB = \left(\frac{A-A^*}{2i}\right)\left(\frac{A+A^*}{2}\right) = \frac{1}{4i}[A^2 + AA^* - A^*A - (A^*)^2]$$

Now if $AA^* = A^*A$, then in both BC and CB the two middle terms cancel giving us $BC = \frac{1}{4i}[A^2 - (A^*)^2] = CB$. Conversely, if $BC = CB$ then

$$\frac{1}{4i}[A^2 - AA^* + A^*A - (A^*)^2] = \frac{1}{4i}[A^2 + AA^* - A^*A - (A^*)^2]$$

or $AA^* = A^*A$.

Exercises 4.4

1 H a Hilbert space $\{x_n\} \subseteq H$ weakly convergent to $x_0 \in H$.

(a) Suppose $\{x_n\}$ is weakly convergent to y_0. Then $\lim \langle x_n, z \rangle = \langle x_0, z \rangle = \langle y_0, z \rangle$ for all $z \in H$. But then $\langle x_0 - y_0, z \rangle = 0$ for all $z \in H$ showing $x_0 - y_0 = 0$. Hence $x_0 - y_0$.

(b) Since $\{x_n\}$ converges weakly to x_0, $\lim\langle x_n, z\rangle = \langle x_0, z\rangle$ for all $z \in H$. Now $\lim\langle Ax_n, z\rangle = \lim\langle x_n, A^*z\rangle = \langle x_0, A^*z\rangle = \langle Ax_0, z\rangle$ for all $z \in H$ because A^*z is just another z.

2 For each fixed $x \in E$ we have $|f(x)| \le \|f\|\|x\| \le M\|x\|$ for all $f \in \mathcal{O}$. Thus $|f(x)| \le K$ for all $f \in \mathcal{O}$ where $K = M\|x\|$.

3 We choose $x_n \in \mathcal{S}_n$ for each n. Then for any m, n with $m \ge n$ we have $x_n, x_m \in \mathcal{S}_n$ and so $\|x_n - x_m\| \le \text{dia}\mathcal{S}_n$. Since the latter tends to zero, $\{x_n\}$ is a Cauchy sequence in the Banach space $(B, \|\cdot\|)$. It follows that $\{x_n\}$ is convergent to $x_0 \in B$. Now $\{x_n\} \subseteq \mathcal{S}_1$ and this is a closed set, so $x_0 = \lim x_n$ is in $\mathcal{S}_1$. Also $\{x_n\}_{n=2}^\infty \subseteq \mathcal{S}_2$ and this, too, is a closed set, so $x_0 = \lim x_n \in \mathcal{S}_2$. Clearly x_0 is in every $\mathcal{S}_m$ because, for a given m, $\{x_n\}_{n=m}^\infty \subseteq \mathcal{S}_m$ and this is a closed set. Thus $x_0 \in \cap_{n=1}^\infty \mathcal{S}_n$ showing that this intersection is not empty.

Exercises 5.1

3 A, C are similar $n \times n$ matrices if there is an invertible matrix B such that $A = BCB^{-1}$.

(a) Suppose that A and C are similar and that A is invertible. Then $A = BCB^{-1}$ so $B^{-1}AB = C$. But then C^{-1} is the matrix $(B^{-1}AB)^{-1} = B^{-1}A^{-1}B$ all three of which exist because A is given to be invertible.

(b) A complex number λ is an eigenvalue of A iff $\det(A - \lambda I) = 0$. Now if $A = BCB^{-1}$ then $A - \lambda I = BCB^1 - \lambda I = BCB^{-1} - \lambda(BIB^{-1}) = B(C - \lambda I)B^{-1}$ and $\det(A - \lambda I) - (\det B)(\det(C - \lambda I))(\det B^{-1})$. Since B is invertible $\det B \ne 0 \ne \det B^{-1}$ and so λ is an eigenvalue of A iff it is an eigenvalue of C.

Exercises 5.2

2 Let T be a linear operator on $\mathbb{R}^n$ or $\mathbb{C}^n$. Then T has a matrix representation $M(T)$ and the eigenvalues of T are the same as those of the matrix $M(T)$. A complex number λ is an eigenvalue of $M(T)$ iff λ satisfies $\det[M(T) - \lambda I] = 0$. But the latter is an nth degree polynomial equation and we know, by the fundamental theorem of algebra, that every such equation has a complex root. Thus $M(T)$ and hence T has an eigenvalue; of course T has n-eigenvalues if we count them according to their multiplicity.

4(a) Let $x_j \in M_j$ for $j = 1, 2, \ldots, n$ and suppose that no x_j is the zero vector. Set $\sum_{j=1}^n \alpha_j x_j = 0$ for unknown scalars α_j which we now determine. We have, for each $k = 1, 2, \ldots, n$,

$$\|x_k\|^2 \alpha_k = \langle \sum_{j=1}^{n} \alpha_j x_j, x_k \rangle = \langle 0, x_k \rangle = 0$$

(b) Recall $\sum_{j=1}^{n} \oplus M_j = \{x \in H | x = \sum_{j=1}^{n} x_j \text{ each } x_j \in M_j\}$. Note that $\|x\|^2 = \langle x, x \rangle = \langle \sum x_j, \sum x_j \rangle = \sum \|x_j\|^2$ by the orthogonality of our spaces. Suppose that $\{y_k\}$ is a sequence in our sum (i.e., in $\sum_{j=1}^{n} \oplus M_j$) and that this sequence converges to y_0. Then $y_k = \sum_{j=1}^{n} x_{kj}$, $x_{kj} \in M_j$ for each $j = 1, 2, \ldots, n$ and every k. Since $y_k - y_0 = \sum_{j=1}^{n}(x_{kj} - x_{\ell j})$, $\sum_{j=1}^{n} \|x_{kj} - x_{\ell j}\|^2 = \|y_k - y_\ell\|^2 \to 0$ as $k, \ell \to \infty$. Thus each $\{x_{kj}\}_{k=1}^{\infty}$ is a Cauchy sequence in the subspace M_j. This subspace is, by definition, closed so $\lim x_{kj} \equiv x_j \in M_j$, $j = 1, \ldots, n$. $\|y_0 - \sum_{j=1}^{n} x_j\| \leq \|y_0 - y_\ell\| + \|y_\ell - \sum x_j\| = \|y_0 - y_\ell\| + \|\sum_{j=1}^{n}(x_{\ell j} - x_j)\| \leq \|y_0 - y_\ell\| + \sum_{j=1}^{n} \|x_{\ell j} - x_j\|$. Given $\varepsilon > 0$ we choose L so that $\|y_0 - y_\ell\| < \varepsilon/2$ and $\|x_{\ell j} - x_j\| < \frac{\varepsilon}{2n}$ for all $d \geq L$ and each $j = 1, \ldots, n$.

(c) See mini-theorem 2 for Exercises 4.2, problem 2(d) and 2(e).

Exercises 5.3

5(a) Clearly $A : M \to M$ because this was given. Now $\|A\|_M = \sup\{\|Ax\| | x \in M, \|x\| = 1\} \leq \sup\{\|Ax\| | x \in H, \|x\| = 1\} = \|A\| < \infty$.

(b) For all x, y in H we have $\langle Ax, y \rangle = \langle x, Ay \rangle$ and this is then true for all $x, y \in M$ showing $A|_M$ is self-adjoint.
Next suppose A is normal. Then $AA^* = A^*A$, where $\langle Ax, y \rangle = \langle x, A^*y \rangle$ for all $x, y \in H$. Since $A|_M$ is a bounded, linear operator on M it has an adjoint A_M^* such that $\langle Ax, y \rangle = \langle x, A_M^* y \rangle$ for all $x, y \in M$. Clearly $A_M^* = A^*|_M$ because $A^*|_M$ is also an adjoint of $A|_M$ and an operator can only have one adjoint. Thus $A|_M A_M^* = A|_M A^*|_M = A^*|_M A|_M$ showing $A|_M$ is normal.

(c) We want to show that $A|_M$ is a compact operator given that A is a compact operator on H. We shall do it as we outlined in the problem: (i) If $\{x_n\} \subseteq M$ converges weakly to $y \in M$ (i.e., if $\lim \langle x_n, z \rangle = \langle y, z \rangle$ all $z \in M$) then $\{x_n\}$ converges weakly to y in H (i.e., $\lim \langle x_n, z \rangle = \langle y, z \rangle$ all $z \in H$). To see this note that $M \oplus M^\perp = H$ so for $z \in H$, $z = z_1 + z_2$ where $z_1 \in M$, $z_2 \in M^\perp$ and this decomposition of z is unique. Hence $\lim \langle x_n, z \rangle = \lim \langle x_n, z_1 + z_2 \rangle = \lim \langle x_n, z_1 \rangle + \lim \langle x_n, z_2 \rangle = \lim \langle x_n, z_1 \rangle = \langle y, z_1 \rangle = \langle y, z_1 \rangle + \langle y, z_2 \rangle = \langle y, z \rangle$ for all $z \in H$ because $\langle x_n, z_2 \rangle = 0$ for all n and $\langle y, z_2 \rangle = 0$.

(ii) If $\{x_n\} \subseteq M$ is weakly convergent to $y \in M$ then $\{x_n\} \subseteq H$ is weakly convergent to y by (a) and so $\{Ax_n\}$ is convergent to Ay. But this just says $\{(A|_M)(x_n)\}$ is convergent to $(A|_M)(y)$ and so $A|_M$ is compact.

2 Let A be a self-adjoint operator on H and let M be a linear subspace of H which is invariant under A. Then M is also invariant under $A^* = A$. By Corollary 1 to Theorem 1, M reduces A. Now go back to page 5.3.1 and do problem 5.

Exercises 5.4

1 $A(\vec{z}) = A(\{z_k\}_{k=1}^\infty) = \{\frac{z_k}{k}\}_{k=1}^\infty$.

(a) Done in Exercises 4.2, problem 5.

(b) Clearly $A\vec{e}_n = \frac{1}{n}\vec{e}_n$ showing $\{\frac{1}{n}\} \subseteq \sigma(A)$. Of course $0 \in \sigma(A)$ since it is a compact operator on an infinite dimensional space (also because $\lim \frac{1}{n} = 0 \in \sigma(A)$). $A\vec{z} - \lambda\vec{z} = 0$ iff $\frac{z_k}{k} = \lambda z_k$ for all k and this impossible unless $\lambda = \frac{1}{k}$ for some fixed k. For each n the eigenspace $M_n = \{\vec{z} | A\vec{z} = \frac{1}{n}\vec{z}\} = \{\alpha\vec{e}_n | \alpha \text{ a scalar}\}$ so $\P_n : \ell^2 \to M_n$ is $P_n(\vec{z}) = \langle \vec{z}, \vec{e}_n\rangle \vec{e}_n$ $A = \sum_{j=1}^\infty \frac{1}{n} P_n$.

(c) $A^n = \sum_{k=1}^\infty (\frac{1}{n})^n P_k$

(d) $p(A) = \sum_{k=1}^\infty p(\frac{1}{k}) P_k$.

2 A is a compact, self-adjoint operator and $\{\lambda_j\}$ are its non-zero, distinct eigenvalues. For each j, M_j is the eigenspace corresponding to λ_j.

(a) First recall that every M_j is finite dimensional and, since A normal, each M_j reduces A. Furthermore, the M_j's are pairwise orthogonal by Theorem 3.3. Hence, by Lemma 1 and the discussion following it, $\sum_{j=1}^\infty \oplus M_j$ is a linear subspace of H. If x is in this subspace then $x = \sum y_j$ where $y_j \in M_j$ for every j and so $Ax = A(\sum y_j) = A(\lim_n \sum_{j=1}^n y_j) = \lim_n A(\sum_{j=1}^n y_j) = \lim \sum_{j=1}^n Ay_j = \sum_{j=1}^\infty Ay_j \in \sum \oplus M_j$ because, for every j, $Ay_j \in M_j$ and $\sum \|Ay_j\|^2 \leq \|A\| \sum \|y_j\|^2 < \infty$ because $\sum y_j$ converges to x (Lemma 1). Thus this subspace is invariant under A but by Section 3, Corollary 1 to Theorem 1, this says our subspace reduces A.
If P_j denotes the orthogonal projection of H onto M_j then $A = \sum_{j=1}^\infty \lambda_j P_j$ by Theorem 3. Thus $Ax = \sum \lambda_j P_j(x)$ shows $\{Ax | x \in H\} \subseteq \sum_{j=1}^\infty \oplus M_j$. But for any $y \in M_j$, $Ay = \lambda_j y$ so $y = A(y/\lambda_j) \in \{Ax | x \in H\}$ showing that $\sum_{j=1}^n \oplus M_j \subseteq \{Ax | x \in\}$ for any n because A is linear. But then $\overline{\{Ax | x \in H\}} = \sum_{j=1}^\infty \oplus M_j$.
Finally, let us show that $\sum_{j=1}^\infty \oplus M_j$ is separable. By Section 1.9, Theorem 1, it suffices to show that this subspace has a countable orthonormal basis. Each M_j has such a basis, call it $\{y_{jk}\}_{k=1}^{n(j)}$ and this is a finite set. Thus $\cup_{j=1}^\infty \{y_{j,k}\}_{k=1}^{n(j)}$ is countable and, since the M_j are pairwise orthogonal, and

orthonormal. Given $x \in \sum_{j=1}^{\infty} \oplus M_j$, $x = \sum_{j=1}^{\infty} y_j$, $y_j \in M_j$ so $x = \sum_{j=1}^{\infty} (\sum_{k=1}^{n(j)} \alpha_{jk} y_{jk})$ showing this is a basis for $\sum \oplus M_j$. Note $\alpha_{jk} = \langle x, y_{jk} \rangle$ for $k = 1, 2, \ldots, n(j)$ and all j. This proves our infinite direct sum is separable and it also proves part (c) below.

(b) We have $\overline{A(H)} = \sum_{j=1}^{\infty} \oplus M_j$. If x is in the orthogonal complement of $\overline{A(H)}$ then $\langle x, y \rangle = 0$ for all $y \in M_j$, $j = 1, 2, \ldots$. Thus $P_j x = 0$ for all j. But then $Ax = \sum \lambda_j P_j(x) = 0$. Thus $\overline{A(H)}^{\perp} \subseteq \{x | Ax = 0\}$. However, if $Ax = 0$ then $P_j x = 0$ for all j, so $\langle x, y \rangle = 0$ for all $y \in M_j$, $j = 1, 2, \ldots$. It follows from this and the continuity of the inner product that x is in the orthogonal complement of $\sum_{j=1}^{\infty} \oplus M_j$ and this is equal to the orthogonal complement of $\overline{A(H)}$.

(c) This was done in the course of our solution to (a); the last paragraph of that solution.

(d) Since $BA = AB$, B commutes with every P_j and so by Corollary 1 to Theorem 2 (Section 5.3) each M_j reduces B. It follows as in part (a) above that $\sum \oplus M_j$ reduces B. But then $[\sum_{j=1}^{\infty} \oplus M_j]^{\perp}$ must also reduce B and this means that B commutes with P_0 (Section 5.3, same corollary).

Exercises 5.5

2 We have $A = \sum \lambda_n P_n$ where convergence is for the norm and $\{\lambda_n\}$ is the set of eigenvalues of A. Then the set of eigenvalues of A^* is $\{\overline{\lambda}\}$. Each P_n is the orthogonal projection of H onto M_n and M_n reduces A. Furthermore, M_n is the eigenspace of A^* corresponding to $\overline{\lambda}_n$ by Corollary 2 to Lemma 1 (Section 5.3). Thus the P_n's are the projections in the spectral decomposition of A^*. Finally $\|A - \sum_{n=1}^{k} \lambda_n P_n\| = \|A^* - \sum_{n=1}^{k} \overline{\lambda}_n P_n\|$ and since the former tends to zero as $k \to \infty$ so also does the latter.

(b) If T commutes with every P_n then T commutes with both A and A^*.

Exercises 6.2

1 Suppose that $f, g \in L[m, M]$. Then $f = f_1 - f_2$ and $g = g_1 - g_2$ f_1, f_2, g_1, g_2 are in $L^+[m, M]$. Clearly $f + g = (f_1 + g_1) - (f_2 + g_2)$ and $f_1 + g_1$, $f_2 + g_2$ are in $L^+[m, M]$. So $f + g \in L[m, M]$.
Next let $\alpha \in \mathbf{R}$. If $\alpha \geq 0$ then $\alpha f = \alpha f_1 - \alpha f_2$ and αf_1, αf_2 are in $L^+[m, M]$. If $\alpha < 0$ then $\alpha f_1 \notin L^+[m, M]$ but $-\alpha f_2$ is. But $0 \in L^+[m, M]$

so the negative of any $L^+[m, M]$ function is in $L[m, M]$. Thus $\alpha f_1 \in L[m, M]$.

5 $\{g_n(t)\}$ continuous functions; (i) $g_{n+1}(t) \leq g_n(t)$ all t and each n; (ii) $\lim g_n(t) = g(t)$ all t; (iii) $g(t)$ continuous on $[m, M]$.

Given $\varepsilon > 0$ and any $t \in [m, M]$ we have an integer $n(t)$ such that $|g(t) - g_n(t)| < \varepsilon$ for all $n \geq n(t)$. By the continuity of g and each g_n and (i) we have a neighborhood $U(t)$ such that $|g(s) - g_n(s)| < \varepsilon$ for all $x \in U(t)$ and $n \geq n(t)$. Now $\cup U(t)$ covers $[m, M]$ and this is a compact set so there is a finite subfamily $U(t_1), \ldots, U(t_k)$ which covers $[m, M]$. Let $n_0 \geq \max\{n(t_j)\}_{j=1}^k$ then $|g(s) - g_n(s)| < \varepsilon$ for $n \geq n_0$ and all $s \in [m, M]$.

Exercises 6.3

3 $\{A_n\}$ converges to B strongly. So $\lim \|A_n x - Bx\| = 0$ for each x. Then $|\langle Bx, x \rangle - \langle A_n x, x \rangle| = |\langle (B - A_n)x, x \rangle| \leq \|(B - A_n)x\| \|x\|$ by the C.S.B. inequality. Our result, that $\lim \langle A_n x, x \rangle = \langle Bx, x \rangle$ for each x, now follows.

Exercises 6.5

1(a) $\{g(x) | a < x < b\}$ is bounded above by $g(b)$ and hence has a least upper bound L. Then given $\varepsilon > 0$ we have $\delta > 0$ such that $L - g(x) < \varepsilon$ for all x satisfying $b - \delta < x < b$.

(b) As we saw in (a) $\lim_{x \to c^-} g(x) = lub\{g(x) | a < x < c\}$. Now we also have $\lim_{x \to c^+} g(x) = glb\{g(x) | c < x < b\}$. Since $g(s) < g(t)$ for any s in the first set and t in the second, $\lim_{t \to c^-} g(t) \leq \lim_{t \to c^+} g(t)$.

(c) Each of the sets D_n must be finite because the total growth of f on (a, b) is $f(b) - f(a) \geq k(\frac{1}{n})$ where k is the number of points in D_n. Clearly then each D_n is finite and $\cup_{n=1}^\infty D_n$ is a countable set containing all of the discontinuities of f on (a, b).

(d) If $h \in BV$ then $h = f - g$ where f, g are monotonically increasing functions. Thus $\lim_{x \to c^-} h = \lim_{x \to c^-} f - \lim_{x \to c^-} g$ and the two on the right exist. We can no longer say that $\lim_{x \to c^-} h \leq \lim_{x \to c^+} h$. Finally, any discontinuity of h or (a, b) must be a discontinuity of either f or g.

C

THE SPECTRAL THEOREM FOR NORMAL OPERATORS

We recall that a bounded, linear operator T on a Hilbert space H is said to be a normal operator if $TT^* = T^*T$. Also, an operator T is normal if, and only if, $\|T^*x\| = \|Tx\|$ for each $x \in H$ (Section 5.3, Corollary 1 to Lemma 1).

Lemma 1

Suppose that T is a normal operator and that λ is a complex number. Then λ is *not* an eigenvalue for T if, and only if, $(T - \lambda I)(H)$ is dense in H.

Proof. Recall that a complex number λ is an eigenvalue for T (remember T is given to be normal here) if, and only if, $\overline{\lambda}$ is an eigenvalue for T^* (Section 5.3, Corollary 2 to Lemma 1).

To prove the lemma suppose that $y \in H$ and that

$$0 = \langle y, (T - \lambda I)x \rangle = \langle (T^* - \overline{\lambda}I)y, x \rangle$$

for all x in H. Then $(T^* - \overline{\lambda}I)y$ must be the zero vector. Now suppose that λ is *not* an eigenvalue for T. Then $\overline{\lambda}$ is *not* an eigenvalue for T^* and hence $(T^* - \overline{\lambda}I)y = 0$ implies that y is the zero vector. We conclude that any vector which is orthogonal to $(T - \lambda I)(H)$ is the zero vector. It follows that this linear manifold is dense in H (Section 4.2, Lemma 1). Now suppose that $(T - \lambda I)(H)$ is dense in H. Then if $(T^* - \overline{\lambda}I)y = 0$ we must have

$$0 = \langle (T^* - \overline{\lambda}I)y, x \rangle = \langle y, (T - \lambda I)x \rangle$$

for all $x \in H$. But clearly then, y must be the zero vector showing that $\overline{\lambda}$ is *not* an eigenvalue of T^*. It follows that λ is *not* an eigenvalue of T.

Theorem 1

The spectrum of a normal operator consists entirely of generalized eigenvalues.

Proof. Suppose that T is a normal operator and that $\lambda \in \sigma(T)$. We may suppose that λ is not an eigenvalue of T. It follows then that $(T - \lambda I)^{-1}$ is defined on a dense, linear manifold, $(T - \lambda I)(H)$ at least, of H. If this operator was bounded then its domain would be a subspace of H and that would mean its domain is all of H. But that would contradict the fact that we chose $\lambda \in \sigma(T)$. Thus $(T - \lambda I)^{-1}$ is unbounded. By Lemma 6 of Section 7.6 we can now assert that λ is a generalized eigenvalue for T.

Now let T be a given normal operator on H. We recall (Exercises 4.3, problem 5(b)) that we can find two self-adjoint operators B, C such that

$$T = B + iC$$

and $BC = CB$; furthermore, these operators are unique. Now we have a spectral theorem for self-adjoint operators. Let $\{P(\lambda)|\lambda \in \mathbb{R}\}$, $\{Q(\mu)|\mu \in \mathbb{R}\}$ be the spectral families of B and C respectively (Section 6.6). Define

$$R(\lambda, \mu) = P(\lambda)Q(\mu)$$

for all λ, μ in $\mathbb{R}$. Now C commutes with B and so every $P(\lambda)$ must commute with C. It follows from this that every $P(\lambda)$ commutes with every $Q(\mu)$ (Corollary 2 to Theorem 1 of Section 6.6). Thus each $R(\lambda, \mu)$ is an orthogonal projection (Exercises 4.2, problem 2(c)).
Recalling the properties of spectral families we see that

1. $R(\lambda,\mu)R(\lambda',\mu') = R(\min\{\lambda,\lambda'\}, \min\{\mu,\mu'\})$ and hence $R(\lambda,\mu)R(\lambda',\mu') = R(\lambda,\mu)$ when $\lambda \leq \lambda'$, $\mu \leq \mu'$.
2. $R(\lambda,\mu) = 0$ when $\lambda < -\|T\|$ or $\mu < -\|T\|$
3. $R(\lambda,\mu) = I$ when $\lambda \geq \|T\|$ and $\mu \geq \|T\|$
4. $\lim_{n\to\infty} \|[R(\lambda+\alpha_n, \mu+\beta_n) - R(\lambda,\mu)]x\| = 0$

for all $x \in H$ whenever $\{\alpha_n\}$, $\{\beta_n\}$ are decreasing sequences of positive numbers converging to zero.
To see this we write

$$\|[R(\lambda,\mu)x - R(\lambda+\alpha_n, \mu+\beta_n)x]\| \leq \|[P(\lambda) - P(\lambda+\alpha_n)]Q(\mu)x\|$$
$$+ \|P(\lambda+\alpha_n)\|\|[Q(\mu) - Q(\mu+\beta_n)]x\|$$

and recall that

$$\lim_{n\to\infty} \|[P(\lambda+\alpha_n)]x\| = 0, \quad \lim_{x\to\infty} \|[Q(\mu+\beta_n) - Q(\mu)]x\| = 0$$

for all x in H.

One can show that the family $\{R(\lambda,\mu) | \lambda,\mu \text{ in } \mathbb{R}\}$ "represents T."

Theorem 2

(Spectral Theorem) To each normal operator T on H there corresponds a family of orthogonal projections $\{R(\lambda,\mu) | \lambda,\mu \text{ in } \mathbb{R}\}$, each of which commutes with T, such that:

1. $R(\lambda,\mu)R(\lambda',\mu') = R(\min\{\lambda,\lambda'\}, \min\{\mu,\mu'\})$;
2. $R(\lambda,\mu) = 0$ for $\lambda < -\|T\|$ or $\mu < -\|T\|$;
3. $R(\lambda,\mu) = I$ for $\lambda \geq \|T\|$ and $\mu \geq \|T\|$;
4. $\lim_{n\to\infty} \|[R(\lambda+\alpha_n, \mu+\beta_n) - R(\lambda,\mu)]x\| = 0$ for all $x \in H$ whenever $\{\alpha_n\}$, $\{\beta_n\}$ are decreasing sequences of positive numbers converging to zero;
5. $T = \int_{-\infty}^{\infty} \int_{-\infty}^{\infty} (\lambda + i\mu) d\, R(\lambda,\mu)$ meaning

$$\left\| T - \sum_{k=1}^{n}\sum_{\ell=1}^{m} \eta_{k\ell}[R(\lambda_k,\mu_\ell) - R(\lambda_k,\mu_{\ell-1}) - R(\lambda_{k-1},\mu_\ell) + R(\lambda_{k-1},\mu_{\ell-1})] \right\| \leq 4\delta$$

whenever

$$\mu_0 < -\|T\| = \mu_1 < \mu_2 < \ldots < \mu_{n-1} < \mu_n = \|T\|,$$

$$\mu_k - \mu_{k-1} \le \delta \text{ for } 1 \le k \le n;$$
$$\lambda_0 < -\|T\| = \lambda_1 < \lambda_2 < \ldots < \lambda_{m-1} < \lambda_m = \|T\|,$$
$$\lambda_\ell - \lambda_{\ell-1} \le \delta \text{ for } 1 \le \ell \le m;$$
$$\eta_{k\ell} \in \{\eta = \lambda + i\mu | \mu_{k-1} \le \mu \le \mu_k, \lambda_{\ell-1} \le \lambda \le \lambda_\ell\}.$$

In the same sense we also have

$$T^* = \int_{-\infty}^{\infty} \int_{-\infty}^{\infty} (\lambda - i\mu) d\, R(\lambda, \mu).$$

BIBLIOGRAPHY

[1] Berezanskii, Ju.M., Expansions in Eigenfunctions of Self-adjoint Operators, Translations of Mathematical Monographs, Vol. 17, Amer. Math. Soc. Providence, R.I. (1968).

[2] Carleson, L., On Convergence and growth of partial sums of Fourier series. Acta Math. 116 (1966), pp. 135–157.

[3] DeVito, C.L., "Functional Analysis" Academic Press, New York, N.Y., (1978).

[4] Enflo, P., A counter-example to the approximation problem in Banach Spaces, Acta Math. 130 (1973), 309–317.

[5] Faris, W., Perturbations and Non-Normalizable Eigenvectors, Helvetica Physica Acta, Vol. 44/7, (1971), 930–936.

[6] Goldberg, R.R., "Fourier Transforms," Cambridge Univ. Press, London and New York (1961).

[7] Helmberg, G., "Introduction to Spectral Theory in Hilbert Space" North-Holland Publishing Co. Amsterdam (1969).

[8] Lindenstrass, J., and Tzafrini, L., On the complemented subspaces problem, Israel J. Math. 9 (1971), 263–269.

[9] Riesz, F., and Nagy, B., "Functional Analysis" 2nd Ed. Frederick Ungar Publishing Co. New York (1955).

[10] Royden, H.L., "Real Analysis" 2nd Ed. Macmillan New York (1968).

[11] Stakgold, I., "Boundary Value Problems of Mathematical Physics Volume 1." The Macmillan Company, New York, N.Y. (1967).

[12] Taylor, A.E., "Advanced Calculus" Ginn and Company New York (1955).

LIST OF SYMBOLS

$C_p[0,1]$, 37
$C_r[0,1]$, 2
$C[0,1]$, 8
$C_0(\mathbb{R})$, 292
E-integrable, 231, 232, 234, 235
$\mathcal{L}(E_1, E_2)$, 70, 71, 81, 82, 86, 107, 112, 125
$L[m, M]$, 206, 207, 211, 218
L^1, 253, 294
$L^+[m, M]$, 206, 207, 211, 218
L^2, 45, 287
ℓ^2, 19
$P^+[m, M]$, 206, 207, 209, 211, 212, 217
S, 212
S^+, 213

INDEX

absolutely continuous function, 254, 255, 304
adherent point, 103, 107, 121
adjoint matrix, 138, 157
adjoint operator, 135, 136, 138, 141, 143–145, 151, 164, 166, 186, 188, 193, 248, 250, 251, 258
algebraic object, 2
algebraic operations, 3, 4, 8, 287
almost everywhere (a.e.), 50, 51, 63, 254
angle, 14, 15
approximation theorem (Weierstrass), 37, 42, 44, 59, 245
Arzela, 100
Ascoli, 100

ball, closed, 73, 75, 80
ball, open, 73, 80
ball, radius ε, 73
ball, unit, 9
Banach algebra, 296
Banach space, 11, 12, 78, 79, 81, 83, 295, 303
Banach, Stefan, 65
Banach–Steinhaus theorem, 146, 152, 252
basis, Hamel, 3, 5, 6, 9, 26, 159
basis, orthonormal, 25, 33, 35, 36, 52, 57, 59–61, 181, 288

Bessel's inequality, 27, 28, 189
bilinear form, 139
bilinear functional, 139
bounded bilinear form, 139, 143, 144
bounded function, 49, 230, 297
bounded linear functional (form), 139, 141, 142, 228, 229
bounded linear mapping, 68, 69, 74, 78, 80, 87, 93
bounded linear mappings (space of), 71, 81, 107, 112
bounded quadratic form, 143, 144
bounded sequence of functionals, 146, 151
bounded set, 72, 86, 100, 105, 112, 146–148, 151, 152
bounded variation, 227–230
Bunyakowski, 16, 17

Caratheodory, 49
Carleson, 63
Cauchy, 16, 17
Cauchy sequence, 11, 14, 72, 97, 107, 110, 131, 152
Cauchy sequence (weak), 145, 147, 148
Cayley transform, 260, 262, 263, 265, 272, 276
Cèsaro convergent, 38, 44
Cèsaro means, 38–40, 42

Cèsaro summable, 38, 41
characteristic value, 123, 124, 187
circled minus sign, 250, 262, 263
closed ball, see ball, closed
closed extension of an operator, 246, 251, 260, 262, 263
closed linear operator, 246, 247, 249, 251, 255, 259, 262
closed set, 48, 74, 80, 86, 89, 90, 97, 103, 104, 110, 112, 117, 131, 133, 137, 152, 162, 246, 249
closure of a set, 103, 107, 114, 134, 137, 181, 188, 193, 201, 247, 289
commutative algebra, 296
community operators, 164, 169, 180, 182, 186, 218, 222, 237, 276, 279
compact linear operator, 101, 105, 112, 113, 116, 117, 121, 122, 127, 150, 151, 170, 172, 179, 181, 184–186
compact set, 101, 104, 106, 110
compact set (relatively), 103–105
complement (orthogonal), 133, 134, 136
complement (set-theoretic), 48, 86
complement (subspace), 123, 129, 130, 131
complete normed space, 11, 12, 55, 78, 79, 81, 83
complete set, 110
completion, 63
complex vector space, 1, 6–10, 15, 19, 44, 52, 55
composition, 23, 72, 82, 83, 88, 106, 137, 143, 144
contained in, 246, 258, 259, 275, 276
continuous function
 of a matrix, 153, 156
 of a bounded operator, 209, 212, 218, 269, 271, 297
 of a bounded self-adjoint operator, 211, 213, 234, 241
 of a unitary operator, 269, 271
 on $[a, b]$, 2, 8, 18, 37, 45, 51, 52, 66, 99, 193, 207, 211, 226, 228, 244
continuous linear mapping, 68, 69, 74, 78, 80, 87, 93
convergent sequence, 10, 13, 14, 64, 77, 82, 103, 104, 107, 122, 179, 185, 186, 217
convergent sequence (strongly), 219, 221
convergent sequence (weakly), 145, 148, 150–152, 169
convergent sequence of functions, 10–12, 41, 42, 44, 54, 63, 100, 189, 206, 211, 212, 245, 299
convergent series, 12, 28, 29, 43, 63, 83, 85, 125, 126, 175, 189, 191, 220
convolution, 295–297
countable orthonormal basis, 33, 35, 36, 61
countable set, 30, 32, 37, 47–49, 56, 64, 113, 285
countably additive, 48
C.S.B. inequality, 16, 17, 139, 140, 145, 214

deficiency indicies, 262, 264
deficiency subspaces, 262, 263
dense linear manifold, 63, 245, 248, 250, 251, 255, 256, 266, 267, 288, 304
dense set, 57, 64, 289, 303, 311
determinant, 156
diagonal matrix, 114, 154, 155
diagonalizable matrix, 155, 156, 158, 163
diagonalizable operator, 158–162, 167, 168

354 INDEX

differential equation, 197, 198, 242, 243, 247, 284, 285
differential operator, 253, 255, 258
differentiation, 244, 252, 254
dimension, 3, 4, 262
dimensional, see finite or infinite dimensional
Dini's theorem, 212, 217
Dirac δ-function, 96
direction, 14
distance, 195
domain, 245, 246, 273–275

eigenfunction, 190, 244, 285
eigenspace, 87, 122, 159, 161, 240
eigenvalue, 86–88, 95, 96, 118, 121, 122, 126, 138, 158, 160–162, 168, 172–174, 179, 181, 184, 190, 201, 204, 205, 265–267, 270, 293
eigenvalue (continuous), 96
eigenvalue (generalized), 96, 98, 170, 203, 268, 293
eigenvector, 87, 88, 114, 158, 159, 204
eigenvector (generalized), 115
E-integrable, 231, 232, 234, 235
equal operators, 246
equation, differential, 197, 198, 242, 243, 247, 284, 285
equation, integral, 45, 65, 66, 99, 100, 102, 103, 108, 123, 124, 126, 186, 187, 191
equicontinuous family, 100
equivalence classes, 55
equivalence relation, 55
equivalent norms, 13, 76, 77, 79, 80
equivalent operators, 306, 307, 312
Euclidean norm, 7–12, 17
extension, 246, 258

finite dimensional, 3, 6, 9, 14, 15, 17, 18, 26, 36, 77, 78, 101, 107, 114, 115, 122, 163, 167–169
finite rank, 107, 109, 110, 113, 179, 185, 186
first kind (integral equation), 123, 186, 191
form, bilinear, 139, 144, 205
form, linear, 139
form, quadratic, 143, 144, 205
Fourier coefficients, 37, 42
Fourier Plancherel operator, 300–302, 304–307
Fourier series, 37, 38, 41, 43, 54, 63
Fourier transform, 286, 292, 296, 297–307
Fredholm, 123, 124, 187, 191
Fredholm alternative, 187
Fubini's theorem, 55, 96
function, bounded variation, 227–230
function, integrable, 50, 51, 54, 56, 295
function, measurable, 49–51, 54, 55
function, monotonic, 226, 227, 229, 230
function, of a matrix, 156
function, of a self-adjoint operator, 217, 218
functional, 139
functional, bounded bilinear, 139, 144
functional, bounded linear, 139, 141, 142, 229

generated by, 2
generalized C.S.B. inequality, 214
generalized eigenvalue, see eigenvalue (generalized)
generalized eigenvector, see eigenvector (generalized)

generalized Pythagorean theorem, 175, 176, 273
Goldberg, R.R., 299
Gram–Schmidt, 25, 36, 59, 60
graph of an operator, 248–252

Hamel, see basis, Hamel
Hausdorff maximal principle, 29
Hermite function, 291
Hermite polynomial, 291, 292
Hilbert, David, 18
Hilbert relations, 88
Hilbert–Schmidt operator, 97, 103, 108, 109, 113, 123, 138, 187, 190, 193
Hilbert–Schmidt theorem, 189
Hilbert space, 18, 21, 26, 29, 30, 33, 36, 46, 52, 57, 59, 61, 62, 130, 132–36, 139, 141, 147, 148, 150, 151, 170, 188, 202, 203, 220, 288, 306

ideal, 296, 303
identity operator, 23, 66, 72, 82, 83, 101, 114, 128, 129, 137, 155, 160, 167, 168, 220, 221, 269
identity, Parseval's, 30
identity, polar, 18, 143, 144, 236
indefinite integral, 253, 254
independent, linearly, 2, 3, 5, 36, 59, 131, 159, 162
indicies, see deficiency indicies
induced operator, 119, 121, 159, 169, 274, 280
inequality, Bessel's, 27, 28
inequality, C.S.B., 16, 17
inequality, C.S.B., (generalized), 214
inequality, triangle, 8, 17

infinite dimensional, 3, 4, 6, 19, 62, 112, 121, 122, 126, 170, 176
infinite interval, 285
infinite sum of subspaces, 174, 175, 181, 182, 273, 280
inner measure, 48, 49
inner product, 14–20, 23, 24, 51, 52
inner product space, 16–20, 24–26, 29, 36, 63
integrable function, Lebesgue, 50, 56
integrable function, Riemann, 50, 56
integral, 224
integral equation (first kind), 123, 191, 192, 193
integral equation (second kind), 123, 187–190
integral operator, 193–195
integration, 194
invariant subspace, 163, 164, 169, 273, 274
inverse, 64, 74, 82, 83, 85, 93, 114, 155, 240, 247, 260, 267, 301–303, 305–307, 312
invertible operator, 82–86, 114, 155–157, 212, 271, 287, 301
isometric isomorphism, 25, 26, 33, 54, 62, 63, 72, 306, 307
isometry, 25, 72, 88, 263
isomorphic, 4, 6
isomorphic, topologically, 78
isomorphism, 4, 6, 9, 24, 25, 64, 78
isomorphism, topological, 78

kernel (integral operator), 97, 108, 123, 124
kernel (null space), 4

law, parallelogram, 18

Lebesgue integral, 45, 46, 48–50
Lebesgue measurable functions, 49–51
Lebesgue measurable sets, 47–50, 55, 56
Lebesgue measure, 48, 307
Lebesgue dominated convergence theorem, 52, 54, 223
left-shift operator, 22, 23
Legendre polynomial, 60
length, 7
L' Hospital, 291
limit, 10, 18, 64
limit, strong in $L(E_1, E_2)$, 219, 220
limit, weak, 145, 148, 150–152
lin, 2, 5, 25, 42, 57, 59, 174, 175
linear combination, 2
linear functional, 139, 141, 142, 228, 229
linear manifold, 2, 5, 103, 107, 116, 119, 123, 128, 134, 245, 246
linear mapping, 4, 93
linear operator, 4
linear subspace, 103, 107, 112, 115, 117, 119, 162, 174, 175, 246
linearly dependent, 2, 5
linearly independent, 2, 3, 5, 19, 25, 36, 87, 136, 159
linearly independent subspaces, 159, 160, 162
Lipschitz, 197

Maclaurin series, 156, 220
manifold, see linear manifold
mapping, bounded-continuous, 67–69
mapping, linear, see linear mapping
matrix, 80, 138, 150, 154–159, 163
matrix representation of a linear operator, 80, 157–159

maximal linearly independent set, 3
maximal orthonormal family, 29, 30, 35, 36
measurable functions, 49–51
measurable sets, 47–50, 55, 56
measure, 47, 51
measure, inner, 48, 49
measure, outer, 47
metric, 195, 197
metric space, 195, 196
monotonically increasing, 226, 227, 229, 230, 239
multiplication operator, 92, 95, 292–294, 298, 299, 302, 303, 305–307
multiplicity, 122
mutually orthogonal, 162, 166, 273, 280

neighborhood, 73, 74
non-linear, 75, 106, 195
non-negative self-adjoint operator, 213–215
non-normalizable eigenvectors, 312
non-trivial linear manifold, 2
norm, 8, 9, 13, 19, 24, 71, 75, 79, 80
norm, Euclidean, 8, 10, 11, 13
norm from an inner product, 16, 17, 18
norm of a bilinear form, 139, 141
norm of a linear functional, 139, 141, 142
norm of a linear map, 71
norm of a linear operator, 71, 144
norm of a quadratic form, 144
norm, sup, 8, 11, 19, 41, 244
normed space, 8, 9, 12–14, 64, 71, 74
normal operator, 144, 165–168
normal operator, compact, 182, 184–186

null space (kernel) of a linear map, 4, 97, 101, 103, 115

open ball, 73, 74, 80
open interval, 47, 56
open set, 47, 55, 74, 75, 80
open-mapping theorem, 93, 203
operator, 69
orthogonal, 24, 88
orthogonal, pairwise, 162, 163, 166, 280
orthogonal complement, 133, 136, 187
orthonormal, 24
orthonormal basis, 33, 36, 52, 54, 57, 59, 61
orthonormal set, 24, 25, 28, 30, 43, 44, 57
orthonormal set (complete), 36
orthonormal set (maximal), 29, 35, 42, 43
orthonormalization (Gram–Schmidt), 25, 59, 60

pairwise disjoint, 47, 48, 55, 56
pairwise orthogonal, 162, 166, 273, 280
Parseval's identity (relation), 30
Peano, 198
Picard, 191, 194
polar identity, 143, 144, 236
polynomial, 3, 42, 44, 60, 126, 206, 209, 211, 291
polynomial, Hermite, 291, 292
polynomial, Legendre, 60
polynomial, trigonometric, 42
polynomial of an operator, 181, 209, 211, 212
positive self-adjoint operator, 213
precompact, 110, 112, 113

product, Cartesian, 55, 56, 64, 66, 98, 99, 130, 248
product, inner, see inner product
projection of a vector, 132
projection operator, 128, 129, 131, 134–137, 160–164, 167, 168, 178–181, 184–186, 219, 221, 231, 238, 239, 269, 270, 273–275, 277, 279, 281, 283
Pythagorean theorem, 7, 18
Pythagorean theorem generalized, 175, 176, 273

quadratic form, 143, 144

range of a linear map, 4, 87, 116, 117, 119
real vector space, 2, 3, 22, 26, 33
reducing subspace, 163, 164, 166, 274, 275
regular point, 86
relatively compact set, 103–105
resolution of the identity, 221
resolvent set, 86
restriction of an operator, 120, 121
Riemann integral, 49, 50, 56, 226
Riemann–Stieltjes integral, 224, 226, 228, 231, 236
Riesz representation theorem, 143, 144, 228, 229
right-shift operator, 23, 91
Russell, B., 7

Schwarz, 16
self-adjoint extension, 258, 260
self-adjoint matrix, 157
self-adjoint operator, 135, 136, 144, 172, 200, 203, 212, 213, 272, 276
separable space, 57, 61–64, 288

sequential continuity, 67
space of functions, 2–6, 45, 205, 244, 253, 287, 294, 295, 308–312
span, linear, 2, 3, 5
spectral bounds, 203, 236, 237, 293
spectral decomposition, 180
spectral family, 231, 237, 269, 270, 281, 282
spectral theorem finite dimensional case, 160, 161, 167, 168
spectral theorem for compact, normal operators, 184
spectral theorem for compact, self-adjoint operators, 179, 180
spectral theorem for normal operators, appendix C
spectral theorem for self-adjoint operators, 232
spectral theorem for unbounded operators, 281, 282
spectral theorem for unitary operators, 269
spectrum, 80, 86, 91, 95, 97, 98, 114, 121, 122, 170, 193, 194, 203, 268, 293
strong limit, see limit, strong
subspace, linear, 103, 112, 117, 130, 133, 134, 137
supplementary linear manifolds, 123, 127–129
symmetric operator, 258, 259, 262, 263

unbounded linear map, 242, 244, 245, 248, 255, 256, 258, 264, 267, 272, 281, 282, 292, 298
uniform continuity, 67, 113, 114, 137, 244, 245
uniformly convergent sequence, 10–12, 212
unitary matrix, 157
unitary operator, 240, 241, 249, 252, 260, 265–267, 268 (spectrum), 269–271, 303 (Fourier-Plancherel)

value, characteristic, see characteristic value
vector, 6, 7, 14, 132
vector space, 1, 2, 4–7
Volterra, V., 65
Volterra integral equation, 123, 124, 126
Von Neumann, J., 248

weak Cauchy, 145–147
weak limit, 145, 147, 150–152, 169
Weierstrass theorem, 37, 42, 44
Weyl, 194

zero element, 3, 71
zero operator, 71, 72
Zorn's lemma, 29

topological isomorphism, 78, 79
topology, 74, 75
total set, 64
transform, Cayley, 260, 263, 265, 272, 276
transform, Fourier, 296–305, 307

The Addison-Wesley **Advanced Book Program** would like to offer you the opportunity to learn about our new mathematics, statistics, and scientific computing titles in advance. To be placed on our mailing list and receive pre-publication notices and special offers, just **fill out this card completely** and return to us, postage paid. Thank you.

Title and Author of this book: _____ **Date purchased:** _____

Name _____
Title _____
School/Company _____
Department _____
Street Address _____
City _____ State _____ Zip _____
Telephone/s( ) _____ ( ) _____

Where did you buy/obtain this book?
☐ Bookstore   ☐ Mail Order   ☐ School (Required for Class)
☐ Campus Bookstore   ☐ Toll Free # to Publisher   ☐ Professional Meeting
☐ Other _____   ☐ Publisher's Representative

What professional mathematics and statistics associations are you an active member of?
☐ AMS (Amer Mathematical Society)   ☐ NCTM (Nat Counc Teachers Math)   ☐ SIAM (Soc Indust Applied Math)
☐ ASA (Amer Statistical Association) ☐ ORSA (Oper Research Soc America) ☐ AAAS (Amer Assoc for the Advancement of Science)
☐ MSA (Math Society of America)
☐ Other _____

Check your areas of interest.

⓺⓪ ✓**Mathematics/Statistics**
61 ☐ Advanced Calculus   69 ☐ Discrete Math   77 ☐ Operations Research
62 ☐ Algebra   70 ☐ Dynamical Systems   78 ☐ Optimization
63 ☐ Analysis   71 ☐ Geometry   79 ☐ Probability Theory
64 ☐ Applied Math   72 ☐ Logic/Probability   80 ☐ Statistical Modelling
65 ☐ Applied Statistics   73 ☐ Math-Biology   81 ☐ Stochastic Processes
66 ☐ Combinatorics   74 ☐ Math-Modelling   82 ☐ Time Series Analysis
67 ☐ Complex Variables   75 ☐ Math-Physics   83 ☐ Topology
68 ☐ Decision Theory   76 ☐ Number Theory   84 ☐ Other _____

Are you more interested in: ☐ pure math ☐ applied math?

Are you currently writing, or planning to write a textbook, research monograph, reference work, or create software in any of the above areas?
☐ Yes ☐ No
Area: _____

(If Yes) **Are you interested in discussing your project with us?**
☐ Yes ☐ No

Mathematics/Statistics

fold and staple

IIlıIııılıııIIIıııııIıIıIıIıIıIııııIIIıIııılıII

No Postage
Necessary
if Mailed in the
United States

BUSINESS REPLY MAIL
FIRST CLASS PERMIT NO. 828   REDWOOD CITY, CA   94065

Postage will be paid by Addressee:

**ADDISON-WESLEY
PUBLISHING COMPANY, INC.®**
Advanced Book Program
390 Bridge Parkway, Suite 202
Redwood City, CA 94065-1522